T0214070

Lecture Notes in Computer Science   12581

More information about this subseries at http://www.springer.com/series/7409

Ladjel Bellatreche · Vikram Goyal ·
Hamido Fujita · Anirban Mondal ·
P. Krishna Reddy (Eds.)

Big Data Analytics

8th International Conference, BDA 2020
Sonepat, India, December 15–18, 2020
Proceedings

 Springer

Editors
Ladjel Bellatreche
ISAE-ENSMA
Chasseneuil, France

Hamido Fujita
Iwate Prefectural University
Takizawa, Japan

P. Krishna Reddy
IIIT Hyderabad
Hyderabad, India

Vikram Goyal
Indraprastha Institute of Information
Technology
New Delhi, India

Anirban Mondal
Ashoka University
Sonepat, India

ǀ

ISSN 0302-9743   ISSN 1611-3349 (electronic)
Lecture Notes in Computer Science
ISBN 978-3-030-66664-4   ISBN 978-3-030-66665-1 (eBook)
https://doi.org/10.1007/978-3-030-66665-1

LNCS Sublibrary: SL3 – Information Systems and Applications, incl. Internet/Web, and HCI

This Springer imprint is published by the registered company Springer Nature Switzerland AG
The registered company address is: Gewerbestrasse 11, 6330 Cham, Switzerland

Preface

The amount of data stored in computer systems has greatly increased in recent years due in part to advances in networking technologies, storage systems, the adoption of mobile and cloud computing and the wide deployment of sensors for data collection. To make sense of this big data to support decision-making, the field of Big Data Analytics has emerged as a key research and application area for industry and other organizations. Numerous applications of big data analytics are found in several important and diverse fields such as e-commerce, finance, healthcare, education, e-governance, media and entertainment, security and surveillance, smart cities, telecommunications, agriculture, astronomy, and transportation.

Analysis of big data raises several challenges such as how to handle massive volumes of data, process data in real-time, and deal with complex, uncertain, heterogeneous and streaming data, which are often stored in multiple remote repositories. To address these challenges, innovative big data analysis solutions must be designed by drawing expertise from recent advances in several fields such as big data processing, data mining, database systems, statistics, machine learning and artificial intelligence. There is also an important need to build data analysis systems for emerging applications such as vehicular networks and social media analysis, and to facilitate the deployment of big data analysis techniques.

The Eighth International Conference on Big Data Analytics (BDA) was held from 15th December to 18th December 2020 at Ashoka University, Sonepat, Haryana, India. This proceedings book includes 15 peer-reviewed research papers and contributions by keynote speakers, invited speakers and tutorial speakers. This year's program covered a wide range of topics related to big data analytics on themes such as: big data analytics, gesture detection, networking, social media, search, information extraction, image processing and analysis, spatial, text, mobile and graph data analysis, machine learning, and healthcare.

It is expected that research papers, keynote speeches, invited talks and tutorials presented at the conference will encourage research on big data analytics and stimulate the development of innovative solutions and their adoption in industry.

The conference received 48 submissions. The Program Committee (PC) consisted of researchers from both academia and industry from 13 countries or territories, namely Australia, Canada, China, Estonia, France, India, Japan, New Caledonia, New Zealand, Norway, Spain, Taiwan and the USA. Each submission was reviewed by at least two, and most by three Program Committee members, and was discussed by PC chairs before taking the decision. Based on the above review process, the Program Committee selected 15 full papers. The overall acceptance rate was about 31%.

We would like to extend our sincere thanks to the members of the Program Committee and external reviewers for their time, energy and expertise in providing support to BDA 2020.

Additionally, we would like to thank all the authors who considered BDA 2020 as the forum to publish their research contributions. The Steering Committee and the Organizing Committee deserve praise for the support they provided. A number of individuals contributed to the success of the conference. We thank Prof. H. V. Jagadish, Prof. M. Kitsuregawa, Prof. Divyakant Agrawal, Prof. Amit Sheth and Prof. Ouri Wolfson for their insightful keynote talks. We also thank all the invited speakers and tutorial speakers. We would like to thank the sponsoring organizations including National Institute of Technology Delhi (NITD), India; Indraprastha Institute of Information Technology Delhi (IIITD), India; International Institute of Information Technology Hyderabad (IIITH), India; The University of Aizu, Japan; University of Delhi, India; Ahmedabad University, India and the Department of Computer Science at Ashoka University, as they deserve praise for the support they provided.

The conference received invaluable support from the management of Ashoka University in hosting and organizing the conference. Moreover, thanks are also extended to the faculty, staff members and student volunteers of the Department of Computer Science at Ashoka University for their constant cooperation and support.

Ladjel Bellatreche
Vikram Goyal
Hamido Fujita
Anirban Mondal
P. Krishna Reddy

Organization

BDA 2020 was organized by Ashoka University, Sonepat, Haryana 131029, India.

Honorary Chairs

Malabika Sarkar   Ashoka University, India
  (Vice-chancellor)
Praveen Kumar (Director)   NIT Delhi, India

General Chair

Anirban Mondal   Ashoka University, India

Steering Committee Chair

P. Krishna Reddy   IIIT Hyderabad, India

Steering Committee

S. K. Gupta   IIT Delhi, India
Srinath Srinivasa   IIIT Bangalore, India
Krithi Ramamritham   IIT Bombay, India
Sanjay Kumar Madria   Missouri University of Science and Technology, USA
Masaru Kitsuregawa   University of Tokyo, Japan
Raj K. Bhatnagar   University of Cincinnati, USA
Vasudha Bhatnagar   University of Delhi, India
Mukesh Mohania   IBM Research, Australia
H. V. Jagadish   University of Michigan, USA
Ramesh Kumar Agrawal   Jawaharlal Nehru University, India
Divyakant Agrawal   University of California, Santa Barbara, USA
Arun Agarwal   University of Hyderabad, India
Subhash Bhalla   The University of Aizu, Japan
Jaideep Srivastava   University of Minnesota, USA
Anirban Mondal   Ashoka University, India
Sharma Chakravarthy   The University of Texas at Arlington, USA
Sanjay Chaudhary   Ahmedabad University, India

Program Committee Chairs

Ladjel Bellatreche   ENSMA, France
Hamido Fujita   Iwate Prefectural University, Japan
Vikram Goyal   IIIT Delhi, India

Organizing Chair

Shelly Sachdeva   NIT Delhi, India

Publication Chair

Subhash Bhalla   The University of Aizu, Japan

Workshop Chairs

Sanjay Chaudhary   Ahmedabad University, India
Vasudha Bhatnagar   University of Delhi, India
Satish Narayana Srirama   University of Tartu, Estonia

Tutorial Chairs

Sanjay Kumar Madria   Missouri University of Science and Technology, USA
Punam Bedi   University of Delhi, India

Publicity Chairs

Manu Awasthi   Ashoka University, India
Sonali Agarwal   IIIT Allahabad, India
Rajiv Ratn Shah   IIIT Delhi, India

Panel Chair

Sharma Chakravarthy   The University of Texas at Arlington, USA

Website Chair

Samant Saurabh   IIM Bodh Gaya, India

Website Administrator

Raghav Mittal   Ashoka University, India

Program Committee

Satish Narayana   University of Tartu, Estonia
Naresh Manwani   IIIT Hyderabad, India
Jun Sasaki   Iwate Prefectural University, Japan
Praveen Rao   University of Missouri, USA
Nazha Selmaoui-Folcher   University of New Caledonia, France
Morteza Zihayat   Ryerson University, Canada
Tin Truong Chi   Dalat University, Vietnam

Ji Zhang	University of Southern Queensland, Australia
Himanshu Gupta	IBM Research, India
Jose Maria Luna	University of Córdoba, Spain
Chun Wei	Western Norway University of Applied Sciences, Norway
Srinath Srinavasa	IIIT Bangalore, India
Engelbert Mephu Nguifo	Université Clermont Auvergne, France
Uday Kiran	The University of Aizu, Japan
Sebastián Ventura	University of Córdoba, Spain
Prem Prakash Jayaraman	Swinburne University of Technology, Australia
Santhanagopalan Rajagopalan	IIIT Bangalore, India
Tzung-Pei Hong	National University of Kaohsiung, Taiwan
Amin Beheshti	Macquarie University, Australia
Andres Gerardo Hernandez-Matamoros	ESIME, Mexico
Rinkle Aggarwal	Thapar Institute of Engineering and Technology, India
Amin Mesmoudi	LIAS-Université de Poitiers, France
Sadok Ben Yahia	Tallinn University of Technology, Estonia
Sangeeta Mittal	JIIT, India
Carlos Ordonez	University of Houston, USA
Alok Singh	University of Hyderabad, India
Samant Saurabh	IIM Bodh Gaya, India

Sponsoring Institutions

National Institute of Technology Delhi, India
Indraprastha Institute of Information Technology Delhi, India
International Institute of Information Technology Hyderabad, India
The University of Aizu, Japan
University of Delhi, India
Ahmedabad University, Ahmedabad, India

Contents

Information Interchange of Web Data Resources

Business Analytics

Data Science: Systems

A Comparison of Data Science Systems

Carlos Ordonez[✉]

University of Houston, Houston, USA
carlos@central.uh.edu

Abstract. Data Science has subsumed Big Data Analytics as an inter-disciplinary endeavor, where the analyst uses diverse programming languages, libraries and tools to integrate, explore and build mathematical models on data, in a broad sense. Nowadays, there exist many systems and approaches, which enable analysis on practically any kind of data: big or small, unstructured or structured, static or streaming, and so on. In this survey paper, we present the state of the art comparing the strengths and weaknesses of the most popular languages used today: Python, R and SQL. We attempt to provide a thorough overview: we cover all processing aspects going from data pre-processing and integration to final model deployment. We consider ease of programming, flexibility, speed, memory limitations, ACID properties and parallel processing. We provide a unifying view of data storage mechanisms, data processing algorithms, external algorithms, memory management and optimizations used and adapted across multiple systems.

1 Introduction

The analytic revolution started with data mining [2], which adapted machine learning and pattern detection algorithms to work on large data sets. In data mining scalability implied fast processing beyond main memory limits, with parallel processing as a secondary aspect. Around the same time data warehousing [2] was born to analyze databases with demanding queries in so-called On-Line Analytical Processing [1]. During the last decade Data Warehouses evolved and went through a disruptive transformation to become Big Data Lakes, where data exploded in the three Vs: Volume, Velocity and Variety. The three Vs brought new challenges to data mining systems. So data mining and queries on Big Data lakes morphed into a catch-all term: Big Data Analytics [3]. This evolution brought faster approaches, tools and algorithms to load data, querying beyond SQL considering semantics, mixing text with tables, stream processing and computing machine learning models (broadly called AI). A more recent revolution brought two more Vs: Veracity in the presence of contradictory information and even Value, given so many development and tool options and the investment to exploit big data. Nevertheless, having so much information in a central repository enabled more sophisticated exploratory analysis, beyond multivariate statistics and queries. Over time people realized that managing so much diverse data required not only database technology, but also a more principled approach

© Springer Nature Switzerland AG 2020
L. Bellatreche et al. (Eds.): BDA 2020, LNCS 12581, pp. 3–11, 2020.
https://doi.org/10.1007/978-3-030-66665-1_1

laying its foundation on one hand in mathematics (probability, machine learning, numerical optimization, statistics) and on the other hand, more abstract, highly analytic, programming (combining multiple languages, pre-processing data, integrating diverse data sources), giving birth to Data Science (DS). This new trend is not another fad: data science is now considered a competing discipline to computer science and even applied mathematics.

In this survey article, we consider popular languages used in Data Science today: Python, R, and SQL. We discuss storage, data manipulation functions, processing and main analytical tasks across diverse systems including their runtimes, the Hadoop stack and DBMSs. We identify their strengths and weaknesses in a modern data science environment. Due to space limitations examples with source code and diagrams are omitted.

2   Infrastructure

2.1   Programming Language

The importance of the specific programming language used for analytic development cannot be overstated: it is the interface for the analyst, it provides different levels of abstraction, it provides different data manipulation mechanisms and it works best with some specific analytic task. There are two main families: interpreted and compiled. DS languages are interpreted, but the systems evaluating them have compiled languages behind, mostly C++ (and C), but also Java (Hadoop stack). Given the need to provide high performance the DS language offers extensibility mechanisms that allow plugging in functions and operators programmed in the compiled language, but which require expertise on the internals of the runtime system. The average analyst does not mess with the compiled language, but systems researchers and developers do. Finally, SQL is the established language to process queries in a database and to compute popular machine learning models and data mining techniques. In a similar manner to DS languages, it is feasible to extend the language with advanced analytic capabilities with UDFs programmed in C++/C and a host language (Java, Python).

2.2   Storage Mechanisms

In this section we attempt to identify the main storage mechanisms used today: arrays, data frames and tables. Notice each programming language offers these mechanism is different flavors as well as with different capabilities and constraints.

An array is a data structure provided by a programming language and it is arrangement of elements of the same data type. An array can represent a table as an array of records and a matrix as a 2-dimensional array of numbers. In most programming languages, like C++ or Java, the array is a data structure in main memory, although there exist systems (SciDB [5]) which allow manipulating arrays on disk. The data frame is another data structure in main memory, somewhat similar to an array. Data frame was introduced in the R language, allowing to assemble rows or columns of diverse data types, which has made its way into Python in the Pandas library. It has similarities to a relational table, but it is not a relational table. When the data type of all entries is a number data frames resemble a matrix, but they are not a matrix either. However, it easy to transform a data frame with numbers into a matrix and vice-versa. A table is a list of records, which can be stored as arrays in main memory or as blocks (small arrays) on disk. By adding a primary key constraint a generic table becomes a relational table [1]. A relational table is a data structure on secondary storage (disk, solid state), allowing manipulation row by row, or block by block. A record is generally understood as physical storage of a tuple, following a specific order and each value taking fixed space. Thus, by design, a relational table is very different from a data frame. Data frames and arrays are somewhat similar, but data frames track row and column names and they do not require contiguous storage in main memory.

The I/O unit is an important performance consideration to process a large data set, which varies depending on the system: line by line or an entire file for a data frame. One line at a time for text files, one record or block of records for tables. DS systems and libraries vary widely on how they read, load and process data sets. Arrays vary according to their content; in 2-dimensional format of numbers they are read in one pass, as one block for a small array or multiple blocks for a large array.

2.3 Physical Primitive Access Operators

We now attempt to identify the main access primitive operators for each storage mechanism.

All systems can process text and binary files, but tend to favor one. DS languages generally work on text files, which can be easily transferred and exchanged with other analytic tools (spreadsheets, editors, word processors). DBMSs use binary files in specific formats, doing a format conversion when records are inserted by transactions or when they are loaded in batch from files. Hadoop system use a combination of both, ranging from plain files to load data to transforming data into efficient block formats as needed in a subsystem.

Large arrays are generally manipulated with blocked access on binary files, where each block is an array. Blocks for dense matrices are generally squared. Few systems attempt to provide subscript-based access on disk because it requires adding special functions and constructs in the programming language.

Data frames are generally loaded with one scan on the input text file, but there exist libraries that allow block by block access (called chunk to distinguish them from database blocks). Therefore, the access is sequential.

Each programming language and system incorporate different access operators, but here we provide a broad classification. In a DS language these are the most common operators in main memory: scan (iterator), sort, merge. Iterators come from object-oriented programming. The merge operator is similar to a relational join, but more flexible to manipulate dirty data. DBMSs feature database operators combining processing in main memory and disk: scan, sort, join; filters are processed with a full scan when there are no indexes, but they can be accelerated with indexes when possible. Sorting can be used at many stages to accelerate further processing.

3 Adding New Data and Updating Old Data

3.1 Input Data Format

It is either text or binary. The most common input file format are text files, with CSV format being a de facto standard. The CSV format allows representing matrices, relational tables and data frames with missing values and strings of varying length. Binary formats are more common in HPC and specific science systems (physics, chemistry, and so on). Streams come in the form of ever growing log files (commonly CSV), where each log record has a timestamp and records that can vary in structure and length. JSON, a new trend, is a more structured text format to import data, which can represent diverse objects, including documents and database records.

3.2 Copying Files vs Loading Data

In most data science languages there is no specific loading phase: any file can be analyzed directly by reading it and loading in the storage mechanisms introduce above in main memory. Therefore, copying a file to the DS server or workstation is all that is required. On the other hand, this phase represents a bottleneck in a database system and a more reasonable compromise in Hadoop systems. In a Hadoop system it can range from copying text files to formatting records in a specific storage format (key-value, Hive, Parquet). In a database system input records are generally encoded into a specific binary format, which allows efficient blocked random access and record indexing. A key aspect in a parallel system, is partitioning the input data set and distributing it, explained below.

3.3 Integrating Diverse Sources

Data science languages (Python, R) provide significant freedom to integrate data sets as needed. They have libraries to merge data sets by a common columns(s), behaving in a similar manner to a relational join. Python is more flexible than R to manipulate text (words, sentences). R provides better defined operators and functions to manipulate matrices. In the absence of columns with overlapping content, data integration is difficult. In a Hadoop system there exist libraries that attempt to match records by content, which is especially challenging with strings. The most principled approaches come from the database world, where records can be matched by columns or by content, but they are generally restricted to tabular data (i.e. relational tables).

3.4   Maintaining Data Integrity: ACID

In data science languages, like Python and R, there is a vague notion of ACID properties, which tends to be ignored, especially with text data (documents, web pages). When there are significant data additions or changes, the data preparation pipeline is re-executed and the analytical tasks repeated with a refreshed data set. Hadoop systems have been gradually strengthened with more ACID properties, especially to explore data sets interactively. It is noteworthy ACID properties have gradually subsumed eventual consistency. DBMSs provide the strongest ACID guarantees, but they are generally considered a second alternative after Hadoop big data systems to analyze large volumes of data with machine learning or graph algorithms. DBMSs forte is query processing. In general, most DBMSs propagate changes on queries recomputing materialized views. As there is more interest in getting near real-time (active) results, but maintaining consistency, the underlying tables are locked or maintained with MVCC.

4   Processing

4.1   A Flexible and Comprehensive Processing Architecture

It is difficult to synthesize all different DS pipelines into a single architecture. Here we list the most common elements: diverse data sources, a data repository, an analytic computer or cluster. The tasks: loading, cleaning, integrating, analyzing, deploying. Data sources include: databases, logs, documents, perhaps with inconsistent and dirty content. The data repository can be a folder in a shared server or a data lake in parallel cluster. Cleaning and integrating big data is generally done combining Hadoop and DBMSs: it just depends where the bulk of information is. Some years ago the standard was to process big data in a local parallel cluster. Cloud computing [6] has changed the landscape, having data lakes in the cloud, which enables analysts to integrate and clean data in the cloud. In general a data set for analysis is much smaller than raw big data, which allow analysts to process data locally, in their own workstation or server.

4.2   Serial Processing

This is the norm in data science languages. It is easier to develop small programs without worrying about low-level details such as multicore CPUs, threads and network communication. In general, each analyst runs in their own space, without worrying about shared memory. In a large project, data integration and preparation is assumed to have been take care of before.

4.3   Parallel Processing

This is the turf of Hadoop and parallel DBMSs, which compete with each other [4]. In most systems parallel processing requires three phases: (1) partitioning data and distributing chunks/blocks; (2) running processing code on each partition; (3) gathering or assembling partial results. Phase 1 may require sorting, automatically or manually and hashing records by a key. In general, the goal is to run independently on each partition and minimize coordination. There is a debate between scaleup (more cores, more RAM) and scale out (more machines, less RAM on each). In DS languages there is automatic parallelism in the object code which is optimized for modern multicore CPUs. Multicore CPUs, larger RAM and SSDs favor doing analysis in one beefy machine. On the other hand, big data, especially "Volume" support distributed processing to overcome the I/O bottleneck. Streams, being sequential data, lead to distributed filtering and summarization, but commonly centralized processing of complex models. Hadoop and DBMSs offer a parallel version of scan, sort and merge/join. However, Hadoop systems trail DBMSs on indexing and advanced join processing, but the gap keeps shrinking. When there are indexes, either in main memory or secondary storage filtering, joining and aggregations can be accelerated.

4.4   Fault Tolerance

There are two main scenarios: fault tolerance to avoid data loss and fault tolerance during processing. This is a contrasting feature with data science languages, which do not provide fault tolerance either to avoid data loss or to avoid redoing work when there are runtime errors. In many cases, this is not seen as a major limitation because analytic data sets can be recreated from source data and because code bugs can be fixed. But they represent a time loss. On the other hand, Hadoop systems provide run-time fault tolerance during processing when one node or machine fail, with automatic data copy/recovery from a backup node or disk. This feature is at the heart of the Google file system (proprietary) and HDFS (open source). Parallel DBMSs provide fault tolerance to avoid losing data, going from secondary copies of each data block to the internal log, where all update operations are recorded, It is generally considered that continuously appending the recovery log is required for transactions, but an overhead for analytics.

5   Analytics

5.1   Definitions

We introduce definitions for the main mathematical objects used in data science. We use matrix, graph and relational table as the main and most general mathematical objects. Graphs can represent cubes and itemsets as a particular case.

The most important one is the matrix, which may be square or rectangular. A matrix is composed of a list of vectors.

A tuple is an n-ary list of values of possibly diverse types, accessed by attribute name. A specific tuple in a relational table is accessed by a key. Finally, bags and sets contain any kind of object without any pre-defined structure, where elements are unique in sets and there may be repetitions in bags.

It is important to emphasize these mathematical objects are different. A matrix is different from a relational table. In a matrix elements are accessed by subscript, whereas in a relational table by key and attribute name. In an analogous manner, a vector is different from a tuple. Relational tables are sets of uniform objects (tuples), but sets in general are not. Then bags represent generic containers for anything, including words, files, images and so on. We will argue different systems tend to be better for one kind of object, requiring significant effort the other two.

A graph $G = (V, E)$ consists of a set of vertices V and a set of edges E connecting them. Graph size is given by $n = |V|$ and $m = |E|$, where $m = O(n)$ for sparse graphs and $m = O(n^2)$ for dense graphs, with a close correspondence to sparse and dense matrices. A graph can be manipulated as a matrix or as a list of edges (or adjacent vertices).

5.2 Exploration

A data set is commonly explored with descriptive statistics and histograms. These statistical mechanisms and techniques give an idea about data distribution. When a hypothesis comes up it is common to run some statistical test whose goal is to reject or accept a user-defined hypothesis. In general, these analyses are fast and they scale reasonably well on large data sets in a data science language, especially when the data sets fits in RAM. Together with statistics Data Science systems provide visualization aids with plots, mesh grids and histograms, both in 2D and 3D. When interacting with a DBMS, analysts explore the data set with queries, or tool that generate queries. In general queries are written in SQL and they combine filters, joins and aggregations. In cube processing and exploration, one query leads to another more focused query. However, the analyst commonly exports slightly pre-processed data sets outside the DBMS due to the ease of use of DS languages (leaving performance as a secondary aspect). Lately, data science languages provide almost equivalent routines to explore the data set with equivalent mechanisms to queries. Out of many systems out there, the Pandas library represents a primitive, but powerful, query mechanism whose flexibility is better than SQL, but lacking important features of query processing.

5.3 Graphs

The most complex exploration mechanism is graphs, which are flexible to represent any set of interconnected objects. Nowadays, they are particularly useful to represent social networks and interconnected computers and devices on the Internet. Their generality allows answering many exploratory questions, which are practically impossible to get with descriptive statistics. Even though graphs represent a mathematical model they can be considered descriptive models rather than predictive. Nevertheless, many algorithms on graphs are computationally challenging, with many of them involving NP-complete problems or exponential time and space complexity. Well-known problems include paths, reachability, centrality, diameter and clique detection, most of which remain open with big data. Graph engines (Spark GraphX, Neo4j) lead, followed by parallel DBMSs (Vertica, Teradata, Tigergraph) to analyze large graphs. DS language libraries do not scale well with large graphs, especially when they do fit in main memory.

5.4   Mathematical Models

Machine learning and statistic are the most prominent mathematical models. We broadly classify models as descriptive and predictive. Descriptive (unsupervised) models are a generalization of descriptive statistics, like the mean and variance, in one dimension. For predictive (supervised) models there is an output variable, which makes the learning problem significantly harder. If the variable is continuous we commonly refer to it as Y, whereas if it a discrete variable we call it G. Nowadays machine learning has subsumed statistics to build predictive models This is in good part due to deep neural networks, which can work on unlabeled data, on text and on images. Moreover, deep neural networks have the capability of automatically deriving features (variables), simplifying data preprocessing. Generally speaking these computation involve iterative algorithms involving matrix computations. The time complexity goes from $O(dn)$ to $O(2^d n)$ and $O(dn^2)$ in practice. But it can go up to $O(2^d n)$ for variable/feature selection or Bayesian variable networks. The number of iterations is generally considered an orthogonal aspect to $O()$, being a challenge for clustering algorithms like K-means and EM. Data science language libraries excel in the variety of models they and the ability to stack them. Models range from simple predictive models like NB and PCA to deep neural networks. It is fair to say Python dominates the landscape in neural networks (Tensorflow, Keras Scikit-learn) and R in advanced statistical models. Hadoop systems offer specialized libraries to compute models like Spark MLlib and Mahout. The trend has been to build wrappers in a DS language, where Python is now the dominant DS language Spark, leaving Scala for expert users. DBMSs offer some algorithms programmed via SQL queries and UDFs, fast cursor interfaces, or internal conversion from relational to matrix format, but they are difficult to use, and they cannot be easily combined and they require importing external data.

Given DBMS rigid architecture, and learning curve, users tend to prefer DS languages, followed by Hadoop with data volume forces the analyst to use such tool. That is, DBMSs have lost ground as an analytical platform, being used mainly for some strategic queries and in a few cases machine learning. In practice, many models are built on samples or highly pre-processed data sets where d and n have been reduced to manageable size. Building predictive models on big data is still an open problem, but it is a moving target towards DS languages.

6   Conclusions

Python and R are now the standard programming languages in Data Science. Clearly, Python and R are not the fastest languages, nor the ones that guarantee

ACID, but they are becoming more popular and their libraries faster. Among the two Python user base is growing faster, but R's vast statistical libraries (CRAN) will maintain its relevance for a long time. However, SQL remains necessary to query databases and extract and pre-process data coming from databases. Transactions remain an important source of new data, but non-transactional data is growing faster. Data warehousing is now an old concept, which has been substituted by so-called data lakes. Cube processing and ad-hoc queries are still used in specialized data warehouses, but they have been subsumed by exploratory statistical analysis in big data. Hadoop big data systems remain relevant for large volume and the cloud. Both Hadoop systems and DBMSs interoperate with Python and R, at different levels depending on the specific analytic system. It is fair to say CSV text files and JSON are now standard formats to exchange big data, leaving behind proprietary formats.

Data Science is opening many possibilities for future research, adapting and extending the state of the art in database systems, programming languages, data mining and parallel computing. Analysts require tools that are reasonably fast, intuitive and flexible. There should be seamless mechanisms to convert data formats, and avoid it when possible. Many research and tool developers are moving beyond the "fastest system" mentality, given hardware advances and the cloud, and they are focusing instead on reducing development time for the analyst. We need analytic algorithms and techniques that can interoperate and exchange data easily with relaxed structure assumptions.

References

1. Garcia-Molina, H., Ullman, J.D., Widom, J.: Database Systems: The Complete Book, 2nd edn. Prentice Hall, Upper Saddle River (2008)
2. Han, J., Kamber, M.: Data Mining: Concepts and Techniques, 2nd edn. Morgan Kaufmann, San Francisco (2006)
3. Ordonez, C., García-García, J.: Managing big data analytics workflows with a database system. In: IEEE/ACM CCGrid, pp. 649–655 (2016)
4. Stonebraker, M., et al.: MapReduce and parallel DBMSs: friends or foes? Commun. ACM **53**(1), 64–71 (2010)
5. Stonebraker, M., Brown, P., Zhang, D., Becla, J.: SciDB: a database management system for applications with complex analytics. Comput. Sci. Eng. **15**(3), 54–62 (2013)
6. Zhang, Y., Ordonez, C., Johnsson, L.: A cloud system for machine learning exploiting a parallel array DBMS. In: Proceedings of the DEXA Workshops (BDMICS), pp. 22–26 (2017)

Architectural Patterns for Integrating Data Lakes into Data Warehouse Architectures

Olaf Herden[(✉)]

Department of Computer Science, Baden-Wuerttemberg Cooperative State University,
Florianstr. 15, 72160 Horb, Germany
`o.herden@hb.dhbw-stuttgart.de`

Abstract. Data Warehouses are an established approach for analyzing data. But with the advent of big data the approach hits its limits due to lack of agility, flexibility and system complexity. To overcome these limits, the idea of data lakes has been proposed. The data lake is not a replacement for data warehouses. Moreover, both solutions have their application areas. So it is necessary to integrate both approaches into a common architecture. This paper describes and compares both approaches, shows different ways of integrating data lakes into data warehouse architectures.

Keywords: Data warehouse · Data lake · Big data · Architecture · Advanced analytics

1 Introduction

Data warehouse architectures have been established as standard for decision support. They work well for reporting, OLAP (Online Analytical Processing) and some other applications like dashboarding or scorecarding. But requirements have changed: In the era of big data large volumes of poly-structured data have to be analyzed quickly. So different modifications respectively extensions of the analytical architecture were proposed. One important suggestion is the concept of data lakes. This concept and its integration into a data warehouse system is handled in detail in this contribution.

The reminder of the paper is organized as follows: After revisiting data warehouse systems in Sect. 2, in Sect. 3 we introduce big data and show the limits of data warehouses to fulfill analytical tasks in this area. After introducing data lakes in Sect. 4, we give an in-depth comparison between data warehouse systems and data lakes in Sect. 5. Different architectural patterns for the integration of both approaches are given in Sect. 6, these approaches are evaluated in Sect. 7. The paper closes with a summary and an outlook.

2 Data Warehouse Systems

A typical DWS (data warehouse system) (or synonym: data warehouse reference architecture) [1–5] is depicted in Fig. 1.

© Springer Nature Switzerland AG 2020
L. Bellatreche et al. (Eds.): BDA 2020, LNCS 12581, pp. 12–27, 2020.
https://doi.org/10.1007/978-3-030-66665-1_2

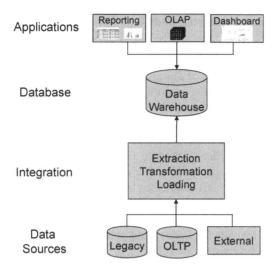

Fig. 1. Data warehouse system

At the lowest layer we can see the data sources. Typically, most of them are internal databases, e.g. legacy systems or inhouse OLTP (Online Transaction Processing) systems. These internal sources are enriched by some external data. All source data is passing through an integration process, often denoted as ETL (extraction – transformation – loading) process to get a uniform way of data.

The persistence layer is realized by the DWH (data warehouse) itself, a subject-oriented, integrated, time-variant, nonvolatile collection of data [3]. The data warehouse has a predefined schema, in most cases a star or snowflake schema [6]. The DWH operates as single point of truth, i.e. every data analytics is based on this database.

On top of the data warehouse the classical analytical tasks of reporting and OLAP (Online Analytical Processing) can be implemented [7]. While reports offer predefined, parametrized analysis, OLAP is more explorative and flexible. It follows the cube paradigm. This means we have a multidimensional view on the data and the user can explore the data by rolling up and drilling down within dimensions, slicing and dicing and executing rotations.

Over the years, the application layer has been enriched by several additional applications, e.g. dashboards and scorecards or data mining applications. For the extended architecture sometimes the term BI (business intelligence) is used [8, 9].

All kind of application can operate directly on the DWH (like depicted in Fig. 1) or with an intermediate server component.

3   Big Data and Limits of Data Warehouses

First, this section describes the term big data and mentions some application scenarios. Afterwards we give reasons why the classical DWS is not sufficient for nowadays data analytics.

3.1 Big Data, Advanced Analytics and Data Scientists

Over the last few years, in organizations a rapid development in data management has been observed. Due to new sources like ubiquitous social media, open data, machine generated data from IoT (Internet of Things) environments the volume of data is increasing steadily. Over 90 percent of the world's data was produced in just the past two years [10]. These new sources deliver data not merely in a structured manner, but rather in many heterogeneous formats. Data can be structured, semi-structured (like emails or JSON (JavaScript Object Notation) documents) or unstructured (like texts) or binary (like photos, music, videos). Finally, modern analytics needs a new speed. The data flow process within a DWS is relatively slow. The data has to be integrated and loaded into the data warehouse. Afterwards, the data can be analyzed and reports can be generated. In modern environments there is a need for faster use of the data.

In summary, these aspects are referred to as big data. Big data is characterized by the three Vs concept for volume, variety and velocity [11, 12]. In the meantime this model has been extended by three more Vs [13]: value (aiming to generate significant value for the organization), veracity (reliability of the processed data) and variability (the flexibility to adapt to new data formats through collecting, storing and processing) [14].

Besides reporting, OLAP and some BI techniques mentioned above, there is a new category of applications, referred to as advanced analytics. Since there is no common definition of the term advanced analytics, in this context we define it as a summary of machine learning techniques [15, 16]. They can either be supervised or unsupervised. Examples for supervised approaches are neural networks, decision trees or regression analysis. Clustering, principal component analysis and association analysis are examples for unsupervised techniques.

While data warehouses are typically used by some data analysts, decision makers and report consumers, these days there are also new professional fields like data or business analysts or data scientists. Moreover, even apps and machines consume the data for further actions.

Data scientists have the job to extract knowledge and give insights into all the data described above. Therefore data scientists need strong skills in statistics, machine learning, programming and algorithms [17, 18].

3.2 Limits of Data Warehouse Systems

All the requirements and challenges of managing big data do DWSs fulfill partially only. The main drawbacks of DWS in this respect are [19]:

- Lack of agility: While reporting is relatively fixed and OLAP applications show a certain flexibility, a typical feature of advanced analytics is the "need for ad-hoc, one time on demand analysis" [19]. This can not be fulfilled by DWSs.
- Structured data only: Most DWHs are realized with relational technology. This allows the storage of structured data only and is not sufficient to serve the variety of big data.
- Hard to extend: If there are new data sources or a change in a known data source, the systems' extension is a comprehensive task. The schema of the DWH has to be altered as well as the ETL processes have to be adopted.

- Performance: Many DWHs have long response times for complex SQL (structured query language) queries. Although various techniques like advanced indexing. in memory computation or column stores have been implemented, the performance problem can only be mitigated but not solved.
- ETL process as bottleneck: Because all data has to pass the complete ETL process there is a latency before the data is available for use.
- High hardware cost: Due to increasing data volume, database servers have to be extended by adding additional hardware, especially hard disks, main memory and processors. In the length of time this is cost intensive.

All in all, we can see that DWSs are not able to fulfill the requirements of handling big data and implement advanced analytics.

4 The Concept Data Lake

This section introduces the term data lake. It follows [20–23]. After giving a definition, we sketch some use cases. In the last subsection we show what kind of tools are necessary and give some examples.

4.1 Definition

Up to now there is no common accepted definition of the term data lake. We define a data lake as a collection of components and tools for storing, administrating and provisioning data for analytic use cases. The data lake stores the data in its original format. Therefore, a data lake can capture data in many formats: structured data from relational databases, semi- and unstructured data like defined in Subsect. 3.1.

Within this contribution, a data lake has an internal structure like depicted Fig. 2.

The first component is the raw data store, also called landing zone, where raw data is stored as an one-to-one copy from its source. The process of transferring data from the sources to the landing zone is called data ingestion. Ingestion should be possible at any rate using arbitrary import methods like initial load, batch processing or streaming of source data [24].

The next stage is the data preparation process (synonyms are standardization zone or data refinery) where the data is profiled. The aim of this process step is the generation of knowledge and producing an understanding of the data with respect to standardization, cleansing and summarization. Even different profiles for the same raw data depending on the analytic task can be produced in the preparation task.

The data profiles are stored in the processed data store. Here it can be either used directly for downstreaming for consumption by advanced analytics applications or can be forwarded to the analytics sandbox.

The analytics sandbox works as a data lab, an environment where data scientists do their work. They do research for creating new models, exploring new ideas, etc. In a sandbox this can be done without impairing productive processes.

Here we can see that the data lake architecture realizes the ELT (extraction – loading – transformation) paradigm. The data is extracted from the sources, is loaded into the

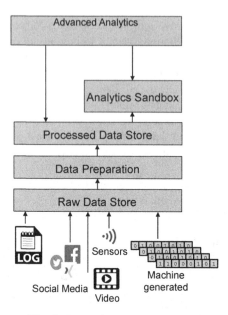

Fig. 2. Internal structure data lake

processed data store and the transformation is done before using the data. This offers more flexibility than the standardized pre-defined ETL process in a DWS.

A very important point is an appropriate management of the data lake. This data lake management should contain data governance, security and metadata management.

Data governance [25–27] is well known since years especially in the context of DWS and should ensure a high data quality by documenting the complete lifecycle of data. In the context of data lakes [31] this especially means that the raw data store needs a data catalog where the origin and transfer date are documented. Moreover, the data lineage should be documented well to understand where the data come from. Also the ELT process described former should be documented well to permit reuse. And finally, the usefulness of the models and analytic results should also be documented by end users and data scientists.

Beside data governance it is important to apply security [28–30] and privacy to the data lake, too. All aspects of privacy and legal concerns have to be considered within the data lake. Especially, security becomes important with respect to the GDPR (General Data Protection Regulation) [32] which was implemented in 2018 and is a regulation on data protection and privacy for all individuals in the European Union. Particularly, the storage of any individual-related data is permitted only for a specific purpose. Moreover, the GDPR implements the right to be forgotten and the principle of data economy. All these aspects are contradictory to the data lake's core idea to store data first and to use them later. Possible solutions for overcoming these problems are anonymization and pseudonymization.

Governance as well as security and data moving processes generate a lot of metadata. Therefore, a data lake should contain a metadata repository. Ideally, this repository is realized centrally spanning over all components of the data lake. In reality, often the metadata is stored decentralized at the different stages.

We can distinguish three metadata categories: technical, operational, and business. Technical metadata describe structure and format of each data set as well as its source. Operational metadata capture the lineage, quality and profile of data. Moreover, information about the transfer time into the data lake or failed loading processes are stored. Business metadata covers all business definitions and contexts.

The absence of metadata management in a data lake keeps some dangers. The use of the lake might be limited because only a small group of users is able to work with the data. Moreover, the integration of the data lake into the enterprise architecture is more complicated by this fact. There is the danger of an isolated existence of the data lake, denoted as data silo. Sometimes, these effects are referred to as data swamp [23] or garbage dump [21].

4.2 Use Cases

There are several use cases for data lakes, e.g. [33] describes multichannel marketing, fraud detection and predictive maintenance and logistics as real world examples. In multichannel marketing a mix of old and new data from websites, call center applications and social media give an insight about customers behavior in different situations. This knowledge can be used for cross-selling, up-selling or acquisition. By combining data from more and different sources we can get larger data sets which can be used for fraud or risk detection. In industrial production sensor data give a lot of information about the machine, certain patterns indicating a breakdown in the nearest future can be detected. This finding can be used for predictive maintenance. Also in the logistics industry sensors can be widely used, e.g. in combination with spatial coordinates delivery routes and delivery times can be reduced.

4.3 Tools

Although the market of software libraries, tools and frameworks is a very volatile one, we want give a brief overview about the tool ecosystem for a data lake.

For the ingestion process step scripting or classical integration tools fail because they do not fulfill the requirement of massive parallel transfer of data, hundreds or even thousands of sources are possible, high scalability and support and integration of many platforms is a necessary requirement. Moreover, the dataflow process should be automated, managed (i.e. documentation and detection of failures during transfer) and visualized. Popular tools for ingestion are Apache Kafka [34–36] for data streams and Apache Sqoop [37–39] for bulk loads.

A raw store has to capture last amounts of data with high throughput and should also be fault tolerant. The most popular system is HDFS (Hadoop Distributed File System) [40, 41], a distributed file system operating on low cost hardware. The last aspect ensures high scalability. Even if HDFS is the quasi standard as raw zone storage there are some complete alternatives like the MapR file system [42] or additions like Ozone [43].

For handling the task of data refinery there are two kind of tools, the class of micro-batch operating and the one of stream-based frameworks. Representatives of the latter class are Apache Flink [44, 45] and Storm [46, 47], for the former one Hadoop MapReduce [38, 48] and Apache Spark [49, 50].

For storing the refined data in the processed data store besides the same storages as in raw data store NoSQL databases (Not only SQL) [51–53] are in line.

Typical tools for data analysts are integrated commercial suites like SAS [54] or open source programming languages or libraries like R [55] or Python [56].

As a summary, we can see that tools from Apache projects dominate the market, but there are also alternatives. A principal alternative is realizing the data lake in the cloud. The most popular solutions in this field are Microsoft Azure [57] and AWS (Amazon Web Services) [58].

5   Comparison

In this section we want to contrast data warehouses with data lakes. This comparison is done by dividing into the topics data, schema, database, users and processes.

In a data warehouse only data necessary for analytics is stored, it is typically structured, integrated and cleansed. In a data lake architecture all data generated by the sources is stored, it can have arbitrary kind of structure and are raw and refined. Data in a data warehouse are stored in a schema designed at systems' built-time. The data lake has no schema, or rather the schema is given by the source. The volume of a data warehouse is large (up to petabyte dimension), the growth is moderate and calculable in most cases. After a certain period data often becomes less interesting for reporting, so it can be archived, often an aggregate is stored. Volumes in a data lake can be orders of magnitude higher, it is rapidly growing and in principle reside in the lake forever. Due to the

Table 1. Data Warehouse vs. Data Lakes (Data)

Criterion	Data warehouse	Data lake
Data	Cleansed, integrated	Raw, refined
Model	At design time (star schema)	No model
Duration of stay	Aggregated after a while, archived	Stay "forever"
Which data?	Necessary for analytics	All
Types	Structured	Structured Semi-structured Unstructured Raw
Volume	"Large"	Quasi unlimited
Growth quantity	Moderate, calculated	Rapidly growing
Quality	High	Not guaranteed
Governance	Well established	Has to be implemented

well defined integration process in data warehouses data governance is well established and guarantees high quality of the data. On the contrary, in data lakes governance has to be realized explicitly and data quality is not guaranteed. Table 1 wraps up these facts.

The data warehouse has a structured schema, usually a star or snowflake schema, this is modelled during building the system. The schema is quality assured but lacks with respect to agility. New requirements mean altering the schema. No modeling is applied for data lakes, the schemas are given by the source systems. For this reason there is a high degree of agility because new or altered sources can be used directly. Table 2 summarizes the schema properties.

Table 2. Data Warehouse vs. Data Lakes (Schema)

Criterion	Data warehouse	Data lake
Time of modeling	During systems built-time	No modeling
Type	Structured (star or snowflake)	Unstructured
Degree of agility	Low	High

Data warehouses typically run on relational DBMS (database management system). With respect to the large data volumes high performance servers are necessary, especially storage costs are high with increasing volumes. Relational database servers follow the scale-up paradigm. This implies middle to high costs for scaling and a principal limit of scaling. In contrast data lakes store the data on Hadoop or NoSQL databases. These systems are based on low cost common hardware. Since these databases realize the scale-out approach, scaling is not expensive and there is quasi no limit in scaling. Security is very mature in data warehouses because relational DBMS offer access roles, in data lakes there is no built-in security mechanism. While data warehouses are accessed by SQL, in most cases generated by reporting or BI tools, the access to data lakes can be done by SQL, arbitrary programs or big data analytic tools. Table 3 shows the database properties.

Table 3. Data Warehouse vs. Data Lakes (Database Properties)

Criterion	Data warehouse	Data lake
Technology	Relational DBMS	Hadoop, NoSQL
Hardware	(High performance) server	Connected common computers
Storage	Expensive for large volumes	Low cost
Scale paradigm	Scale-Up	Scale-Out
Scale Cost	Middle/High	Low
Access	SQL, BI tools	SQL, programs, big data analytic tools
Security	Very mature	Less mature

Users in a DWS are business analysts and decision makers. They access the data via predefined reports or in an explorative manner by using OLAP tools. There exists an integrated view on the data and the access paradigm is on-write. In data lakes there is the additional user role data scientist, data are used for advanced analytics and explorative research. Users can see all (raw or refined) data, the access is on-read.

Table 4 summarizes the user properties.

Table 4. Data Warehouse vs. Data Lakes (User Properties)

Criterion	Data warehouse	Data lake
Roles	Business analysts, decision makers	Business analysts, decision makers, data scientists
Usage	Standard reports, explorative (OLAP)	Advanced analytics, explorative, research
View	Integrated data	All (raw) data
Access paradigm	Access limited by DWH structures (on-write)	Original format allows transformation after storing (on-read)

Ingestion of data lakes occurs permanently, loading the data warehouse periodically. While the latter follows the ETL process, the former is organized in ELT manner. As a consequence, the most complex process step in a DWS is the integration, in a data lake the processing of data. This offers high agility, data can be configured and reconfigured as needed, while the data warehouse is less agile, the integration process is fixed.

Table 5 shows the process properties.

Table 5. Data Warehouse vs. Data Lakes (Process Properties)

Criterion	Data warehouse	Data lake
Ingestion/Loading	Periodically	Permanently
Integration	ETL	ELT
Agility	Less agile, fixed configuration	Highly agile, (re)configure as needed
Most complex process step	Integration	Processing

All in all, we can see that both approaches have different aims and different pros and cons. So the data lake can not work as replacement of the data warehouse. It rather is an important addition for building a modern analytics architecture.

6   Integrating a Data Lake into the Data Warehouse Architecture

As we have seen in the last section, both approaches have different orientation and both are necessary in a modern analytical information system architecture. Therefore, we discuss different feasibilities how to integrate data lakes into the DWS.

6.1 Sequential Approach

The first model is the sequential approach. The data lake and the data warehouse are used in sequence. In other words everything has to pass the data lake and can then be used directly or serves as input for the ETL process. Figure 3 shows this approach.

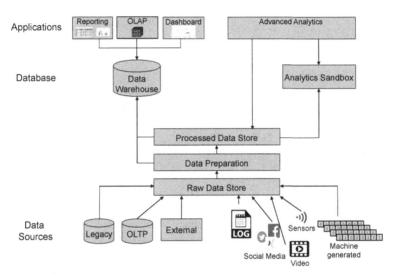

Fig. 3. Sequential architecture

In this architecture the data lake replaces the integration layer of the data warehouse architecture and serves as the single point of truth. The schema-on-read paradigm can be hold. The drawback of this approach is the reimplementation of all integration processes into the data warehouse. In established data warehouse environments this is very cost intensive and therefore not very feasible.

6.2 Parallel Approach

In the parallel approach the data warehouse and the data lake both use the same sources and work in parallel independently. Figure 4 sketches this approach.

The advantage in this pattern is the fact that the DWS keeps untouched. The implementation effort is restricted to the realization of the data lake, the right part of Fig. 4. This architectures' big drawback is the missing integration of DWS and data lake. On the one hand, the applications on top of the data warehouse do not profit from the data

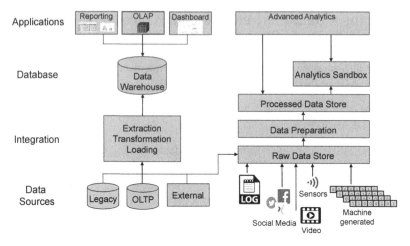

Fig. 4. Parallel architecture

lake. On the other hand, all data from the data warehouse is also in the data lake, but the copy process is placed before the integration step. So results of e.g. reports from data warehouse and an advanced analytics tool consuming data from the data lake are hardly comparable.

6.3 DWH Offloading

The architectural pattern data warehouse offloading is characterized by the process step of transferring data from the data warehouse back to the data lake's raw data store. This process is called offload following the word usage in logistics.

Typical data for the offloading process is cold data defined as data that is not frequently accessed or actively used and consolidated master data like account systems or product catalogues.

This architecture improves the parallel approach by a unidirectional integration of DWS and data lake. But there is still no single point of truth, outputs from DWS and advanced analytic tools can still show different results and give rise to questions (Fig. 5).

6.4 Hybrid Approaches

Hybrid approaches are a combination of the data warehouse offloading architecture from the former subsection and either the sequential approach from 6.1 or the parallel approach from 6.2.

The first combination possibility is depicted in Fig. 6. The main characteristic of the sequential approach (the core data lake replaces the ETL process) is extended by the backcoupling step of offloading the data warehouse into the raw data store.

Alternatively, the data warehouse offloading can be combined with the parallel architectural pattern. This strategy is depicted in Fig. 7. Beside the data warehouse offloading into the raw data store, we also have an onloading process from the data lakes' core as

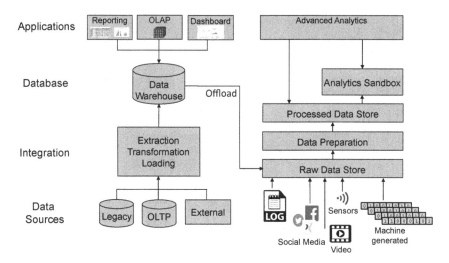

Fig. 5. DWH offloading

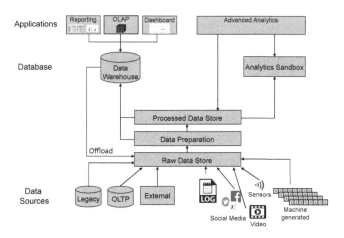

Fig. 6. Hybrid approach: combination sequential and offload

input for the ETL process of the data warehouse. This architecture combines the advantages of both basic approaches and delivers a complete integration of DWS and data lake.

Both hybrid approaches combine the features of their basic models. The combination of the sequential approach with offloading brings a single point of truth but has still the drawback of losing the established DWS architecture and integration process. Combining the parallel architecture with offloading overcomes this disadvantage. On the other hand, this approach does not fulfill the single point of truth idea completely because it depends on the realization of the offloading and onloading steps.

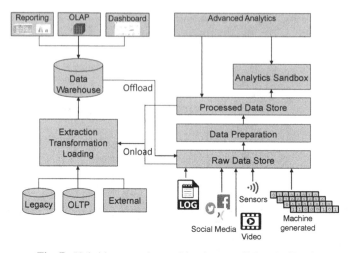

Fig. 7. Hybrid approach: combination parallel and offload

7 Evaluation

This section summarizes the evaluations made in the former section. Table 6 shows the main criteria and different approaches. The architectural patterns are referred to the number of the subsection where they are introduced.

Table 6. Evaluation of architectural patterns

Criterion	6.1	6.2	6.3	6.4.1	6.4.2
Single point of truth	+	−	−	+	o
Schema-on-read	+	+	+	+	+
Preserving DWS	−	+	+	−	+
Cost	High	Low	Middle	High	Middle
Integration of DWS and data lake	+	−	o	+	++

Each criterion is valuated by a value from the scale ++/+/o/−. The criterion cost is rated by the relative scale high/middle/low.

As we can see solution 6.4.2 (combining parallel and offload from Fig. 7) has the most advantages. It preserves the existing DWS structure, especially the complex integration process, and offers the agility of the data lake approach. Both worlds are well integrated on the condition that the offload and onload job are realized properly.

But there might also be use cases where one of the other solutions has it right to exist. E.g. when data for reporting and data for advanced analytics are completely disjoint, the parallel approach can be the best solution.

8   Summary and Outlook

This contribution handled data warehouse, data lakes and their integration. After a short summary, the established concept of DWS was revisited. After describing big data, we showed the limits of DWS for advanced analytics. Then we presented the concept of data lakes and made a detailed comparison of data warehouses and data lakes. Afterwards we showed and discussed several architectural patterns for the integration of DWS and data lakes.

As a result, we have different approaches to handle this integration. All approaches have different pros and cons. Therefore, the specific situation must be considered and analyzed to come to a decision.

As one future task, we want to build prototypes for different architectural patterns and develop a list of criteria for tools and frameworks for the different components of the architecture.

But there are even application fields where the architectural patterns are not a sufficient solution. E.g. news coverages or stock market updates or some social media posts update in very short intervals, some of them even in seconds. For these use cases we need a special kind of analytics called (near) real time analytics. In other words, data has to be processed before they are stored. This paradigm is referred to as stream analytics [59, 60]. Adding stream processing into the architecture might be another future task.

References

1. Devlin, B.: Data Warehouse: From Architecture to Implementation. SEI Series in Software Engineering. Addison Wesley, Boston (1996)
2. Gardner, S.R.: Building the data warehouse. CACM **41**(9), 52–60 (1998)
3. Inmon, W.H.: Building the Data Warehouse, 4th edn. Wiley, New York (1996)
4. Kimball, R.: The Data Warehouse Toolkit, 3rd edn. Wiley, New York (2013)
5. Vaisman, A., Zimányi, E.: Data Warehouse Systems. DSA. Springer, Heidelberg (2014). https://doi.org/10.1007/978-3-642-54655-6
6. Kimball, R., Reeves, L., Ross, M., Thornthwaite, W.: The Data Warehouse Life Cycle Toolkit. Wiley, New York (1998)
7. Thomsen, E.: OLAP Solutions 2E w/WS: Building Multidimensional Information Systems, 2nd edn. Wiley, New York (2002)
8. Golfarelli, M., Rizzi, S.: Data Warehouse Design: Modern Principles and Methodologies. McGraw-Hill, New York (2009)
9. Hartenauer, J.: Introduction to Business Intelligence: Concepts and Tools, 2nd edn. AV Akademikerverlag, Riga (Latvia) (2012)
10. Gudivada, V., Baeza-Yates, R., Raghavan, V.: Big data: promises and Problems. IEEE Comput. **48**(3), 20–23 (2015)
11. Laney, D.: 3D Data Management: Controlling Data Volume, Velocity, and Variety. https://blogs.gartner.com/douglaney/files/2012/01/ad949-3D-Data-Management-Controlling-Data-Volume-Velocity-and-Variety.pdf. Accessed 31 Aug 2020
12. Siewert, S.: Big data in the cloud: data velocity, volume, variety, veracity. IBM Developer, 9 July 2013. https://www.ibm.com/developerworks/library/bd-bigdatacloud/index.html. Accessed 31 Aug 2020

13. Flouris, I., Giatrakos, N., Deligiannakis, A., Garofalakis, M., Kamp, M., Mock, M.: Issues in complex event processing: status and prospects in the big data era. J. Syst. Softw. **127**, 217–236 (2017)
14. Orenga-Rogla, S., Chalmeta, R.: Framework for implementing a big data ecosystem in organizations. Commun. ACM **62**(1), 58–65 (2019)
15. Shalev-Shwartz, S., Ben-David, S.: Understanding Machine Learning: From Theory to Algorithms. Cambridge University Press, Cambridge (2014)
16. Bonaccorso, G.: Mastering machine learning algorithms: expert techniques to implement popular machine learning algorithms and fine-tune your models. Packt Publishing, Birmingham (2018)
17. Cohen, J., Dolan, B., Dunlap, M., Hellerstein, J., Welton, C.: MAD skills: new analysis practices for big data. PVLDB **2**(2), 1481–1492 (2009)
18. Dhar, V.: Data science and prediction. Commun. ACM **56**(12), 64–73 (2013)
19. Deshpande, K., Desai, B.: Limitations of datawarehouse platforms and assessment of hadoop as an alternative. IJITMIS **5**(2), 51–58 (2014)
20. Pasupuleti, P., Purra, B.: Data Lake Development with Big Data. Packt Publishing, Birmingham (2015)
21. Inmon, W.H.: Data Lake Architecture: Designing the Data Lake and Avoiding the Garbage Dump. Technics Publications, New Jersey (2016)
22. John, T., Misra, P.: Data Lake for Enterprises: Lambda Architecture for Building Enterprise Data Systems. Packt Publishing, Birmingham (2017)
23. Gupta, S., Giri, V.: Practical Enterprise Data Lake Insights: Handle Data-Driven Challenges in an Enterprise Big Data Lake. Apress, New York (2018)
24. Mathis, C.: Data lakes. Datenbank-Spektrum **17**(3), 289–293 (2017)
25. Ladley, J.: Data Governance: How to Design, Deploy and Sustain an Effective Data Governance Program. The Morgan Kaufmann Series on Business Intelligence. Morgan Kaufmann, Burlington (2012)
26. Seiner, R.S.: Non-Invasive Data Governance: The Path of Least Resistance and Greatest Success. Technics Publications, New Jersey (2014)
27. Soares, S.: The Chief Data Officer Handbook for Data Governance. MC Press LLC, Boise (2015)
28. Talabis, M.: Information Security Analytics: Finding Security Insights, Patterns, and Anomalies in Big Data. Syngress, Rockland (2014)
29. Spivey, B., Echeverria, J.: Hadoop Security: Protecting Your Big Data Platform. O'Reilly, Newton (2015)
30. Dunning, T., Friedman, E.: Sharing Big Data Safely: Managing Data Security. O'Reilly Media, Newton (2016)
31. Ghavami, P.: Big Data Governance: Modern Data Management Principles for Hadoop, NoSQL Big Data Analytics. CreateSpace Independent Publishing Platform, Scotts Valley (2015)
32. Regulation (eu) 2016/679 of the european parliament and of the council of 27 April 2016 on the protection of natural persons with regard to the processing of personal data and on the free movement of such data, and repealing directive 95/46/ec (general data protection regulation). https://eur-lex.europa.eu/eli/reg/2016/679/oj. Accessed 31 Aug 2020
33. Russom, P., (eds.).: Data lakes: purposes, practices, patterns, and platforms. best practice report Q1/2017, TDWI (2017)
34. Bejek Jr., W.P.: Kafka Streams in Action. Manning, New York (2017)
35. Narkhede, N., Shapira, G., Palino, T.: Kafka: The Definitive Guide: Real-Time Data and Stream Processing at Scale. O'Reilly, Newton (2017)
36. Apache Kafka Project homepage. https://kafka.apache.org/. Accessed 31 Aug 2020

37. Ting, K., Cecho, J.: Apache Sqoop Cookbook. O'Reilly, Newton (2013)
38. White, T.: Hadoop: The Definitive Guide. O'Reilly, Newton (2015)
39. Apache sqoop Project homepage. https://sqoop.apache.org/. Accessed 31 Aug 2020
40. Alapati, S.: Expert Hadoop Administration: Managing, Tuning, and Securing Spark, YARN, and HDFS. Addison Wesley, Boston (2016)
41. HDFS. http://hadoop.apache.org/hdfs/. Accessed 31 Aug 2020
42. MapR. https://mapr.com/. Accessed 31 Aug 2020
43. Ozone. https://hadoop.apache.org/ozone/. Accessed 31 Aug 2020
44. Ellen, M.D., Tzoumas, K.: Introduction to Apache Flink: Stream Processing for Real Time and Beyond. O'Reilly, Newton (2016)
45. Apache Flink. https://flink.apache.org/. Accessed 31 Aug 2020
46. Allen, S., Pathirana, P., Jankowski, M.: Storm Applied: Strategies for Real-Time Event Processing. Manning, New York (2015)
47. Apache Storm. https://storm.apache.org/. Accessed 31 Aug 2020
48. Dean, J., Ghemawat, S.: MapReduce: simplified data processing on large clusters. In: Brewer, E.A., Chen, P. (eds), 6th Symposium on Operating System Design and Implementation (OS-DI 2004), San Francisco, California, USA, 6–8 December 2004, pp. 137–150. USENIX Association (2004)
49. Chambers, B., Zaharu, M.: Spark: The Definitive Guide: Big data processing made simple. O'Reilly, Newton (2018)
50. Apache Spark. https://spark.apache.org/. Accessed 31 Aug 2020
51. Sadalage, P., Fowler, M.: NoSQL Distilled: A Brief Guide to the Emerging World of Polyglot Persistence. Addison-Wesley, Boston (2012)
52. Harrison, G.: Next Generation Databases: NoSQL and Big Data. Apress, New York (2015)
53. Harrison, G.: Seven NoSQL Databases in a Week: Get Up and Running with the Fundamentals and Functionalities of Seven of the Most Popular NoSQL Databases. Packt Publishing, Birmingham (2018)
54. SAS Institure. https://www.sas.com/. Accessed 31 Aug 2020
55. The R Project for Statistical Computing. https://www.r-project.org/. Accessed 31 Aug 2020
56. Python Software Foundation. https://www.python.org/. Accessed 31 Aug 2020
57. Microsoft Azure. https://azure.microsoft.com/. Accessed 31 Aug 2020
58. AWS. https://aws.amazon.com/. Accessed 31 Aug 2020
59. Andrade, H., Gedik, B., Turaga, B.: Fundamentals of Stream Processing: Application Design, Systems, and Analytics. Cambridge University Press, Cambridge (2014)
60. Basak, A., Venkataraman, K., Murphy, R., Singh, M.: Stream Analytics with Microsoft Azure: Real-Time Data Processing for Quick Insights using Azure Stream Analytics. Packt Publishing, Birmingham (2017)

Study and Understanding the Significance of Multilayer-ELM Feature Space

Rajendra Kumar Roul[(✉)] [iD]

Department of Computer Science, Thapar Institute of Engineering and Technology,
Patiala 147004, Punjab, India
raj.roul@thapar.edu

Abstract. Multi-layer Extreme Learning Machine (Multi-layer ELM) is one of the most popular deep learning classifiers among other traditional classifiers because of its good characteristics such as being able to manage a huge volume of data, no backpropagation, faster learning speed, maximum level of data abstraction etc. Another distinct feature of Multi-layer ELM is that it can be able to make the input features linearly separable by mapping them non-linearly to an extended feature space. This architecture shows acceptable performance as compared to other deep networks. The paper studies the high dimensional feature space of Multi-layer ELM named as *MLELM-HDFS* in detail by performing different conventional unsupervised and semi-supervised clustering techniques on it using text data and comparing it with the traditional *TF-IDF* vector space named as *TFIDF-VS* in order to show its importance. Results on both unsupervised and semi-supervised clustering techniques show that *MLELM-HDFS* is more promising than the *TFIDF-VS*.

Keywords: Classification · Deep learning · Feature space · Multi-layer ELM · Semi-supervised · Unsupervised

1 Introduction

The digital web consists of a huge number of resources which include a vast amount of text documents, and a large number of end users as active participants. In order to personalize the large amount of information to the needs of each individual end user, systematic ways to organize data and retrieve useful information in real time are the need of the day. *Clustering* is one of the most robust techniques which can address these issues to a large extend by grouping similar documents into one place. It has been used for various applications such as organizing, browsing, classification, summarization etc. [1–4]. Clustering technique can either *hard* such as DBSCAN, k-Means, Agglomerative etc., or *soft* such as Fuzzy C-means, Rough C-means, Rough Fuzzy C-means etc. Many studies have been conducted on hard and soft clusterings [5]. In terms of efficiency and effectiveness, different clustering algorithms have different tradeoffs, and these algorithms are compared based on their experimental results [6].

© Springer Nature Switzerland AG 2020
L. Bellatreche et al. (Eds.): BDA 2020, LNCS 12581, pp. 28–48, 2020.
https://doi.org/10.1007/978-3-030-66665-1_3

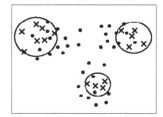   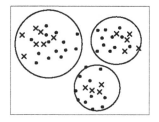

Fig. 1. Semi-supervised technique (before)   **Fig. 2.** Semi-supervised technique (after)

Clustering technique can be either *unsupervised* or *semi-supervised*. In unsupervised one, the entire clustering process takes place without the presence of any class labels (also called as supervised or labeled data), whereas in supervised learning, the labeled data are present. But if the labeled data can be mixed with the unlabeled data to guide the clustering process then we called it as semi-supervised clustering as shown in Figs. 1 and 2 respectively where '×' indicates the labeled data and '•' represents the unlabeled data. Researchers believe that the mixture of labeled and unlabeled data can improve the performance of the clustering process [7]. The semi-supervised clustering is initiated with a small amount of labeled data, and then the centroids are formed based on the number of clusters we want. Using a distance or similarity mechanism, the unlabeled data are assigned to their respective cluster. The problem with any clustering techniques is that visualizing the data becomes more difficult when the dataset size increases, and it needs a large storage size. To handle such problem, *feature selection* technique is essential which selects the important features from a large corpus and removes the unnecessary features during the model construction [8–11]. The main aim of the feature selection technique is:

- to reduce the storage size and training time of the classifier.
- to remove the multicollinearity so that the performance of the system gets improved.
- to make the data visualization more clear by reducing the large feature space to a dimension of two or three.

Based on the subset selection, the entire feature election technique is divided into three categories.

– *Filter methods* where no involvement of any classifier occurs instead, features are selected as per their statistical properties [12].
– *wrapper methods* where feature selection is done based on the decision of the classifier. Because of the involvement of the classifier, these methods are costlier compared to the earlier methods and have practically less use [13].
– *Embedded methods* combine both the earlier approaches [14].

Another complex problem that is faced by clustering technique is the exponential growth of the input data. This raises the difficulty for separating the data items in the vector space of *TF-IDF* due to low dimension. Although feature selection can reduce the curse of dimensionality, but cannot handle separation of the data items in the low dimensional feature space. To handle such problem, kernel methods [15] are generally used by any clustering or classification process. In the past, kernel methods have been used for clustering and classifications techniques, and better results are obtained [16,17].

But kernel methods are expensive because they use inner product among the features to compute the similarity in the higher dimensional space. To reduce the cost in higher dimensional space, Huang et al. [18] identified that by using feature mapping technique of ELM, if the feature vector non-linearly transfers to a higher dimensional space, then the data become more simpler and linearly separable. They have acknowledged that ELM feature mapping can outperform the kernel techniques, and any complex non-linear mappings can be estimated directly from the training data [19] by ELM. This shows that if we cluster or classify the text data using ELM feature space, then it costs less compared to the kernel techniques. In spite of many advantages of ELM in comparisons with other machine learning classifiers, it also has some limitations which cannot be ignored such as:

- having a single layer, it is unable to achieve a high level data abstraction.
- it needs a huge network which is difficult to implement for fitting perfectly the highly-modified input data.
- it is quite difficult to handle large-scale learning tasks using ELM due to memory constraints and huge cost of computation involved in evaluating the inverse of larger matrices.

Aiming in this direction, this paper studied the feature space of Multilayer-ELM (ML-ELM) [20,21], a deep learning network architecture which handles the above two limitations of ELM by taking the advantages of *ELM feature mapping* [22] extensively. An important and distinct feature of ML-ELM compared to other existing deep learning models is that it can be able to make the input features linearly separable by mapping them non-linearly to an extended space with less cost [18]. The main objective of this study is to test extensively the *MLELM-HDFS* for unsupervised and semi-supervised clustering of text documents in comparison with the *TFIDF-VS*. The main advantages and the major contributions of the paper are as follows:

i. The architecture and feature space of ML-ELM are studied in this paper and the proposed approach extensively test the unsupervised and semi-supervised clustering using text data on *MLELM-HDFS* (without using any kernel techniques while transferring the feature vector from lower to extended space), and on *TFIDF-VS* to justify the suitability and importance of ML-ELM feature space.
ii. From the past literature, it can be seen that no research works on unsupervised and semi-supervised clustering using text data have been done on the extended feature space of ML-ELM. Hence, this work can be the

first work to test the suitability and importance of higher dimensional feature space of Multi-layer ELM for unsupervised and semi-supervised clustering.

iii. PageRank based feature selection (*PRFS*), a novel feature selection technique is introduced to select the important features from the corpus which can improve the performances of unsupervised and semi-supervised clustering algorithms using *MLELM-HDFS*.

iv. The performance of ML-ELM using the proposed *PRFS* technique is compared with state-of-the-art deep learning classifiers to justify its competency and effectiveness.

Empirical results on three benchmark datasets show that both unsupervised and semi-supervised clustering results are promising on *MLELM-HDFS*.

2 Basics Preliminaries

2.1 Extreme Learning Machine

Extreme Learning Machine (ELM), where 'extreme' indicates 'faster' is a single hidden layer feed forward network [23,24]. If X is the input feature vector having n number of nodes, ELM output function Y can be defined using Eq. 1.

$$Y_j = \sum_{i=1}^{L} \beta_i g(w_i \cdot x_j + b_i),\ j = 1, \cdots N \tag{1}$$

where,

- N is independent and identically distributed samples, and L is the number of nodes in the hidden layer H.
- $w_i = [w_{i1}, w_{i2}, w_{i3}, \cdots, w_{in}]^T$ is the weight vector and b_i is the bias of the ith hidden node which are randomly generated.
- g is the activation function and β is the output weight vector that connects each hidden node to every output node.

Equation 2 represents the relationship between the hidden and the output feature vectors.

$$H\beta = Y \tag{2}$$

This indicates $\beta = H^{-1}Y$, where H^{-1} is the Moore-Penrose inverse [25]. The architecture of ELM and its subnetwork are shown in the Figs. 3 and 4 respectively. The feature space of ELM is shown in Fig. 5.

2.2 ML-ELM

ML-ELM is an artificial neural network having multiple hidden layers [20,26] and it is shown in Fig. 6. Equations 3, 4 and 5 are used for computing β (output weight vector) in ELM Autoencoder where H represents the hidden layer, X is the input layer, n and L are number of nodes in the input and hidden layer respectively.

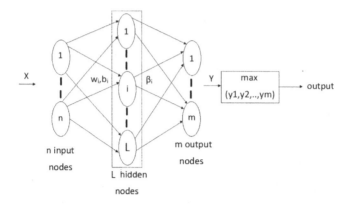

Fig. 3. Architecture of ELM

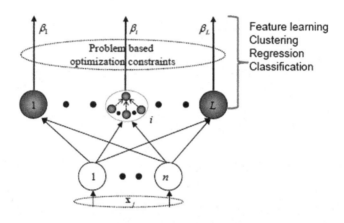

Fig. 4. Subnetwork of ELM

i. $n = L$

$$\beta = H^{-1}X \tag{3}$$

ii. $n < L$

$$\beta = H^T \left(\frac{I}{C} + HH^T\right)^{-1} X \tag{4}$$

iii. $n > L$

$$\beta = \left(\frac{I}{C} + H^T H\right)^{-1} H^T X \tag{5}$$

Here, $\frac{I}{C}$ generalize the performance of ELM [22,27] and is known as regularization parameter. Using Eq. 6, ML-ELM transfers the data between the hidden layers.

$$H_i = g((\beta_i)^T H_{i-1}) \tag{6}$$

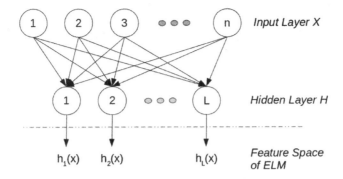

Fig. 5. ELM feature space

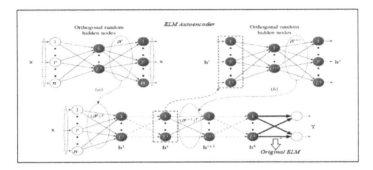

Fig. 6. Architecture of ML-ELM

2.3 ML-ELM Feature Mapping

ML-ELM uses Eq. 7 to map the features to higher dimensional space and thereby makes them linearly separable [22].

$$
h(\mathbf{x}) =
\begin{bmatrix}
h_1(\mathbf{x}) \\
h_2(\mathbf{x}) \\
\cdot \\
\cdot \\
\cdot \\
h_L(\mathbf{x})
\end{bmatrix}^T
=
\begin{bmatrix}
g(w_1, b_1, \mathbf{x}) \\
g(w_2, b_2, \mathbf{x}) \\
\cdot \\
\cdot \\
\cdot \\
g(w_L, b_L, \mathbf{x})
\end{bmatrix}^T
\tag{7}
$$

where, $h_i(\mathbf{x}) = g(w_i.\mathbf{x} + b_i)$. $h(\mathbf{x}) = [h_1(\mathbf{x})...h_i(\mathbf{x})...h_L(\mathbf{x})]^T$ can directly use for feature mapping [28,29]. w_i is the weight vector between the input nodes and the hidden nodes and b_i is the bias of the ith hidden node, and $\mathbf{x}$ is the input feature vector with g as the activation function. No kernel technique is required here and that reduces the cost at higher dimensional space because kernel function uses dot product very heavily in the higher dimensional space to find the similarity between the features. ML-ELM takes the benefit of the ELM feature mapping [22] extensively as shown in Eq. 8.

$$\lim_{L \to +\infty} ||y(\mathbf{x}) - y_L(\mathbf{x})|| = \lim_{L \to +\infty} ||y(\mathbf{x}) - \sum_{i=1}^{L} \beta_i h_i(\mathbf{x})|| = 0 \qquad (8)$$

The ML-ELM feature mapping is shown in Fig. 7.

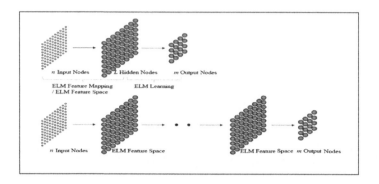

Fig. 7. ML-ELM feature mapping

2.4   Seeded k-Means Algorithm

Seeded kMeans is a semi-supervised algorithm suggested by Basu et al. [30]. The main difference between Seeded kMeans and the traditional kMean is, deciding the initial number of clusters and the initial centroid calculation. The rest of the steps are same on both clustering algorithms. The following steps discuss about the initial centroid calculation of the Seeded k-Means algorithm.

i. Consider a dataset D having x number of documents from n categories.
ii. Select a subset S having k documents along with their label (a label represents the category from which the document is selected) such that atleast one document should be selected from each of the n categories.
iii. The remaining $(x\text{-}k)$ documents of D are marked as unlabeled.
iv. Now the number of clusters to be formed is same as the number of categories we have, which is equal to n. The initial centroids sc_i of a cluster c_i is computed by finding the average of the number of p documents (i.e., vectors) of S belonging to that category and is shown in Eq. 9 where d represents document.

$$sc_i = \sum_{i=1}^{p} \frac{d_i}{p} \qquad (9)$$

v. The remaining unlabeled documents of D are assigned to any of these initial centroid based on Euclidean distance mechanism. The rest part of this algorithm is same as the traditional kMeans clustering algorithm [31].

3   Methodology

The following steps discuss the PageRank based Feature selection ($PRFS$) technique to generate the feature vector that is used for unsupervised and semi-supervised clustering algorithms on $MLELM\text{-}HDFS$ and $TFIDF\text{-}VS$. Overview of the methodology is shown in the Fig. 10.

1. *Document Pre-processing*:
   Consider a corpus P having a set of classes $C = \{c_1, c_2, \cdots, c_n\}$. At the beginning, the documents of all the classes of P are merged into a large document set called $D_{large} = \{d_1, d_2, \cdots, d_b\}$. Then all documents of D_{large} of P are pre-processed using lexical-analysis, stop-word elimination, removal of HTML tags, stemming[1], and then index terms are extracted using Natural Language Toolkit[2]. The term-document matrix of P is illustrated in Table 1.

Table 1. Term-document matrix

	d_1	d_2	d_3	$\cdots$	d_q
t_1	t_{11}	t_{12}	t_{13}	$\cdots$	t_{1q}
t_2	t_{21}	t_{22}	t_{23}	$\cdots$	t_{2q}
t_3	t_{31}	t_{32}	t_{33}	$\cdots$	t_{3q}
$\vdots$	$\vdots$	$\vdots$	$\vdots$	$\ddots$	$\vdots$
t_r	t_{r1}	t_{r2}	t_{r3}	$\cdots$	t_{rq}

2. *Computing semantic similarity between each pair of terms:*
   To compute the semantic similarity between each pair of terms of the corpus P, a semantic knowledge database is generated using WordNet that consists of a hierarchical structure of the general English terms [32]. The terms are organized into sets of synonyms called synsets and each synset represents a node in the hierarchical structure (Fig. 8). To compute the similarity between two terms, the length of the path between the two synsets which containing these two words is need to be find out. In this case, there can be three possible situations for path length between two terms:

   i. both terms belong to the same synset.
   ii. terms does not belong to the same synset, but there are one or more common terms between the respective synsets of the two terms.
   iii. terms are not belong to the same synset and there are no common terms between the respective synsets of the two terms.

In the first case, both terms are synonyms to each other, hence a path length of 0 is assigned. In the second case, although both terms are not belong to the same synset, but they share some common features, and in this case a path

---

[1] https://pythonprogramming.net/lemmatizing-nltk-tutorial/.
[2] https://www.nltk.org/.

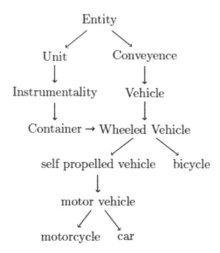

Fig. 8. Hierarchical structure from WordNet

length of 1 is assigned. In third case, neither they belong to same synset nor they share common features, and hence we compute the shortest path length between the respective synsets containing each term [33, 34]. It is also taken into consideration that terms have more general concepts if they close to the root of the hierarchical structure compared to the terms that are away from the root. Hence, it is necessary to determine the depth of the subsumer of both the terms. Subsumer is the node which is closest to the root of the hierarchical structure, on the path between two synsets. The semantic similarity between two words t_p and t_q is computed using Eq. 10.

$$sim(t_p, t_q) = e^{\theta l} * \frac{e^{\phi h} - e^{-\phi h}}{e^{\phi h} + e^{-\phi h}} \qquad (10)$$

where 'θ' represents how much path length contributes to the overall term similarity in the corpus P and lies in the range of [0, 1]. 'ϕ' represents the contribution of subsumer length and lies in the range of (0,1]. 'h' is the depth of the subsumer of both terms, and 'l' represents the path length between t_p and t_q. The value of $sim(t_p, t_q)$ lies between [0, 1]. For WordNet, the optimal values of θ and ϕ are 0.2 and 0.45 respectively [35]. For 'word sense disambiguation', 'max similarity' algorithm [36] is used and is implemented in Pywsd[3] (Python library based on NLTK[4]).

3. Generating Graph:
   An undirected graph is created where each term represents a node and the similarity value $(sim(t_p, t_q))$ is the edge between two terms t_p and t_q of the corpus P. Now all the term-pairs (i.e., edges) are arranged in the descending order of their similarity value as calculated in the previous step. Of this, the

---

[3] https://github.com/alvations/pywsd.
[4] https://www.nltk.org/.

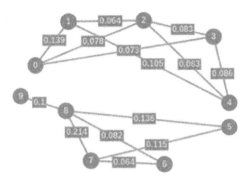

Fig. 9. Graph created on Classic4 dataset (Top 10 rows)

top $n\%$ similarity values are selected. If a term-pair (both inclusive) is not present in the top $n\%$ similarity values then there will be no edge between that term-pair. To understand this technique, we have shown a graph (Fig. 9) after running the algorithm on Clasic4 dataset and the details are discussed in Sect. 4.

4. PageRank calculation:
   PageRank [37,38] is an algorithm used to determine the relevance or importance of a web page, and rank it in the search results. The rank value indicates the importance of a particular page. We have modified the existing PageRank algorithm so that it can be used on terms where each term is assumed to be a web page in the real PageRank algorithm. Assume that term T has incoming edges from terms T_1, T_2, T_3, $\cdots$, T_n. The PageRank of a term T is given by Eq. 11 where 'd' represents the damping factor and its value lies between 0 and 1 and $C(T)$ represents the number of edges that leaving out from the term T. The PageRanks form a probability distribution over the terms, hence the sum of the PageRank of all the terms $PR(T)$ values will be one.

$$PR(T) = (1 - d) + d\left(\frac{PR(T_1)}{C(T_1)} + \cdots + \frac{PR(T_n)}{C(T_n)}\right) \tag{11}$$

   If there are a lot of terms that link to the term T then there is a common belief that the term T is an important term. The importance of the terms linking to T is also taken into account, and using PageRank terminology, it can be said that terms T_1, T_2, T_3, $\cdots$, T_n transfer their importance to T, albeit in a weighted manner. Thus, it is possible to iteratively assign a rank to each term, based on the ranks of the terms that point to it. In the proposed approach, the PageRank algorithm is run on the graph created in the previous step, with damping factor $d = 0.5$.

5. *Generating the input (or training) feature vector:*
   All the terms of the corpus P are sorted in descending order based on their PageRank score and out of these top $k\%$ terms are selected. The *TF-IDF* values of these terms will act as the values of the features for the documents.

Finally, all these top $k\%$ terms are put into a new list L_{new} to generate the input feature vector.

6. *Mapping L_{new} into MLELM-HDFS*:
   Equation 7 is used to transfer L_{new} to *MLELM-HDFS*. Before the transformation, to make the features of L_{new} much simpler and linearly separable on *MLELM-HDFS*, the number of nodes in hidden layer of ML-ELM are set as higher compare to the nodes of input layer (i.e., $L > n$).

7. *Unsupervised and semi-supervised clusterings on MLELM-HDFS and TFIDF-VS*: Different state-of-the-art unsupervised clustering algorithms using L_{new} are run separately both on *MLELM-HDFS* and *TFIDF-VS* respectively as illustrated in Algorithm 1. For semi-supervised clustering, the conventional kMeans and seeded-kMeans are run on *MLELM-HDFS* and *TFIDF-VS* separately to generate required number of clusters as illustrated in Algorithm 2.

Algorithm 1: ML-ELM feature space (Unsupervised clustering)

1: **Input:** L_{new}, number of clusters c.
2: **Output:** Result set R of *MLELM-HDFS* and results set V of *TFIDF-VS*.
3: Initially, $R \leftarrow \phi$ and $V \leftarrow \phi$
4: L_{new} is transferred to *MLELM-HDFS*.
5: The unsupervised clustering algorithm is run on the *MLELM-HDFS* by passing c, and the results is stored in R.
6: Similarly, the unsupervised clustering algorithm is run on the *TFIDF-VS* by passing c, and the results is stored in V.
7: Returns R and V.

Algorithm 2: ML-ELM feature space (Semi-supervised clustering)

1: **Input:** New list L_{new}, number of clusters c'.
2: **Output:** Generated clusters.
3: L_{new} is transferred to *MLELM-HDFS*.
4: The kMeans algorithm is run on *MLELM-HDFS*, and on *TFIDF-VS* by passing c' separately.
5: Retain $n\%$ of labels of L_{new} and a subset S is built as discussed in section 2.4. The Seeded-$kMeans$ algorithm is run on the *MLELM-HDFS* by passing, c' and S, and on *TFIDF-VS* by passing c'.

---

4   Experimental Analysis

To conduct the experiment, three benchmark datasets (Classic4[5], 20-Newsgroups[6], and Reuters[7]) are used. The details about these datasets are shown in the Table 2.

---

[5] http://www.dataminingresearch.com/index.php/2010/09/classic3-classic4-datasets/.

[6] http://qwone.com/~jason/20Newsgroups/.

[7] http://www.daviddlewis.com/resources/testcollections/reuters21578/.

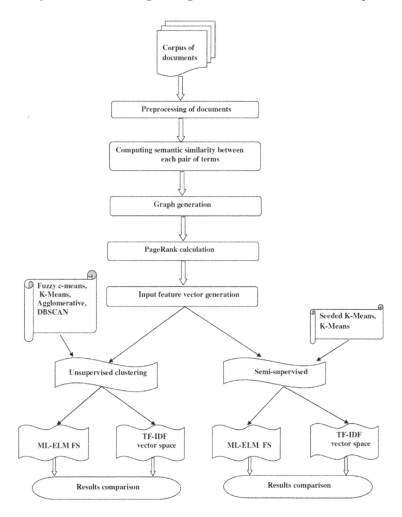

Fig. 10. Overview of the proposed methodology

Table 2. The details of different datasets

Dataset	No. of categories	No. of documents	Length of L_{new} (top 10% terms)
20-NG	7	19728	16272
Classic4	4	7095	5972
Reuters	8	7674	6532

Table 3. Parameters settings of different deep learning classifiers

Classifiers	No. of hidden layers	Activation function	Dropout	Optimizers	Epochs
ANN	3	sigmoid	0.1	SGD	500
CNN	3	relu	0.2	SGD	500
RNN	3	tanh	0.2	SGD	400
ML-ELM	20-NG=4, others=3	sigmoid	0.1	SGD	450

4.1   Parameters Setting:

Parameter settings of different deep learning classifiers used for experimental purposes are depicted in Table 3. The values of the parameters of deep learning classifiers are decided based on the experiment for which the best results are obtained.

4.2   Comparison of the Performance ML-ELM Using *PRFS* with Other Deep Learning Classifiers

The existing deep learning techniques have high computational cost, require massive training data and training time. The F-measure and accuracy comparisons of ML-ELM along with other deep learning classifiers such as Recurrent Neural Network (RNN) [39], Convolution Neural Network (CNN) [40], and Artificial Neural Network (ANN) using *PRFS* technique on top 10% features are depicted in Figs. 11 and 12 respectively. Results show that the performance of ML-ELM outperforms existing deep learning classifiers.

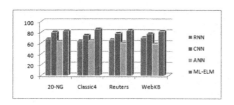

Fig. 11. Deep learning (F-measure)

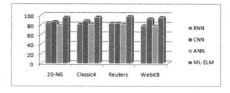

Fig. 12. Deep learning (Accuracy)

4.3   Advantages of ML-ELM over the Deep Learning Classifiers

All such complex problems of existing deep learning classifiers can be handled by ML-ELM in the following ways:

i. No fine tuning of hidden node parameters, no back propagations, and the number of parameters used are less in ML-ELM. Hence, increasing multiple hidden nodes and hidden layers would not cost much to ML-ELM in terms of computation as compared to other deep learning architectures. This saves lots of training time and increases the learning speed during the clustering process, and hence, GPUs are not required.
ii. Excellent performance is achieved when the size of the dataset increases. In other ways, it can manage a large volume of dataset.
iii. Using the universal approximation and classification capabilities of ELM, it is able to map a huge volume of features from a low to high dimensional feature space and can separate those features linearly in higher dimensional space without using kernel techniques.
iv. Majority of the training in ML-ELM is unsupervised in nature except the last level where supervised training takes place.
v. A high level of data abstraction is achieved by having multiple hidden layers where each layer learns a different form of input which enhances the performance of ML-ELM.
vi. Less complex, easy to understand, and simple to implement.

4.4   Discussion on Performance Evaluation of Unsupervised and Semi-supervised Clustering

After generating the new list (L_{new}) from a corpus (i.e., dataset), the L_{new} is used for different clustering algorithms. It is necessary to identify the documents of the generated cluster comes from which class of the corpus. While running the clustering algorithm on *MLELM-HDFS* and *TFIDF-VS* respectively, to make the process easy, the number of clusters to be formed on any dataset is considered as equivalent to the total number of classes present in that dataset. The ith cluster is assigned to j^{th} class iff i^{th} cluster contains maximum documents of jth class. For example: 20-Newsgroups has 20 classes, and hence the number of clusters formed on *MLELM-HDFS* and *TFIDF-VS* will be considered as 20. The implementation details are illustrated in Algorithm 3.

Algorithm 3: Assigning a cluster to its target class

1: **Input:** n generated clusters and k given classes
2: **Output:** Cluster and its corresponding class
3: $c \leftarrow 0$ // count variable
4: $cos\text{-}sim \leftarrow$ cosine-similarity
5: **for** r in 1 to n **do**
6:    **for** s in 1 to k **do**
7:       $Vcount \leftarrow 0$
8:       **for all** $d \in i$ **do**
9:          //d represent the document
10:          **for all** $d' \in j$ **do**
11:             //D' represent the document
12:             $cs \leftarrow cos\text{-}sim(d, d')$
13:             **if** $cs = 1$ **then**
14:                $c \leftarrow c + 1$
15:             **end if**
16:          **end for**
17:       **end for**
18:       $A[clust\_no][class\_no] \leftarrow c$
19:    **end for**
20:    $max \leftarrow A[r][p]$ //examine the class which gets large no. of documents of rth cluster
21:    **for** p in 2 to k **do**
22:       **if** $A[r][p] > max$ **then**
23:          $max \leftarrow A[r][p]$
24:          $m \leftarrow p$ // assign the class which gets large no. of documents of rth cluster
25:       **end if**
26:    **end for**
27:    $c[p] \leftarrow r$ // allocate the rth cluster to p^{th} class because p^{th} class gets large no. of documents of rth cluster
28: **end for**

Parameters Used for Clustering Algorithms:

Two well known supervised techniques *Purity* and *Entropy* are used to measure the performance of the unsupervised and semi-supervised clustering algorithms.

i. *Purity* indicates the extent to which clusters contain a single class and can be defined using Eq. 12.

$$purity = \frac{1}{N} \sum_{i=1}^{C} n_i * \max_j p_{ij}$$ (12)

where N and C are total number of documents and clusters in the corpus respectively. $p_{ij} = \frac{n_{ij}}{n_i}$ where n_{ij} represents the number of documents of class j belongs to cluster i and n_i is the number of documents belongs to i. High purity value indicates that the cluster is good.

ii. *Entropy* define the goodness of a clustering as shown in Eq. 13.

$$entropy = \sum_{i=1}^{Cl} \frac{n_i}{N} e_i$$ (13)

where $e_i = -\sum_{j=1}^{Cl} p_{ij} log_2 p_{ij}$. Here, Cl is the total number of classes, and all other variables are same as defined in Eq. 12. A small value of entropy indicates a better cluster.

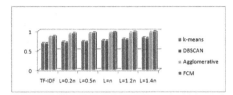

Fig. 13. 20-NG (Purity)

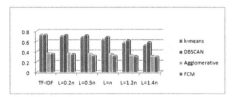

Fig. 14. 20-NG (Entropy)

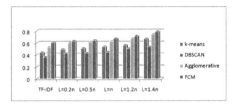

Fig. 15. Classic4-Purity

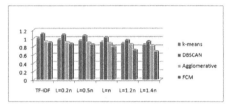

Fig. 16. Classic4 (Entropy)

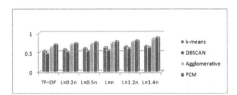

Fig. 17. Reuters (Purity)

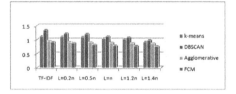

Fig. 18. Reuters (Entropy)

Performance Comparisons of Unsupervised Clustering:
Among different traditional unsupervised clustering algorithms, four algorithms such as Agglomerative (complete-linkage), DBSCAN, Fuzzy C-means (FCM), and k-Means are compiled on *MLELM-HDFS* and *TFIDF-VS* using three benchmark datasets separately (Figs. 13–18). From the results, the following points can be observed:

i. The empirical results of all unsupervised clustering algorithms are more promising on the *MLELM-HDFS* in comparison with *TFIDF-VS*.

ii. Empirical results obtained using Fuzzy C-means outperform other clustering algorithms on both spaces. This can be due to the non-overlapping of
features in a higher dimensional space where FCM works well [41], and it is a
soft clustering technique where the fuzziness can be exploited to generate a
more crisp behaving technique that produces better results in higher dimensional feature space [42]. But the empirical results of DBSCAN algorithm
are not good enough. This can be due to handling with clusters of varying
densities and this is one of the major drawbacks of DBSCAN. Agglomerative
is better compared to k-means in all datasets.

iii. It is also seen that in the *MLELM-HDFS*, while increasing the L values
compare to n (discussed in Eq. 4), after a certain stage, further increment
of L has either no effect or it decreases the performance of clustering algorithms. The reason may be due to the feature vector which becomes more
spare. Those results are not included as they are not point of interest of the
proposed work.

Performance Comparisons of Semi-supervised Clustering:

i. Both kMeans and seeded kMeans clustering techniques are compiled on
MLELM-HDFS and *TFIDF-VS* for many iterations (only third and fifth
iterations results are shown for experimental purposes) and the results are
shown in the Tables 4, 5, 6, 7, 8 and 9. Here, L ← 60%(n) indicates $L = 60\%$
of n. Here, L has started from 60% and extended till 140% and beyond this
the results become unchanged.

ii. For seeded kMeans only 10% label data are selected from the corpus and
remaining 90% of data are unlabeled. From the results, it can be concluded
that the results of seeded kMeans is better compared to kMeans clustering
on both *MLELM-HDFS* and *TFIDF-VS*, and the results of *MLELM-HDFS*
is better than *TFIDF-VS*.

iii. For experimental purposes, we have shown the execution time (using Clas-
sic4 and Reuters datasets) of seeded kMeans and kMeans in Figs. 19 and
20 on both vector (without using ELM) and feature space. Time taken by
MLELM-HDFS is less compared to *TFIDF-VS*, but the increase in L value
compared to n also increases the execution time of *MLELM-HDFS*, and it
is obvious, because the number of nodes in L get increased compared to n.

Table 4. Purity (20-NG)

Method	TFIDF	L ← 60%(n)	L ← 80%(n)	L ← n	L ← 1.2n	L ← 1.4n
kMeans (3 iterations)	0.693	0.711	0.726	0.757	0.791	0.763
Seeded k-Means (3 iterations)	0.857	0.948	0.948	0.952	0.962	0.966
kMeans (5 iterations)	0.695	0.732	0.747	0.768	0.821	0.837
Seeded k-Means(5 iterations)	0.857	0.953	0.957	0.958	0.967	0.978

Table 5. Entropy (20-NG)

Method	TFIDF	L ← 60%(n)	L ← 80%(n)	L ← n	L ← 1.2n	L ← 1.4n
kMeans (3 iterations)	0.734	0.723	0.719	0.688	0.622	0.587
Seeded k-Means (3 iterations)	0.365	0.360	0.358	0.353	0.331	0.313
kMeans (5 iterations)	0.739	0.702	0.68	0.637	0.578	0.522
Seeded k-Means (5 iterations)	0.361	0.353	0.323	0.321	0.311	0.302

Table 6. Purity (Classic4)

Method	TFIDF	L ← 60%(n)	L ← 80%(n)	L ← n	L ← 1.2n	L ← 1.4n
kMeans (3 iterations)	0.451	0.537	0.541	0.569	0.618	0.660
Seeded k-Means (3 iterations)	0.765	0.806	0.811	0.82	0.837	0.856
kMeans (5 iterations)	0.475	0.586	0.602	0.632	0.642	0.669
Seeded k-Means(5 iterations)	0.799	0.867	0.868	0.872	0.885	0.896

Table 7. Entropy (Classic4)

Method	TFIDF	L ← 60%(n)	L ← 80%(n)	L ← n	L ← 1.2n	L ← 1.4n
kMeans (3 iterations)	1.355	1.354	1.32	1.271	1.071	0.981
Seeded k-Means (3 iterations)	0.963	0.911	0.908	0.906	0.880	0.868
kMeans (5 iterations)	1.354	1.275	1.252	1.271	1.012	0.976
Seeded k-Means (5 iterations)	0.944	0.861	0.840	0.832	0.812	0.804

Table 8. Purity (Reuters)

Method	TFIDF	L ← 60%(n)	L ← 80%(n)	L ← n	L ← 1.2n	L ← 1.4n
kMeans (3 iterations)	0.541	0.577	0.597	0.669	0.678	0.710
Seeded k-Means (3 iterations)	0.763	0.810	0.821	0.865	0.887	0.898
kMeans (5 iterations)	0.565	0.594	0.612	0.682	0.702	0.721
Seeded k-Means (5 iterations)	0.769	0.837	0.869	0.888	0.910	0.928

Table 9. Entropy (Reuters)

Method	TFIDF	L ← 60%(n)	L ← 80%(n)	L ← n	L ← 1.2n	L ← 1.4n
kMeans (3 iterations)	1.255	1.224	1.202	1.171	1.145	0.910
Seeded k-Means (3 iterations)	0.863	0.811	0.808	0.806	0.780	0.768
kMeans (5 iterations)	1.234	1.175	1.152	1.081	1.072	0.864
Seeded k-Means (5 iterations)	0.844	0.761	0.740	0.732	0.712	0.704

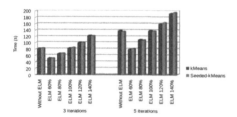

Fig. 19. Execution time (Classic4)

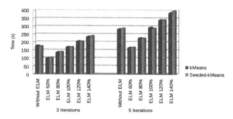

Fig. 20. Execution time (Reuters)

5   Conclusion

The suitability and usage of the extended space of ML-ELM is extensively studied by the proposed approach. To justify that the unsupervised and semi-supervised clustering techniques using text data can be more effective on the *MLELM-HDFS* in comparison with *TFIDF-VS*, exhaustive empirical analysis has been conducted on three benchmark datasets to show the effectiveness of the proposed approach. Before the clustering process starts, a novel feature selection technique *PRFS* is applied on the corpus in order to remove the redundant features from the corpus, which enhances the performance of both unsupervised and semi-supervised clustering methods. After doing a detailed analysis of the experimental work, it is concluded that ML-ELM feature space is more suitable and much useful for unsupervised and semi-supervised clustering of text data in comparison to *TFIDF-VS*. It is also noticed that Fuzzy C-means clustering shows better results on *MLELM-HDFS*. Similarly, seeded k Means gave better results compared to the traditional kMeans on both *MLELM-HDFS* and *TFIDF-VS*. The future works of the proposed approach are as follows:

i. to reduce the searching time of a desired document in the cluster, ranking can be done on the cluster.
ii. text summarization can be done on those generated clusters which will help the user to get the gist of such huge cluster in a summary form.
iii. further, labeling can be done on those clusters so that the user can be able to know the internal content of a cluster before using them for different applications.

References

1. Curiskis, S.A., Drake, B., Osborn, T.R., Kennedy, P.J.: An evaluation of document clustering and topic modelling in two online social networks: Twitter and reddit. Inf. Process. Manage. **57**(2), 102034 (2020)
2. Zhao, Y., Karypis, G.: Empirical and theoretical comparisons of selected criterion functions for document clustering. Mach. Learn. **55**(3), 311–331 (2004)
3. Roul, R.K., Arora, K.: A nifty review to text summarization-based recommendation system for electronic products. Soft Comput. **23**(24), 13183–13204 (2019)
4. Roul, R.K.: Topic modeling combined with classification technique for extractive multi-document text summarization. Soft Comput. **24**(22), 1–15 (2020)

5. Kim, H., Kim, H.K., Cho, S.: Improving spherical k-means for document clustering: fast initialization, sparse centroid projection, and efficient cluster labeling. Expert Syst. Appl. **150**, 113288 (2020)
6. Steinbach, M., Karypis, G., Kumar, V., et al.: A comparison of document clustering techniques. In: KDD Workshop on Text Mining, vol. 400, pp. 525–526, Boston (2000)
7. Basu, S., Bilenko, M., Mooney, R.J.: Comparing and unifying search-based and similarity-based approaches to semi-supervised clustering. In: Proceedings of the ICML-2003 Workshop on the Continuum from Labeled to Unlabeled Data in Machine Learning and Data Mining, pp. 42–49 (2003)
8. Roul, R.K., Gugnani, S., Kalpeshbhai, S.M.: Clustering based feature selection using extreme learning machines for text classification. In: 2015 Annual IEEE India Conference (INDICON), pp. 1–6. IEEE (2015)
9. Sayed, G.I., Hassanien, A.E., Azar, A.T.: Feature selection via a novel chaotic crow search algorithm. Neural Comput. Appl. **31**(1), 171–188 (2019)
10. Qian, W., Long, X., Wang, Y., Xie, Y.: Multi-label feature selection based on label distribution and feature complementarity. Appl. Soft Comput. **90**, 106167 (2020)
11. Roul, R.K., Sahoo, J.K.: Text categorization using a novel feature selection technique combined with ELM. In: Sa, P.K., Bakshi, S., Hatzilygeroudis, I.K., Sahoo, M.N. (eds.) Recent Findings in Intelligent Computing Techniques. AISC, vol. 709, pp. 217–228. Springer, Singapore (2018). https://doi.org/10.1007/978-981-10-8633-5_23
12. Kira, K., Rendell, L.A.: The feature selection problem: traditional methods and a new algorithm. AAAI **2**, 129–134 (1992)
13. Kohavi, R., John, G.H.: Wrappers for feature subset selection. Artif. Intell. **97**(1), 273–324 (1997)
14. Lal, T.N., Chapelle, O., Weston, J., Elisseeff, A.: Embedded methods. In: Guyon, I., Nikravesh, M., Gunn, S., Zadeh, L.A. (eds.) Feature Extraction. Studies in Fuzziness and Soft Computing, vol. 207, pp. 137–165. Springer, Heidelberg (2006). https://doi.org/10.1007/978-3-540-35488-8_6
15. Da Jiao, Z.L.Z.W., Cheng, L.: Kernel clustering algorithm. Chin. J. Comput. **6**, 004 (2002)
16. Kang, Z., Wen, L., Chen, W., Xu, Z.: Low-rank kernel learning for graph-based clustering. Knowl.-Based Syst. **163**, 510–517 (2019)
17. Hu, G., Du, Z.: Adaptive kernel-based fuzzy c-means clustering with spatial constraints for image segmentation. Int. J. Pattern Recognit. Artif. Intelli. **33**(01), 1954003 (2019)
18. Huang, G.-B., Ding, X., Zhou, H.: Optimization method based extreme learning machine for classification. Neurocomputing **74**(1), 155–163 (2010)
19. Huang, G.-B., Chen, L.: Enhanced random search based incremental extreme learning machine. Neurocomputing **71**(16), 3460–3468 (2008)
20. Kasun, L.L.C., Zhou, H., Huang, G.-B., Vong, C.M.: Representational learning with extreme learning machine for big data. IEEE Intell. Syst. **28**(6), 31–34 (2013)
21. Roul, R.K., Asthana, S.R., Kumar, G.: Study on suitability and importance of multilayer extreme learning machine for classification of text data. Soft Comput. **21**(15), 4239–4256 (2017)
22. Huang, G.-B., Zhou, H., Ding, X., Zhang, R.: Extreme learning machine for regression and multiclass classification. IEEE Trans. Syst. Man Cybern. B (Cybern.) **42**(2), 513–529 (2012)
23. Huang, G.-B., Zhu, Q.-Y., Siew, C.-K.: Extreme learning machine: theory and applications. Neurocomputing **70**(1), 489–501 (2006)

24. Gugnani, S., Bihany, T., Roul, R.K.: Importance of extreme learning machine in the field of query classification: a novel approach. In: 2014 9th International Conference on Industrial and Information Systems (ICIIS), pp. 1–6. IEEE (2014)
25. Weisstein, E.W.: Moore-penrose matrix inverse (2002). https://mathworld.wolfram.com/
26. Roul, R.K.: Detecting spam web pages using multilayer extreme learning machine. Int. J. Big Data Intell. **5**(1–2), 49–61 (2018)
27. Roul, R.K.: Deep learning in the domain of near-duplicate document detection. In: Madria, S., Fournier-Viger, P., Chaudhary, S., Reddy, P.K. (eds.) BDA 2019. LNCS, vol. 11932, pp. 439–459. Springer, Cham (2019). https://doi.org/10.1007/978-3-030-37188-3_25
28. Huang, G.-B., Chen, L., Siew, C.K., et al.: Universal approximation using incremental constructive feedforward networks with random hidden nodes. IEEE Trans. Neural Networks **17**(4), 879–892 (2006)
29. Roul, R.K., Sahoo, J.K., Goel, R.: Deep learning in the domain of multi-document text summarization. In: Shankar, B.U., Ghosh, K., Mandal, D.P., Ray, S.S., Zhang, D., Pal, S.K. (eds.) PReMI 2017. LNCS, vol. 10597, pp. 575–581. Springer, Cham (2017). https://doi.org/10.1007/978-3-319-69900-4_73
30. Basu, S., Banerjee, A., Mooney, R.: Semi-supervised clustering by seeding. In: In Proceedings of 19th International Conference on Machine Learning (ICML-2002), Citeseer (2002)
31. Hartigan, J.A., Wong, M.A.: Algorithm as 136: a k-means clustering algorithm. J. Roy. Stat. Soc. Ser. C (Appl. Stat.) **28**(1), 100–108 (1979)
32. Miller, G.A.: Wordnet: a lexical database for English. Commun. ACM **38**(11), 39–41 (1995)
33. Li, Y., McLean, D., Bandar, Z.A., O'shea, J.D., Crockett, K.: Sentence similarity based on semantic nets and corpus statistics. IEEE Trans. Knowl. Data Eng. **18**(8), 1138–1150 (2006)
34. Gugnani, S., Roul, R.K.: Triple indexing: an efficient technique for fast phrase query evaluation. Int. J. Comput. Appl. **87**(13), 9–13 (2014)
35. Erkan, G., Radev, D.R.: LexRank: graph-based lexical centrality as salience in text summarization. J. Artif. Intell. Res. **22**, 457–479 (2004)
36. Pedersen, T., Banerjee, S., Patwardhan, S.: Maximizing semantic relatedness to perform word sense disambiguation, Technical report
37. Page, L., Brin, S., Motwani, R., Winograd, T.: The PageRank citation ranking: bringing order to the web. In: Proceedings of the 7th International World Wide Web Conference (Brisbane, Australia), pp. 161–172 (1998)
38. Roul, R.K., Sahoo, J.: A novel approach for ranking web documents based on query-optimized personalized PageRank. Int. J. Data Sci. Anal. **10**(2), 1–19 (2020)
39. Williams, R.J., Zipser, D.: A learning algorithm for continually running fully recurrent neural networks. Neural Comput. **1**(2), 270–280 (1989)
40. Fukushima, K.: Neocognitron. Scholarpedia **2**(1), 1717 (2007). revision #91558
41. Bezdek, J.C., Ehrlich, R., Full, W.: FCM: the fuzzy c-means clustering algorithm. Comput. Geosci. **10**(2), 191–203 (1984)
42. Winkler, R., Klawonn, F., Kruse, R.: Fuzzy c-means in high dimensional spaces. Int. J. Fuzzy Syst. Appl. **1**, 1–16 (2013)

Spectral Learning of Semantic Units in a Sentence Pair to Evaluate Semantic Textual Similarity

Akanksha Mehndiratta$^{(\boxtimes)}$ (iD) and Krishna Asawa

Jaypee Institute of Information Technology, Noida, India
mehndiratta.akanksha@gmail.com, krishna.asawa@jiit.ac.in

Abstract. Semantic Textual Similarity (STS) measures the degree of semantic equivalence between two snippets of text. It has applicability in a variety of Natural Language Processing (NLP) tasks. Due to the wide application range of STS in many fields, there is a constant demand for new methods as well as improvement in current methods. A surge of unsupervised and supervised systems has been proposed in this field but they pose a limitation in terms of scale. The restraints are caused either by the complex, non-linear sophisticated supervised learning models or by unsupervised learning models that employ a lexical database for word alignment. The model proposed here provides a spectral learning-based approach that is linear, scale-invariant, scalable, and fairly simple. The work focuses on finding semantic similarity by identifying semantic components from both the sentences that maximize the correlation amongst the sentence pair. We introduce an approach based on Canonical Correlation Analysis (CCA), using cosine similarity and Word Mover's Distance (WMD) as a calculation metric. The model performs at par with sophisticated supervised techniques such as LSTM and BiLSTM and adds a layer of semantic components that can contribute vividly to NLP tasks.

Keywords: Semantic Textual Similarity · Natural Language Processing · Spectral learning · Semantic units · Canonical Correlation Analysis · Word Mover's Distance

1 Introduction

Semantic Textual Similarity (STS) determines the similarity between two pieces of texts. It has applicability in a variety of Natural Language Processing (NLP) tasks including textual entailment, paraphrase, machine translation, and many more. It aims at providing a uniform structure for generation and evaluation of various semantic components that, conventionally, were considered independently and with a superficial understanding of their impact in various NLP applications.

The SemEval STS task is an annual event held as part of the SemEval/*SEM family of workshops. It was one of the most awaited events for STS from 2012 to

© Springer Nature Switzerland AG 2020
L. Bellatreche et al. (Eds.): BDA 2020, LNCS 12581, pp. 49–59, 2020.
https://doi.org/10.1007/978-3-030-66665-1_4

2017 [1–6], that attracted a large number of teams every year for participation. The dataset is available publicly by the organizers containing up to 16000 sentence pairs for training and testing that is annotated by humans with a rating between 0–5 with 0 indicating highly dissimilar and 5 being highly similar.

Generally, the techniques under the umbrella of STS can be classified into the following two categories:

1. **Supervised Systems:** The techniques designed in this category generate results after conducting training with an adequate amount of data using a machine learning or deep-learning based model [9,10]. Deep learning has gained a lot of popularity in NLP tasks. They are extremely powerful and expressive but are also complex and non-linear. The increased model complexity makes such models much slower to train on larger datasets.
2. **Unsupervised Systems:** To our surprise, the basic approach of plain averaging [11] and weighted averaging [12] word vectors to represent a sentence and computing the degree of similarity as the cosine distance has outperformed LSTM based techniques. Examples like these strengthen the researchers that lean towards the simpler side and exploit techniques that have the potential to process a large amount of text and are scalable instead of increased model complexity. Some of the techniques under this category may have been proposed even before the STS shared task [19,20] whiles some during. Some of these techniques usually rely on a lexical database such as paraphrase database (PPDB) [7,8], wordnet [21], etc. to determine contextual dependencies amongst words.

The technique that is proposed in this study is based on spectral learning and is fairly simple. The idea behind the approach stems from the fact that the semantically equivalent sentences are dependent on a similar context. Hence goal here is to identify semantic components that can be utilized to frame context from both the sentences. To achieve that we propose a model that identifies such semantic units from a sentence based on its correlation from words of another sentence. The method proposed in the study, a spectral learning-based approach for measuring the strength of similarity amongst two sentences based on Canonical Correlation Analysis (CCA) [22] uses cosine similarity and Word Mover's Distance (WMD) as calculation metric. The model is fast, scalable, and scale-invariant. Also, the model is linear and have the potential to perform at par with the non-linear supervised learning architectures such as such as LSTM and BiL-STM. It also adds another layer by identifying semantic components from both the sentences based on their correlation. These components can help develop a deeper level of language understanding.

2   Canonical Correlation Analysis

Given two sets of variables, canonical correlation is the analysis of a linear relationship amongst the variables. The linear relation is captured by studying the latent variables (variables that are not observed directly but inferred) that represent the direct variables. It is similar to correlation analysis but multivariate. In the statistical analysis, the term can be found in multivariate discriminant analysis and multiple regression analysis. It is an analog to Principal Component Analysis (PCA), for a set of outputs. PCA generates a direction of maximal covariance amongst the elements of a matrix, in other words for a multivariate input on a single output, whereas CCA generates a direction of maximal covariance amongst the elements of a pair of matrices, in other words for a multivariate input on a multivariate output.

Consider two random multivariable x and y. Given C_{xx}, C_{yy}, C_{yx} that represents the within-sets and between-sets covariance matrix of x and y and C_{xy} is a transpose of C_{yx}, CCA tries to generate projections CV_1 and CV_2, a pair of linear transformations, using the optimization problem given by Eq. 1.

$$\max_{CV_1, CV_2} \frac{CV_1^T C_{xy} CV_2}{\sqrt{CV_1^T C_{xx} CV_1} \sqrt{CV_2^T C_{yy} CV_2}} \tag{1}$$

Given x and y, the canonical correlations are found by exploiting the eigenvalue equations. Here the eigenvalues are the squared canonical correlations and the eigenvectors are the normalized canonical correlation basis vectors. Other than eigenvalues and eigenvectors, another integral piece for solving Eq. 1 is to compute the inverse of the covariance matrices. CCA utilizes Singular value decomposition (SVD) or eigen decomposition for performing the inverse of a matrix. Recent advances [24] have facilitated such problems with a boost on a larger scale. This boost is what makes CCA fast and scalable.

More specifically, consider a group of people that have been selected to participate in two different surveys. To determine the correlation between the two surveys CCA tries to project a linear transformation of the questions from survey 1 and questions from survey 2 that maximizes the correlation between the projections. CCA terminology identifies the questions in the survey as the variables and the projections as variates. Hence the variates are a linear transformation or a weighted average of the original variables. Let the questions in survey 1 be represented as $x_1, x_2, x_3 .... x_n$ similarly questions in survey 2 are represented as $y_1, y_2, y_3 .... y_m$. The first variate for survey 1 is generated using the relation given by Eq. 2.

$$CV_1 = a_1 x_1 + a_2 x_2 + a_3 x_3 + ...a_n x_n \tag{2}$$

And the first variate for survey 2 is generated using the relation given by Eq. 3.

$$CV_1 = b_1 y_1 + b_2 y_2 + b_3 y_3 + ...b_m y_m \tag{3}$$

Where $a_1, a_2, a_3 ..... a_n$ and $b_1, b_2, b_3 .... b_m$ are weights that are generated in such a way that it maximizes the correlation between CV_1 and CV_2. CCA can generate the second pair of variates using the residuals of the first pair of variates and many more in such a way that the variates are independent of each other i.e. the projections are orthogonal.

When applying CCA the following fundaments are needed to be taken care of:

1. Determine the minimum number of variates pair be generated.
2. Analyze the significance of a variate from two perspectives – one being the magnitude of relatedness between the variate and the original variable from which it was transformed and the magnitude of relatedness between the corresponding variate pair.

2.1 CCA for Computing Semantic Units

Given two views $X = (X^{(1)}, X^{(2)})$ of the input data and a target variable Y of interest, Foster [23] exploits CCA to generate a projection of X that reduces the dimensionality without compromising on its predictive power. Authors assume, as represented by Eq. 4, that the views are independent of each other conditioned on a hidden state h, i.e.

$$P(X^{(1)}, X^{(2)}|h) = P(X^{(1)}|h)P(X^{(2)}|h) \tag{4}$$

Here CCA utilizes the multi-view nature of data to perform dimensionality reduction.

STS is an estimate of the prospective of a candidate sentence to be considered as a semantic counterpart of another sentence. Measuring text similarity has had a long-serving and contributed widely in applications designed for text processing and related areas. Text similarity has been used for machine translation, text summarization, semantic search, word sense disambiguation, and many more. While making such an assessment is trivial for humans, making algorithms and computational models that mimic human-level performance poses a challenge. Consequently, natural language processing applications such as generative models typically assume a Hidden Markov Model (HMM) as a learning function. HMM also indicates a multi-view nature. Hence, two sentences that have a semantic unit(s) c with each other provide two natural views and CCA can be capitalized, as shown in Eq. 5, to extract this relationship.

$$P(S_1, S_2|c) = P(S_1|c)P(S_2|c) \tag{5}$$

Where S_1 and S_2 mean sentence one and sentence two that are supposed to have some semantic unit(s) c. It has been discussed in the previous section that CCA is fast and scalable. Also, CCA neither requires all the views to be of a fixed length nor have the views to be of the same length; hence it is scale-invariant for the observations.

3   Model

3.1   Data Collection

We test our model in three textual similarity tasks. All three of which were published in SemEval semantic textual similarity (STS) tasks (2012–2017). The first dataset considered for experimenting was from SemEval -2017 Task 1 [6], an ongoing series of evaluations of computational semantic analysis systems with a total of 250 sentence pairs. Another data set was SemEval textual similarity dataset 2012 with the name "OnWN" [4]. The sentence pair in the dataset is generated from the Ontonotes and its corresponding wordnet definition. Lastly, SemEval textual similarity dataset 2014 named "headlines" [2] that contains sentences taken from news headlines. Both the datasets have 750 sentence pairs. In all the three datasets a sentence pair is accompanied with a rating between 0–5 with 0 indicating highly dissimilar and 5 being highly similar. An example of a sentence pair available in the SemEval semantic textual similarity (STS) task is shown in Table 1.

Table 1. A sample demonstration of sentence pair available in the SemEval semantic textual similarity (STS) task publically available dataset.

	Example - 1	Example - 2
Sentence 1	Birdie is washing itself in the water basin	The young lady enjoys listening to the guitar
Sentence 2	The bird is bathing in the sink	The woman is playing the violin
Similarity Score	5 (The two sentences mean the same thing hence are completely equivalent)	1 (The two sentences may be around the same topic but are not equivalent)

3.2   Data Preprocessing

It is important to pre-process the input data to improve the learning and elevate the performance of the model. Before running the similarity algorithm the data collected is pre-processed based on the following steps.

1. **Tokenization -** Processing one sentence at a time from the dataset the sentence is broken into a list of words that were essential for creating word embeddings.
2. **Removing punctuations -** Punctuations, exclamations, and other marks are removed from the sentence using regular expression and replaced with empty strings as there is no vector representation available for such marks.
3. **Replacing numbers -** The numerical values are converted to their corresponding words, which can then be represented as embeddings.

4. **Removing stop words** - In this step the stop words from each sentence are removed. A stop word is a most commonly used word (such as "the", "a", "an", "in") that do not add any valuable semantic information to our sentence. The used list of stop words is obtained from the nltk package in python.

3.3   Identifying Semantic Units

Our contribution to the STS task adds another layer by identifying semantic units in a sentence. These units are identified based on their correlation with the semantic units identified in the paired sentence. Each sentence s_i is represented as a list of the word2vec embedding, where each word is represented in the m -dimensional space using Google's word2vec. $s_i = (w_{i1}, w_{i2}, ..., w_{im})$, $i = 1, 2, ...,$ m, where each element is the embedding counterpart of its corresponding word. Given two sentences s_i and s_j, CCA projects variates as linear transformation of s_i and s_j. The number of projections to be generated is limited to the length, i.e. no. of words, of the smallest vector between s_i and s_j. E.g. if the length of s_i and s_j is 8 and 5 respectively, the maximum number of correlation variates outputted is 5. Conventionally, word vectors were considered independently and with a superficial understanding of their impact in various NLP applications. But these components obtained can contribute vividly in an NLP task. A sample of semantic units identified on a sentence pair is shown in Table 2.

Table 2. A sample of semantic units identified on a sentence pair in the SemEval dataset.

Sentence	The group is eating while taking in a breath-taking view.	A group of people take a look at an unusual tree.
Pre-processed tokens	['group', 'eating', 'taking', 'breath-taking', 'view']	['group', 'people', 'take', 'look', 'unusual', 'tree']
Correlation variates	['group', 'taking', 'view', 'breathtaking', 'people']	['group', 'take', 'look', 'unusual', 'people']

3.4   Formulating Similarity

The correlation variates projected by CCA are used to generate a new representation for each sentence s_i as a list of the word2vec vectors, $s_i = (w_{i1}, w_{i2}, ..., w_{in})$, $i = 1, 2, ..., n$, where each element is the Google's word2vec word embedding of its corresponding variate identified by CCA.

Given a range of variate pairs, there are two ways of generating a similarity score for sentence s_i and s_j:

1. Cosine similarity: It is a very common and popular measure for similarity. Given a pair of sentence represented as $s_i = (w_{i1}, w_{i2}, ..., w_{im})$ and $s_j = (w_{j1}, w_{j2}, ..., w_{jm})$, cosine similarity measure is defined as Eq. 6

$$sim(s_i, s_j) = \frac{\sum_{k=1}^{m} w_{ik} w_{jk}}{\sqrt{\sum_{k=1}^{m} w_{ik}^2} \sqrt{\sum_{k=1}^{m} w_{jk}^2}} \tag{6}$$

Similarity score is calculated by computing the mean of cosine similarity for each of these variate pairs.

2. Word Mover's Distance (WMD): WMD is a method that allows us to assess the "distance" between two documents in a meaningful way. It harnesses the results from advanced word –embedding generation techniques like Glove [13] or Word2Vec as embeddings generated from these techniques are semantically superior. Also, with embeddings generated using Word2Vec or Glove it is believed that semantically relevant words should have similar vectors. Let $T = (t_1, t_2, ..., t_m)$ represents a set with m different words from a document A. Similarly $P = (p_1, p_2, ..., p_n)$ represents a set with n different terms from a document B. The minimum cumulative distance traveled amongst the word cloud of the text document A and B becomes the distance between them.

A min-max normalization, given in Eq. 7, is applied on the similarity score generated by cosine similarity or WMD to scale the output similarity score to 5.

$$x_{scaled} = \frac{x - x_{min}}{x_{max} - x_{min}} \tag{7}$$

4   Results and Analysis

The key evaluation criterion is the Pearson's coefficient between the predicted scores and the ground-truth scores. The results from the "OnWN" and "Headlines" dataset published in SemEval semantic textual similarity (STS) task 2012 and 2014 respectively is shown in Table 3. The first three results are from the official task rankings followed by seven models proposed by Weintings [11]. The last two column indicate the result from the model proposed with cosine similarity and WMD respectively. The dataset published in SemEval semantic textual similarity (STS) tasks 2017 is identified as Semantic Textual Similarity Benchmark (STS-B) by the General Language Understanding Evaluation (GLUE) benchmark [16]. The results of the official task rankings for the task STS-B are shown in Table 4. Table 5 indicate the result from the model proposed with cosine similarity and WMD respectively. Since the advent of GLUE, a lot models have been proposed for the STS-B task, such as XLNet [17], ERNIE 2.0 [18] and many more, details of these models are available on the official website of GLUE[1], that produces result above 90% in STS-B task. But the increased model complexity

---

[1] https://gluebenchmark.com/leaderboard.

Table 3. Results on SemEval -2012 and 2014 textual similarity dataset (Pearson's r x 100).

Dataset	50%	75%	Max	PP	proj	DAN	RNN	iRNN	LSTM (output gate)	LSTM	CCA (CoSim)	CCA (WMD)
OnWN	60.8	65.9	72.7	70.6	70.1	65.9	63.1	70.1	65.2	56.4	60.5	37.1
Headlines	67.1	75.4	78.4	69.7	70.8	69.2	57.5	70.2	57.5	50.9	62.5	55.8

Table 4. Results on STS-B task from GLUE Benchmark (Pearson's r x 100).

Model	STS-B
Single task training	
BiLSTM	66.0
+ELMo [14]	64.0
+CoVe [15]	67.2
+Attn	59.3
+Attn, ELMo	55.5
+ATTN, CoVe	57.2
Multi-task training	
BiLSTM	70.3
+ELMo	67.2
+CoVe	64.4
+Attn	72.8
+Attn, ELMo	74.2
+ATTN, CoVe	69.8
Pre-trained sentence representation models	
CBow	61.2
Skip-Thought	71.8
Infersent	75.9
DisSent	66.1
GenSen	79.3

Note. Adapted from "Glue: A multi-task benchmark and analysis platform for natural language understanding" by Wang, A., Singh, A., Michael, J., Hill, F., Levy, O., Bowman, S. R.(2019), In: International Conference on Learning Representations (ICLR).

makes such models much slower to train on larger datasets. The work here focuses on finding semantic similarity by identifying semantic components using an approach that is linear, scale-invariant, scalable, and fairly simple.

Table 5. Results of proposed spectral learning-based model on the SemEval 2017 dataset (Pearson's r x 100).

Model	STS-B
CCA (Cosine similarity)	73.7
CCA (WMD)	76.9

5 Conclusion

We proposed a spectral learning based model namely CCA using cosine Similarity and WMD, and compared the model on three different datasets with various other competitive models. The model proposed utilizes a scalable algorithm hence it can be included in any research that is inclined towards textual analysis. With an added bonus that the model is simple, fast and scale-invariant it can be an easy fit for a study.

Another important take from this study is the identification of semantic units. The first step in any NLP task is providing a uniform structure for generation and evaluation of various semantic units that, conventionally, were considered independently and with a superficial understanding of their impact. Such components can help in understanding the development of context over sentence in a document, user reviews, question-answer and dialog session.

Even though our model couldn't give best results it still performed better than some models and gave competitive results for others, which shows that there is a great scope for improvement. One of the limitations of the model is its inability to identify semantic units larger than a word for instance, a phrase. It will also be interesting to develop a model that is a combination of this spectral model with a supervised or an unsupervised model. On further improvement the model will be helpful in various ways and can be used in applications such as document summarization, word sense disambiguation, short answer grading, information retrieval and extraction, etc.

References

1. Agirre, E., et al.: SemEval-2015 task 2: semantic textual similarity, English, Spanish and pilot on interpretability. In: Proceedings of the 9th International Workshop on Semantic Evaluation (SemEval 2015), pp. 252–263. Association for Computational Linguistics, June 2015
2. Agirre, E., et al.: SemEval-2014 task 10: multilingual semantic textual similarity. In: Proceedings of the 8th International Workshop on Semantic Evaluation (SemEval 2014), pp. 81–91. Association for Computational Linguistics, August 2014
3. Agirre, E., et al.: SemEval-2016 task 1: semantic textual similarity, monolingual and cross-lingual evaluation. In: SemEval 2016, 10th International Workshop on Semantic Evaluation, San Diego, CA, Stroudsburg (PA), pp. 497–511. Association for Computational Linguistics (2016)

4. Agirre, E., Bos, J., Diab, M., Manandhar, S., Marton, Y., Yuret, D.: *SEM 2012: The First Joint Conference on Lexical and Computational Semantics-Volume 1: Proceedings of the Main Conference and the Shared Task, and Volume 2: Proceedings of the Sixth International Workshop on Semantic Evaluation (SemEval 2012), pp. 385–393. Association for Computational Linguistics (2012)
5. Agirre, E., Cer, D., Diab, M., Gonzalez-Agirre, A., Guo, W.: *SEM 2013 shared task: semantic textual similarity. In: Second Joint Conference on Lexical and Computational Semantics (*SEM), Volume 1: Proceedings of the Main Conference and the Shared Task: Semantic Textual Similarity, pp. 32–43. Association for Computational Linguistics, June 2013
6. Cer, D., Diab, M., Agirre, E., Lopez-Gazpio, I., Specia, L.: SemEval-2017 task 1: semantic textual similarity-multilingual and cross-lingual focused evaluation. In: Proceedings of the 11th International Workshop on Semantic Evaluation (SemEval 2017), Vancouver, Canada, pp. 1–14. Association for Computational Linguistics (2017)
7. Sultan, M.A., Bethard, S., Sumner, T.: DLS@CU: sentence similarity from word alignment and semantic vector composition. In: Proceedings of the 9th International Workshop on Semantic Evaluation, pp. 148–153. Association for Computational Linguistics, June 2015
8. Wu, H., Huang, H.Y., Jian, P., Guo, Y., Su, C.: BIT at SemEval-2017 task 1: using semantic information space to evaluate semantic textual similarity. In: Proceedings of the 11th International Workshop on Semantic Evaluation (SemEval 2017), pp. 77–84. Association for Computational Linguistics, August 2017
9. Rychalska, B., Pakulska, K., Chodorowska, K., Walczak, W., Andruszkiewicz, P.: Samsung Poland NLP team at SemEval-2016 task 1: necessity for diversity; combining recursive autoencoders, WordNet and ensemble methods to measure semantic similarity. In: Proceedings of the 10th International Workshop on Semantic Evaluation (SemEval 2016), pp. 602–608. Association for Computational Linguistics, June 2016
10. Brychcín, T., Svoboda, L.: UWB at SemEval-2016 task 1: semantic textual similarity using lexical, syntactic, and semantic information. In: Proceedings of the 10th International Workshop on Semantic Evaluation (SemEval 2016), pp. 588–594. Association for Computational Linguistics, June 2016
11. Wieting, J., Bansal, M., Gimpel, K., Livescu, K.: Towards universal paraphrastic sentence embeddings. In: International Conference on Learning Representations (ICLR) (2015)
12. Arora, S., Liang, Y., Ma, T.: A simple but tough-to-beat baseline for sentence embeddings. In: International Conference on Learning Representations (ICLR) (2016)
13. Pennington, J., Socher, R., Manning, C.D.: GloVe: global vectors for word representation. In: Proceedings of the 2014 Conference on Empirical Methods in Natural Language Processing, pp. 1532–1543, October 2014
14. Peters, M.E., et al.: Deep contextualized word representations. In: Proceedings of the North American Chapter of the Association for Computational Linguistics: Human Language Technologies (2018)
15. McCann, B., Bradbury, J., Xiong, C., Socher, R.: Learned in translation: contextualized word vectors. In: Advances in Neural Information Processing Systems, pp. 6297–6308 (2017)
16. Wang, A., Singh, A., Michael, J., Hill, F., Levy, O., Bowman, S.R.: GLUE: a multi-task benchmark and analysis platform for natural language understanding. In: International Conference on Learning Representations (ICLR) (2019)

17. Yang, Z., Dai, Z., Yang, Y., Carbonell, J., Salakhutdinov, R.R., Le, Q.V.: XLNet: generalized autoregressive pretraining for language understanding. In: Advances in Neural Information Processing Systems, pp. 5753–5763 (2019)
18. Sun, Y., et al.: ERNIE 2.0: a continual pre-training framework for language understanding. In: AAAI, pp. 8968–8975 (2020)
19. Islam, A., Inkpen, D.: Semantic text similarity using corpus-based word similarity and string similarity. ACM Trans. Knowl. Discov. Data (TKDD) **2**(2), 1–25 (2008)
20. Li, Y., McLean, D., Bandar, Z.A., O'shea, J.D., Crockett, K.: Sentence similarity based on semantic nets and corpus statistics. IEEE Trans. Knowl. Data Eng. **18**(8), 1138–1150 (2006)
21. Wu, H., Huang, H.: Sentence similarity computational model based on information content. IEICE Trans. Inf. Syst. **99**(6), 1645–1652 (2016)
22. Hotelling, H.: Canonical correlation analysis (CCA). J. Educ. Psychol. **10** (1935)
23. Foster, D.P., Kakade, S.M., Zhang, T.: Multi-view dimensionality reduction via canonical correlation analysis (2008)
24. Golub, G.H., Reinsch, C.: Singular value decomposition and least squares solutions. In: Bauer, F.L. (ed.) Linear Algebra, pp. 134–151. Springer, Heidelberg (1971). https://doi.org/10.1007/978-3-662-39778-7_10

Data Science: Architectures

i-Fence: A Spatio-Temporal Context-Aware Geofencing Framework for Triggering Impulse Decisions

Jaiteg Singh[1], Amit Mittal[2(✉)], Ruchi Mittal[1], Karamjeet Singh[3], and Varun Malik[1]

[1] Institute of Engineering and Technology, Chitkara University, Punjab, India
[2] Chitkara Business School, Chitkara University, Punjab, India
amit.mittal@chitkara.edu.in
[3] Department of CSE, Thapar University, Punjab, India

Abstract. Unlike traditional recommender systems, temporal and contextual user information has become indispensable for contemporary recommender systems. The internet-driven Context-Aware Recommender Systems (CARS) have been used by big brands like Amazon, Uber and Walmart for promoting sales and boosting revenues. This paper argues about the relevance of contextual information for any recommender system. A smartphone-based CARS framework was proposed to record temporal information about user actions and location. The framework was deployed within a geofenced area of five hundred meters around a popular food joint to record contextual and temporal information of seventy-two volunteers for ninety consecutive days. The framework was augmented with an integrated analytics engine to generate task-specific, intrusive, location-based personalized promotional offers for volunteers based upon their gained temporal and contextual knowledge. There were a total of 1261 promotional offers sent under different categories to registered smartphones of volunteers out of which 1016 were redeemed within the stipulated period. The food joint experienced a rise of 6.45% in its average weekly sales and proved the versatility of such a system for profitable business ventures and capital gain. The significance of the rise in sales was also confirmed statistically.

Keywords: Context-aware recommender systems · Geofencing · Mobile promotions · Sales promotion · Temporal and contextual information

1 Introduction

A variety of traditional recommender systems merely focused upon recommending relevant items to individual users without taking into account their contextual information such as their geographical location, time, and their likes and dislikes as a community or target group. Temporal contextual information has become indispensable for contemporary recommender systems to recommend personalized content to any prospective consumer. The use of such contextual information, over the years, to anticipate consumer preferences has been duly supported by the outcomes of consumer decision

© Springer Nature Switzerland AG 2020
L. Bellatreche et al. (Eds.): BDA 2020, LNCS 12581, pp. 63–80, 2020.
https://doi.org/10.1007/978-3-030-66665-1_5

making research [1, 2]. Therefore, correct contextual information is indispensable for a recommender system to anticipate precise consumer preferences. The recommender systems considering contextual information of users before recommending any item are called Context-Aware Recommender Systems (CARS). Such CARS have become versatile since they became capable of sharing contextual information through the internet. The Internet has resulted in a more connected and interactive community. The internet revolution has permanently changed the way people used to communicate among themselves [3].

The internet initially connected personal computers then notebook computers and eventually connected smartphones and palmtop computers for sharing resources and information. Contemporary smartphones have virtually substituted desktop and laptop computers for browsing the internet. The smartphones are not only powerful computing devices but are also power efficient. Mobile applications played a major role to augment the popularity and success of the mobile internet era [4]. The number of smartphone users is nearing 3 billion, and 90% of the world's population is expected to own a mobile phone by the year 2020 [5]. Given the growing popularity of smartphones, their use as a potential channel for context-aware product promotion, branding, customer relationship management, and customer retention cannot be ignored. Brands have already started exploiting the potential of mobile technology for promoting their products. Mobile promotion platforms offer unprecedented openings to marketers to get connected with consumers by exploiting the geographical data and consumer behavior of smartphone users. Mobile promotions are defined as information delivered on the mobile device to offer an exchange of value, with the sole objective of driving a specific behavior in the short term [6]. The most prominent way adopted by corporations to send mobile promotions to target consumers is the use of geofencing.

Geofencing being a fusion of internet, mobile and geo-positioning technology is gaining attention because of its potential to promote context-based customized services and promotions [7, 8]. A geofence can be described as a virtual boundary marking any geographic area to monitor any predefined activity. Geofence is created using Global Positioning System (GPS) satellite consortium or any other Wi-Fi/ Bluetooth enabled radio frequency identifiers. The geofence logic is then paired with some hardware and software platform(s), which would respond to the geofencing boundary as required by defined hardware and software parameters. A few of the prominent geofencing based applications include Internet of Things (IoT) based device control, security, family tracking, and even marketing devices. Location-based marketing is considered a promising tool to augment the immersion level of mobile users. The contextual and temporal user information could be profitably exploited by marketers to promote products and services. Prospective customers, on the other hand, would get customized promotional offers from marketers according to their interests, needs, and behavior. Brands like Amazon, Uber, Voucher cloud and Walmart are actively using geofencing to boost sales and revenues. A generic geofencing deployment is shown in Fig. 1.

Such promotional strategies are not novel but have eventually evolved over decades through other channels such as in-store flyers, loyalty card discounts and mailers to stimulate impulse buying decisions among consumers. There are several existing overlapping theories explaining the effectiveness of short-term promotions and impulse buying. The

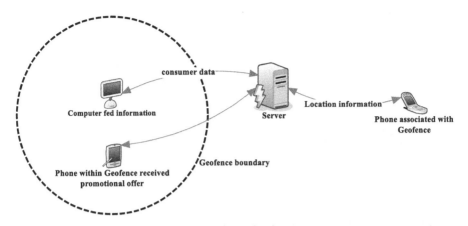

Fig. 1. Generic geofencing setup

theory of impulse buying defends the logic of purchases happening as a result of impulse or last-minute decision to buy an item that consumers had not planned [9]. Such impulse purchase can be triggered from events ranging from financial discounts or cashbacks, family combo-pack offers to events concerning store atmospherics, social concern or charity. Such unprecedented changes are expected to happen near the point of sale/ purchase terminal, where customers would have less time to deliberate on it. This temporal condition of impulse buying is extended by the theory of consumer construal level, it states that consumers would likely consider details associated with any promotional offer as soon as consumption occurs [10]. The time dimension cannot be isolated from the spatial context, where the location and timing of mobile promotion offer would influence consumers to redeem it [11]. If the mobile promotional offer is to be consumed at a farther place, then the promotion must provide more time to lead to consumers [12]. Furthermore, beyond considering only the spatiotemporal perspective, marketers must also may require to evaluate the contexts in which consumers may consider any mobile promotion.

Though there are applications like Google Now for collecting user opinions/ reviews for its clients, yet the data collected is kept strictly confidential and is never shared for research purposes. Traditional marketing strategies have a limited impact on resulting impulse purchase decisions. Such marketing strategies don't consider the spatiotemporal context of users in product promotion and sales optimization. The promising fusion of geofencing, context-aware mobile technology, and the absence of any relevant data encouraged the authors to experiment with it. This paper intends to evaluate the impact of contextual user data for triggering impulse purchases through custom mobile promotions. A context-aware recommender system was deployed within a geofenced area to evaluate its effectiveness in converting impulse consumer decisions to actual sales. About the chosen problem domain, the prime objectives of this paper are:

- To design and deploy a framework for managing geofence within a geographical area
- To collect temporal consumer behavior and location data

- To evaluate the potential of temporal consumer behavior and location data for triggering impulse decisions

The rest of this paper is organized as follows. Section 2 includes a discussion on work done so far in the design and usage of context-aware systems. The Sect. 3 is focused upon the experimental framework. Section 4 includes results and discussions and Sect. 5 concludes the experimental results and provides briefing about the future scope of the proposed framework.

2 Related Work

The research approach followed for the design and deployment of the proposed framework is guided from three different research domains namely CARS, geo-positioning, and mobile promotions. This section consists of three parts. The first part talks about the context-aware systems and their categories as suggested by various researchers. The second part is focused upon fixing the GPS position of mobile devices and the third part talks about the effect of mobile decisions on impulse decisions.

The use of recommender systems as a channel to suggest any product or service to the potential consumer has been a common practice. A recommender system can be defined as an application used to filter out useful information about a user for recommending any product to him. A reference to a personalized recommender system for tracking user interests and past buying behavior for making recommendations is suggested by many researchers [13, 14]. Recommender systems can be classified into Content based, personalized, Collaborative, and hybrid recommender systems and discussed the prime attributes of the aforesaid classifications [15]. The content-based recommender systems are recommended to adhere to information retrieval fundamentals and build a logical model based upon product attributes and historical user data, to recommend any product or service to users [16].

The relevance of moment detection function on the power consumption of mobile devices has always been a favorite area of research. The studies revealed that the motion detection function plays a vital role in reducing power consumption. A motion detection method to reduce the frequency of motion detection by considering the distance to the target spot instead of using the acceleration sensor of the mobile has been proposed [17, 18]. User-defined error bounds around the boundary of the target spot may affect the accuracy of the Geo Positioning System. Though speed can be used to identify and define the time for next position measurement yet no significant reduction in battery consumption of the mobile device can be attained [19]. The use of a state machine to control the interval of position measurement could provide an efficient location tracking mechanism [20]. Tracking the habitual activities of an individual too can be used to realize its location. Subsequently, it was found that location detection is highly dependent on the accuracy of predicting habitual activities [21]. A switch positioning method for position measurement made use of cell-ID for tracking user information. In this technique, the server notifies a cell-ID to a terminal so that the terminal need not measure position while it is away from the target spot [22]. A hybrid localization method to combine data from Global System for Mobile communications (GSM) and Global Positioning System

(GPS) experimented with Kalman filter. The objective of such fusion is to guess location within congested urban and indoor scenarios [23].

Owing to the attention given by people to their mobile devices, mobile promotions have superseded radio and print promotional channels. The increase in consumer touch-points as an outcome of mobile promotions cannot be ignored [24]. Mobile promotions are considered to have the potential for boosting unplanned, unrelated spending causing high margin sales [25]. There are direct financial benefits associated with mobile promotions. Mobile coupons can be received irrelevant to the location of the consumer. Further, consumers can also get non-financial benefits like free access to mobile data associated with any purchase they made following a mobile promotion [26]. Mobile intermediaries have empowered retailers to track point progress for their customers and make use of such data to build lasting relations with their consumers [27]. The value extraction and intrusiveness tradeoff are associated with mobile usage and mobile promotions of high value, yet there is an associated concern about personal consumer data being available over public and quasi-public networks. The case of AT&T and Version facing criticism for allowing partner organizations to track digital mobile footprints without consumers' knowledge for profits was also referred [28]. The mobile advertising augments retailers, service providers, and manufacturers to offer surprisingly relevant offers to consumers. The success of such campaigns is however claimed to be dependent on the understanding of environmental, consumer and technological context user variables. A framework to synthesize current findings in mobile advertising along with a research agenda to stimulate additional work to be explored in this field too was proposed by a few researchers [29]. Mobile devices are considered to be highly individualized and important communication tool which has allowed marketers to reach consumers more directly and consistently. The use of mobiles for surfing the internet and other custom mobile applications may help marketers understand the contemporary needs or interests of consumers and offer relevant advertisement content [30]. The ability of mobile devices to support location-based applications can be of great use for marketers. The use of such applications for finding the nearest point of interest like restaurants cannot be ignored [31]. The effects of hyper-contextual targeting according to physical crowdedness on consumer responses to mobile advertisements have already been examined. An experiment was carried out in collaboration with a renowned telecom service provider, who gauged physical crowdedness in teal time to send targeted mobile ads to individual users, measured purchase rates and provided a survey on purchasers and non-purchasers [32]. In another study, an attempt was made to bridge the gap between the theory and practice of mobile promotions and proposed a framework to benefit various stakeholders in the mobile promotion ecosystem. Further prevalent research questions regarding mobile promotions, privacy-value tradeoff, return on investment along with spatiotemporal targeting were also discussed [6]. A wide variety of research methods has been adopted by researchers to analyze consumer behavior. A research topology based on divergent behavioral research was introduced. Later it was found that mobile research has the potential to cause highly valid conclusions with a minimum effort [33]. The intersection of mobile and shopper marketing was termed as the mobile shopper market. The mobile shopper market was formally defined as the planning and execution of mobile-based activities which may influence any consumer for prospective purchase. The various challenges and opportunities were also discussed [34].

From the literature studied, it was concluded that contemporary literature did not discuss the promising association between location-based spatiotemporal services and mobile promotions. To the best of our knowledge, there is no empirical study published so far to evaluate the scope of context-aware spatiotemporal services like geofencing for measuring the impact of mobile promotions on consumers' impulsive purchase decisions.

3   Proposed Framework

The proposed framework is termed as Spatio-temporal context-driven framework for impulse decisions (SCFID). SCFID makes use of consumer demographics to generate custom promotional offers based on their purchase patterns and group behavior. The actual deployment of the said framework requires a dedicated server for managing geofence around the target area. The geofence server is equipped with an integrated analytics engine for observing actions, location and temporal information of registered users. Based upon the collected demographics the analytics engine is programmed to generate task-specific, intrusive, location-based personalized offers. Figure 2 corresponds to the proposed framework for deploying geofence and recording impulse decisions made by consumers concerning custom mobile promotions sent to them through the geofence server.

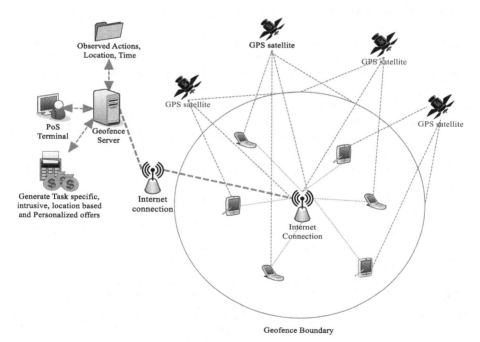

Fig. 2. SCFID framework to deploy geofencing for mobile promotions

Geofence server was fed by sales data relevant to registered users through the point of sale terminal. Geofence and mobile promotions were managed using a custom-built android application named Intelligent Fence (i-Fence). The said application operates in a distributed fashion and is controlled through a geofence server. The geofence server demarks a geofence around the point of sale location using its longitude and latitude coordinates. The circumference of the geofence can be decided during the initial system configuration and can be altered as required. Internet-enabled android mobile phones running client variant of i-Fence would keep on reporting the location of the mobile phone user to the geofence server. The geofence server would also be running an analytics engine to find out the most suitable promotional offer for any registered consumer based upon his/her past purchases. During this experiment, i-Fence monitoring was deployed on a food joint and was designed to record the following consumer demographics.

3.1 Capture Real-Time Feedback

Real-time feedback has proven to be a vital source of consumer retention and sales promotion. i-Fence was designed to record real-time consumer feedback while the order was served to any consumer at a food joint. A few simple questions were made to appear on a consumer's smartphone as soon as they walk out of the venue to get responses in real-time. The intention was to target and engage respondents instantly upon entrance or exit of the fenced location. To minimize false positives, the feedback was sought from only those i-Fence users who spent a predefined time within a fenced location, not from those who have just broken the fence. To ensure a robust consumer follow up mechanism, users who responded within a defined timeframe were allowed to answer feedback questions. Further, A mechanism to identify single or repeated visits within the fence was deployed to send one feedback survey invitation for n-number of visits instead of seeking feedback after every visit within the fenced area.

3.2 Offer Personalized Service

Every consumer was assigned a score concerning any purchase they have made. If a consumer has made a benchmark purchase or beyond within a stipulated period, it would earn him an additional loyalty score. Customized promotional offers were made to consumers according to their loyalty scores.

3.3 Hyper-Localized Deals

Consumers were offered promotional deals/ coupon on their mobile phones as soon as they enter the fenced area, instead of sending them through email. The coupon would be valid for a limited timeframe. When the user enters the geofence, the i-Fence would send the user a welcome message along with the promotional offer. At the same time, a message to the point of sale terminal would also be sent to expect the consumer's visit anytime soon.

3.4 Purchase Based Coupons

The consumers were prompted to visit the food joint, when they enter the fenced area. Besides, it was used to send information about available options to eat or drink. If they're a returning customer, then it would remind them of a previously ordered meal or beverage. The coupons were offered based upon the purchasing habits of consumers. About the past purchases made by consumers, Combo offers were designed for products generally sold together and found to have a strong association among them.

3.5 Tracking Key Metrics

Location data of consumers were used to track how long the consumers spend time at the food joint, the time of their visit, and the frequency and regularity of their visit. It was used to understand not only their purchase practices but also to design promotional coupons.

3.6 Tailored Festive/Special Offers

During special seasonal celebrations or festivals, special discount coupons were offered depending upon the age, gender, purchase habits, and other prerecorded demographics.
  The pseudocode for aforesaid logic is declared as under:

P denotes a set of uniquely identified persons {**P1......Pn**}
Each person **Pi** in set **P** is having the following attributes:
Gender-> Gender of the person i.e. male or female.
Age-> Age of the person in years.
In_time-> Time stamp at which the person entered the fence.
No_of_visits-> Number of times the person has visited the fence.
Score-> Cumulative points earned over every purchase in the past initially set to 0.
Past_Purchases-> Set of items purchased in the past by the person.
Location-> GPS co-ordinates of the person.
Current_Purchase-> Latest set of items purchased by the person.
Time_Threshold-> Minimum time in minutes to be spent in i-Fence to be eligible to fill questionnaire.
Score_Threshold-> Minimum score to be earned to be eligible for promotional offers.
Each coupon is generated based upon the following attributes:
"Coupon_expiry"- Expiry date of the coupon.
"C_Gender" - Gender to which coupon is to be sent.
"C_Age_Group" - Range of age group to which coupon is to be sent.
"C_product" - Product category on which coupon can be applied.

1. If Location (**x, y**) of person **Pi** is within (i-Fence):

    Calculate Time_spent = Minutes (Current_time - In_time);
    If Time_spent > = Time_Threshold
    No_of_visits +=1;

2. If no_of_visits 1 or > = 10:

   Send feedback form on Mobile of **Pi**.

3. If Location (**x, y**) of person **Pi** is within Area (i-Fence) and Current_Purchase! = Null

   Update Score based on current_purchase.
   If Score > Threshold_score and Current_Date <=Coupon_expiry:
   If Gender == C_Gender and Age lies between C_age_group and C_product matches with past_purchases:
   Send combo offers/ promotional offers/ discount coupons on Mobile.

4. If current_day = festival

   Send combo offers/ promotional offers/ discount coupons to most frequently visiting consumers.
   If current_day = Special day (like birthday of registered user(s))
   Send combo offers/ promotional offers/ discount coupons to specific consumers.

There were two algorithms deployed within i-Fence setup, one was programmed to run on server side (i-Fence Server), while the other was taking care of client-side (i-Fence Client). i-Fence server was deployed on the geofence server and was responsible to define and control the fenced area. Further, it kept on referring to the analytics engine for any suitable offer to be sent to any registered user. If there is an eligible candidate to receive a promotional offer, then i-Server would forward the same to the registered user. The i-Fence client was installed on smartphones running on Android 4.2 or above. Its primary objective was to record and send Spatio-temporal user demographics to the Analytics engine through i-Fence server. The algorithms used to deploy geofence and tracking mobile locations is as under:

Algorithm:1   i-Fence *Server*

Input: Ask for latitude and longitude values for PoS terminal
Constant: Latitude (Lat1) and Longitude (Long1) values for PoS and Geofence radius (in Meters/ Kilometers)
Begin:
Echo Latitude, Longitude and radius to i-Fence client mobile application
Get i-Fence client location;
Refer analytics engine to find suitable offer;
If eligible=yes;
Get coupon details;
Send coupon;
Else
Keep monitoring;
Endif
Exit

Algorithm:2   i-Fence *Client*
Input: Lat1 and Long1 as broadcasted by geofence server, Geofence radius, Real time Latitude (lat2) and Longitude (long2) values of consumer as supplied by smartphone's Geo Positioning System (GPS) receiver
Output: Firing a Hypertext Transfer Protocol Request to geoserver with authentication details and distance from PoS

Begin
Read lat1, long1 and lat2, long2 values;
Double R= 6371; (Radius of earth)
dlat= to radians(lat1-lat2); (Difference of latitudes in degrees of user and PoS)
dlong= to radians(long1-long2); (Difference of longitudes in degrees of user and PoS)
lat1= to radians(lat1);
lat2= to radians(lat2);
a=pow(sin(dlat/2),2) + pow(sin(dlong/2),2) * cos(lat1) * cos(lat2); [35]
C= 2 * asin(sqrt(a));
Total distance= R*C;
If Total distance <= Geofence radius;
Generate alert for Geofence server;
Else
Repeat;
Endif
Exit

4   Results and Discussion

Seventy-two participants voluntarily agreed to install i-Fence within their mobile phones and were frequent visitors of the food joint. The i-Fence application was developed to be used on the android platform only and the geofence server ran from 0800 h till 2100 h only. Android platform was chosen owing to the expertise of the team in developing android applications and because of the dominating market share of Android devices within India ("Mobile OS market share in India (Android, iOS) 2012–2016 I Statistic," n.d.). The participating android devices were connected to the geofence server through the internet and every registered device/ consumer was given a unique identification. The experiment lasted for ninety days from December 2016 till February 2017 around a fenced circular area with a radius of five hundred meters across the Point of Sale terminal (PoS) at the food joint.

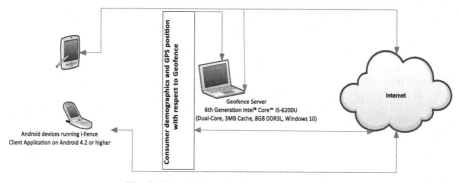

Fig. 3. Testbed for the proposed framework

The testbed deployed to evaluate the functionality of the proposed framework is shown in Fig. 3. The geofence server was hosted on a Dell machine with 6th generation Intel Core i5-6200U dual-core processor, 3 MB cache, 8 GB DDR3L random access memory and running windows 10 operating system. i-Fence client was installed on mobile devices running Android 4.2 or higher. Geofence server running i-Fence server and android devices running i-Fence client, were made to communicate through the internet. There was a total of seventy-two regular/ frequent customers who were convinced to install i-Fence client application on their phone. Among the seventy-two registered i-Fence client users, seven were proven to be dormant and did not provide any feedback or availed any coupon. There was a total of 4192 visits made by i-Fence client users and spent 127938 min within food joint.

Based upon the detailed analysis of consumer demographics, a total of 1152 promotional coupons were sent to registered users out of which 927 coupons were redeemed at the food joint, with a success percentage of 80.46%. A total of 72 coupons were sent on festivals like Christmas, New year eve, Lohri (a prominent festival of north India) out of which 54 were redeemed with a success ratio of 75%. Further a total of 37 promotional offers were sent on special days like Republic day of India, Birthdays of most frequent customers, and customers with highest loyalty score and there were 35 such coupons encashed thus providing a success ratio of 94% as shown in Fig. 7. Cumulative details of the experiment are provided in Table 1 and screenshots of i-Fence server and i-Fence client applications are shown in Figs. 4 and 5 respectively.

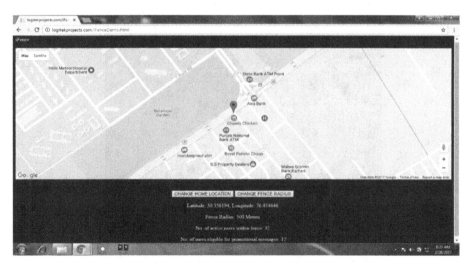

Fig. 4. i-Fence Server application hosted at Geoserver and controlled from PoS terminal

There were a few prominent limitations associated with this experiment. The most relevant limitation was that i-Fence client application was not published at publicly accessible platforms like Google Playstore, hence the proposed framework was tested in highly controlled environment. Secondly, the consumers who agreed to install i-Fence client were mostly employees of a nearby organization who were habitual to visit

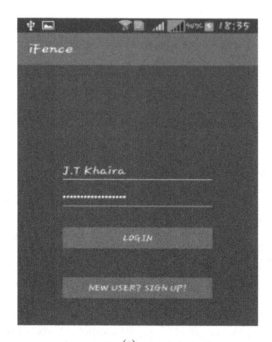

(a)

(b)

Fig. 5. i-Fence Client interface (a), (b), (c)

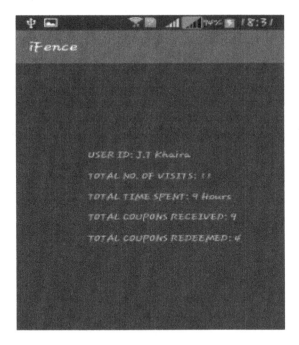

(c)

Fig. 5. (*continued*)

Table 1. Cumulative details of experiment

Total Volunteers	72
Dormant	7
Total Active	65
Total Number of Visits by i-Fence users	4192
Total Time Spent at Food Joint by i-Fence users	127938 min
Total Generic Promotional Coupons/offers sent	1152
Total coupons redeemed	927
Total coupons sent on Festive days	72
Total coupons redeemed	54
Total coupons sent on special days	37
Total coupons redeemed	35
Average Daily sales previous year for the same period	52,543 INR
Average Daily sales after i-Fence deployment	55,933 INR
Find Percentage Rise:	6.45%

food joint during lunch hours. Thirdly, owing to the limited infrastructure and resource availability, the experiment was undertaken for a limited period of ninety days. The sort of data generated and the count of promotional offers may not be sufficient to develop an intelligent system offering end to end support and solution to the chosen domain. The sole aim of this experiment was to evaluate the potential of temporal consumer and location data for triggering impulse decisions. As shown in Fig. 6, the results have shown a 6.45 percent increase in average daily sales of food joint within a short period of ninety days only, and seems promising if deployed as large scale full-time system to exploit margins resulting from impulse consumer decisions.

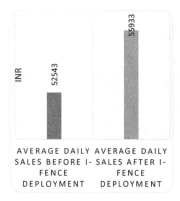

Fig. 6. Average daily sales before and after i-Fence deployment

Fig. 7. Comparison of coupons sent and redeemed under various categories

To validate the significance of the proposed framework, we assumed null hypothesis

H_0– There is no significant difference in the pre-and post SCIFD deployment sales.
H_1– There is a significant difference between pre-and post SCIFD deployment sales.

A two-tailed t-test was performed on the sales data belonging to month December 2015 till February 2016 (Pre deploying SCIFD framework) and from December 2016 till February 2017 (Post deploying SCIFD framework). The results have shown a significant difference between the two samples. Table 2 presents Paired Samples Statistics, Table 3 corresponds to paired samples correlation and Table 4 presents a briefing about the paired samples t-test, showing a significant difference. As the observed *p*-value is less than .05 hence H_0 is rejected and the H_1 is accepted suggesting a significant difference in Pre and Post SCIFD sales of the food joint. Furthermore, given that the mean score of sales post SCIFD deployment (52543) are more than sales pre SCIFD deployment (55933), it shows a significant increase in sales that can be attributed to the deployment of SCIFD. This proves the effectiveness of SCIFD.

Table 2. Paired samples statistics

		Mean	N	Std. deviation	Std. error mean
Pair 1	Sales 1 (Dec 2015-Feb 2016)	52543.0	90	18173.45	1915.65
	Sales 2 (Dec 2016-Feb 2017)	55933.0	90	19414.33	2046.45

Table 3. Paired samples correlations

		N	Correlation	Sig.
Pair 1	Sales 1 & Sales 2	90	.877	.000

Table 4. Paired Samples test

	Paired differences					t	df	Sig. (2-tailed)
	Mean	Std. Deviation	Std. Error Mean	95% Confidence Interval of the Difference				
				Lower	Upper			
Pair 1 Sales 1 - Sales 2	−3423.77778	9387.27259	989.50541	−5389.90396	−1457.65160	−3.460	89	.001

5 Conclusion and Future Scope

Location-based marketing is among the most promising methodologies associated with smartphones [36]. Due to the availability of affordable equipment, internet and smart devices, such technologies are now used for a wide array of purposes. The fusion of smartphone devices, applications, geofencing and other location-aware technologies has the potential to benefit almost every business in almost real-time. The experiment undertaken was able to push the sales of a moderate-sized local food joint by 6.45% within ninety days only, proving the yet underexplored potential of such technology in facilitating business ventures. In the future, similar geofencing establishment s could be augmented with the use of smart sensors recording the physical activity of consumers through the Internet of Things (IoT). It could result in smart promotional messages depending upon the physical activity, routine, dietary habits and according to the physical health standards of any potential consumer, for instance, a person suffering from obesity and hypertension may prefer to visit a store on the ground floor instead of visiting one on the seventh floor of a multiplex, or may even like to get things delivered at his/ her doorstep as per his requirements, or could be offered a sugar-free dessert after a meal.

To conclude, the geofencing based frameworks and applications can prove to be a major success factor for business ventures. Owing to the immense customization potential they offer; custom applications of such a framework can be deployed for almost every type of business venture.

References

1. Saeed, T.U., Burris, M.W., Labi, S., Sinha, K.C.: An empirical discourse on forecasting the use of autonomous vehicles using consumers' preferences. Technol. Forecast. Soc. Change **158**, 120130.—5.846 (2020)
2. Warlop, L., Ratneshwar, S.: The Role of Usage Context in Consumer Choice: a Problem Solving Perspective. NA - Advances in Consumer Research Volume 20 (1993). http://www.acrwebsite.org/volumes/7478/volumes/v20/NA-20
3. Touir, G.: Digitalization and Civic Engagement for the Environment: New Trends. Digitalization of Society and Socio-political Issues 2: Digital, Information and Research, 93–102 (2020)
4. Mittal, A., Aggarwal, A., Mittal, R.: Predicting university students' adoption of mobile news applications: the role of perceived hedonic value and news motivation. Int. J. E-Services Mob. Appl. (IJESMA) **12**(4), 42–59 (2020)
5. Jonsson, P.: Mobility Report, 4–7 February (2014). https://doi.org/10.3103/S0005105510050031
6. Andrews, M., Goehring, J., Hui, S., Pancras, J., Thornswood, L.: Mobile promotions: a framework and research priorities. J. Interact. Mark. **34**, 15–24 (2016). https://doi.org/10.1016/j.intmar.2016.03.004
7. Bareth, U., Kupper, A.: Energy-efficient position tracking in proactive location-based services for smartphone environments. In: 2011 IEEE 35th Annual Computer Software and Applications Conference, pp. 516–521. IEEE (2011). https://doi.org/10.1109/COMPSAC.2011.72
8. Martens, J., Bareth, U.: A declarative approach to a user-centric markup language for location-based services. In: Proceedings of the 6th International Conference on Mobile Technology, Application & Systems - Mobility 2009, pp. 1–7. ACM Press, New York (2009). https://doi.org/10.1145/1710035.1710073
9. Rook, D.W.: The buying impulse. J. Consum. Res. **14**(2), 189–199 (1987). https://doi.org/10.1086/209105
10. Liberman, N., Trope, Y.: The psychology of transcending the here and now. Science **322**(5905), 1201–1205 (2008). https://doi.org/10.1126/science.1161958
11. Danaher, P.J., Smith, M.S., Ranasinghe, K., Danaher, T.S.: Where, when, and how long: factors that influence the redemption of mobile phone coupons. J. Mark. Res. **52**(5), 710–725 (2015). https://doi.org/10.1509/jmr.13.0341
12. Luo, X., Andrews, M., Fang, Z., Phang, C.W.: Mobile targeting. Manage. Sci. **60**(7), 1738–1756 (2014). https://doi.org/10.1287/mnsc.2013.1836
13. Bag, S., Ghadge, A., Tiwari, M.K.: An integrated recommender system for improved accuracy and aggregate diversity. Comput. Ind. Eng. **130**, 187–197 (2019)
14. Ansari, A., Essegaier, S., Kohli, R.: Internet recommendation systems. J. Mark. Res. **37**(3), 363–375 (2000). https://doi.org/10.1509/jmkr.37.3.363.18779
15. Bouneffouf, D.: Mobile Recommender Systems Methods: An Overview (2013). http://arxiv.org/abs/1305.1745
16. Bobadilla, J., Ortega, F., Hernando, A., Gutiérrez, A.: Recommender systems survey. Knowl.-Based Syst. **46**, 109–132 (2013). https://doi.org/10.1016/j.knosys.2013.03.012

17. Lee, C.O., Lee, M., Han, D.: Energy-efficient location logging for mobile device. In: Proceedings - 2010 10th Annual International Symposium on Applications and the Internet, SAINT 2010, pp. 84–90 (2010). https://doi.org/10.1109/SAINT.2010.30

18. Chon, J., Cha, Hojung: LifeMap: a smartphone-based context provider for location-based services. IEEE Pervasive Comput. **10**(2), 58–67 (2011). https://doi.org/10.1109/MPRV.2011.13

19. Farrell, T., Cheng, R., Rothermel, K.: Energy-efficient monitoring of mobile objects with uncertainty-aware tolerances. In: 11th International Database Engineering and Applications Symposium (IDEAS 2007), pp. 129–140. IEEE (2007). https://doi.org/10.1109/IDEAS.2007.4318097

20. Taylor, I.M., Labrador, M.A.: Improving the energy consumption in mobile phones by filtering noisy GPS fixes with modified Kalman filters. In: 2011 IEEE Wireless Communications and Networking Conference, pp. 2006–2011. IEEE (2011). https://doi.org/10.1109/WCNC.2011.5779437

21. Constandache, I., Gaonkar, S., Sayler, M., Choudhury, R.R., Cox, L.: EnLoc: energy-efficient localization for mobile phones. In: IEEE INFOCOM 2009 - The 28th Conference on Computer Communications, pp. 2716–2720. IEEE (2009). https://doi.org/10.1109/INFCOM.2009.5062218

22. Bareth, U.: Privacy-aware and energy-efficient geofencing through reverse cellular positioning. In: 2012 8th International Wireless Communications and Mobile Computing Conference (IWCMC), pp. 153–158. IEEE (2012). https://doi.org/10.1109/IWCMC.2012.6314194

23. Fritsche, C., Klein, A.: Nonlinear filtering for hybrid GPS/GSM mobile terminal tracking. Int. J. Navig. Obs. **2010**, 1–17 (2010). https://doi.org/10.1155/2010/149065

24. Thompson, D.: Facebook, Google, and the Economics of Time. The Atlantic (2015). https://www.theatlantic.com/business/archive/2015/03/facebook-google-and-the-economics-of-time/387877/

25. Ramanathan, S., Dhar, S.K.: The effect of sales promotions on the size and composition of the shopping basket: regulatory compatibility from framing and temporal restrictions. J. Mark. Res. **47**(3), 542–552 (2010). https://doi.org/10.1509/jmkr.47.3.542

26. Barris, M.: McDonald's uses mobile data as currency in new promotion - Mobile Commerce Daily – Merchandising (2015). http://www.mobilecommercedaily.com/mcdonalds-offers-free-sms-messaging-with-food-purchases-in-philippines. Accessed 1 March 2017. Mobile OS market share in India (Android, iOS) 2012–2016|Statistic. (n.d.). https://www.statista.com/statistics/262157/market-share-held-by-mobile-operating-systems-in-india/. Accessed 1 March 2017

27. Pancras, J., Venkatesan, R., Li, B.: Investigating the Value of Competitive Mobile Loyalty Program Platforms for Intermediaries and Retailers, (15) (2015). http://www.msi.org/reports/investigating-the-value-of-competitive-mobile-loyalty-program-platforms-for/. Accessed

28. Singer, N., Chen, B.X. (n.d.): Verizon's Mobile "Supercookies" Seen as Threat to Privacy - The New York Times. https://www.nytimes.com/2015/01/26/technology/verizons-mobile-supercookies-seen-as-threat-to-privacy.html?_r=3. Accessed 1 Mar 2017

29. Grewal, D., Bart, Y., Spann, M., Zubcsek, P.P.: Mobile advertising: a framework and research agenda. J. Interact. Mark. **34**, 3–14 (2016). https://doi.org/10.1016/j.intmar.2016.03.003

30. Bacile, T.J., Ye, C., Swilley, E.: From firm-controlled to consumer-contributed: consumer co-production of personal media marketing communication. J. Interact. Mark. **28**(2), 117–133 (2014). https://doi.org/10.1016/j.intmar.2013.12.001

31. Grewal, D., Levy, M. (n.d.): Marketing (5th ed.). McGraw-Hill/Irwin, Burr Ridge

32. Andrews, M., Luo, X., Fang, Z., Ghose, A.: Mobile ad effectiveness: hyper-contextual targeting with crowdedness. Mark. Sci. **35**(2), 218–233 (2015). https://doi.org/10.1287/mksc.2015.0905

33. Cooke, A.D., Zubcsek, P.P.: The Promise and Peril of Behavioral Consumer Research Using Mobile Devices. Working Paper, pp. 1–33 (2016). http://ssrn.com/abstract=2835408%0A. Accessed

34. Shankar, V., Kleijnen, M., Ramanathan, S., Rizley, R., Holland, S., Morrissey, S.: Mobile shopper marketing: key issues, current insights, and future research avenues. J. Interact. Mark. **34**, 37–48 (2016). https://doi.org/10.1016/j.intmar.2016.03.002

35. Sharp, T.: How Big is Earth? (2012). http://www.space.com/17638-how-big-is-earth.html. Accessed 1 Mar 2017

36. Usman, M., Asghar, M.R., Ansari, I.S., Granelli, F., Qaraqe, K.A.: Technologies and solutions for location-based services in smart cities: Past, present, and future. IEEE Access **6**, 22240–22248 (2018)

Face Mask Detection Using Transfer Learning of InceptionV3

G. Jignesh Chowdary[1]([✉]), Narinder Singh Punn[2], Sanjay Kumar Sonbhadra[2], and Sonali Agarwal[2]

[1] Vellore Institute of Technology Chennai Campus, Chennai, India
guttajignesh.chowdary2018@vitstudent.ac.in
[2] Indian Institute of Information Technology Allahabad, Allahabad, India
{pse2017002,rsi2017502,sonali}@iiit.ac.in

Abstract. The world is facing a huge health crisis due to the rapid transmission of coronavirus (COVID-19). Several guidelines were issued by the World Health Organization (WHO) for protection against the spread of coronavirus. According to WHO, the most effective preventive measure against COVID-19 is wearing a mask in public places and crowded areas. It is very difficult to monitor people manually in these areas. In this paper, a transfer learning model is proposed to automate the process of identifying the people who are not wearing mask. The proposed model is built by fine-tuning the pre-trained state-of-the-art deep learning model, InceptionV3. The proposed model is trained and tested on the Simulated Masked Face Dataset (SMFD). Image augmentation technique is adopted to address the limited availability of data for better training and testing of the model. The model outperformed the other recently proposed approaches by achieving an accuracy of 99.9% during training and 100% during testing.

Keywords: Transfer learning · SMFD dataset · Mask detection · InceptionV3 · Image augmentation

1 Introduction

In view of the transmission of coronavirus disease (COVID-19), it was advised by the World Health Organization (WHO) to various countries to ensure that their citizens are wearing masks in public places. Prior to COVID-19, only a few people used to wear masks for the protection of their health from air pollution, and health professionals used to wear masks while they were practicing at hospitals. With the rapid transmission of COVID-19, the WHO has declared it as the global pandemic. According to WHO [3], the infected cases around the world are close to 22 million. The majority of the positive cases are found in crowded and overcrowded areas. Therefore, it was prescribed by the scientists that wearing a mask in public places can prevent transmission of the disease [8]. An initiative was

---

All authors have contributed equally.

© Springer Nature Switzerland AG 2020
L. Bellatreche et al. (Eds.): BDA 2020, LNCS 12581, pp. 81–90, 2020.
https://doi.org/10.1007/978-3-030-66665-1_6

started by the French government to identify passengers who are not wearing masks in the metro station. For this initiative, an AI software was built and integrated with security cameras in Paris metro stations [2].

Artificial intelligence (AI) techniques like machine learning (ML) and deep learning (DL) can be used in many ways for preventing the transmission of COVID-19 [4]. Machine learning and deep learning techniques allow to forecast the spread of COVID-19 and helpful to design an early prediction system that can aid in monitoring the further spread of disease. For the early prediction and diagnosis of complex diseases, emerging technologies like the Internet of Things (IoT), AI, big data, DL and ML are being used for the faster diagnosis of COVID-19 [19–21, 24, 26].

The main aim of this work is to develop a deep learning model for the detection of persons who are not wearing a face mask. The proposed model uses the transfer learning of InceptionV3 to identify persons who are not wearing a mask in public places by integrating it with surveillance cameras. Image augmentation techniques are used to increase the diversity of the training data for enhancing the performance of the proposed model. The rest of the paper is divided into 5 sections and is as follows. Sections 2 deals with the review of related works in the past. Section 3 describes the dataset. Section 4 describes the proposed model and Sect. 5 presents the experimental analysis of the proposed transfer-learning model. Finally, the conclusion of the proposed work is presented in Sect. 6.

2   Literature Review

Over the period there have been many advancements in the deep learning towards object detection and recognition in various application domains [18, 27]. In general, most of the works focus on image reconstruction and face recognition for identity verification. But the main aim of this work is to identify people who are not wearing masks in public places to control the further transmission of COVID-19. Bosheng Qin and Dongxiao Li [22] have designed a face mask identification method using the SRCNet classification network and achieved an accuracy of 98.7% in classifying the images into three categories namely "correct facemask wearing", "incorrect facemask wearing" and "no facemask wearing". Md. Sabbir Ejaz et al. [7] implemented the Principal Component Analysis (PCA) algorithm for masked and un-masked facial recognition. It was noticed that PCA is efficient in recognizing faces without a mask with an accuracy of 96.25% but its accuracy is decreased to 68.75% in identifying faces with a mask. In a similar facial recognition application, Park et al. [12] proposed a method for the removal of sunglasses from the human frontal facial image and reconstruction of the removed region using recursive error compensation.

Li et al. [14] used YOLOv3 for face detection, which is based on deep learning network architecture named darknet-19, where WIDER FACE and Celebi databases were used for training, and later the evaluation was done using the FDDB database. This model achieved an accuracy of 93.9%. In a similar research, Nizam et al. [6] proposed a GAN based network architecture for the removal of the face mask and the reconstruction of the region covered by the mask. Rodriguez et al. [17] proposed a system for the automatic detection of the presence or absence of the mandatory surgical mask in operating rooms. The objective of this system is to trigger alarms when a staff is not wearing a mask. This system achieved an accuracy of 95%. Javed et al. [13] developed an interactive model named MRGAN that removes objects like microphones in the facial images and reconstructs the removed region's using a generative adversarial network. Hussain and Balushi [11] used VGG16 architecture for the recognition and classification of facial emotions. Their VGG16 model is trained on the KDEF database and achieved an accuracy of 88%.

Following from the above context it is evident that specially for mask detection very limited number of research articles have been reported till date whereas further improvement is desired on existing methods. Therefore, to contribute in the further improvements of face mask recognition in combat against COVID-19, a transfer learning based approach is proposed that utilizes trained inceptionV3 model.

3 Dataset Description

In the present research article, a Simulated Masked Face Dataset (SMFD) [1] is used that consists of 1570 images that consists of 785 simulated masked facial images and 785 unmasked facial images. From this dataset, 1099 images of both categories are used for training and the remaining 470 images are used for testing the model. Few sample images from the dataset are shown in Fig. 1.

4 Proposed Methodology

It is evident from the dataset description that there are a limited number of samples due to the government norms concerning security and privacy of the individuals. Whereas deep learning models struggle to learn in presence of a limited number of samples. Hence, over-sampling can be the key to address the challenge of limited data availability. Thereby the proposed methodology is split into two phases. The first phase deals with over-sampling with image augmentation of the training data whereas the second phase deals with the detection of face mask using transfer learning of InceptionV3.

4.1 Image Augmentation

Image augmentation is a technique used to increase the size of the training dataset by artificially modifying images in the dataset. In this research, the training images are augmented with eight distinct operations namely shearing, contrasting, flipping horizontally, rotating, zooming, blurring. The generated dataset is then rescaled to 224 × 224 pixels, and converted to a single channel greyscale representation. Figure 2 shows an example of an image that is augmented by these eight methods.

Images without mask

Images with mask

Fig. 1. Sample images from the dataset.

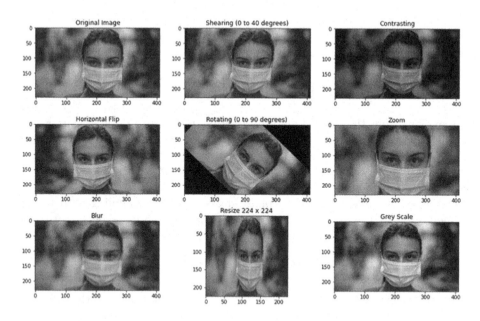

Fig. 2. Augmentation of training images.

4.2 Transfer Learning

Deep neural networks are used for image classification because of their better performance than other algorithms. But training a deep neural network is expensive because it requires high computational power and other resources, and it is time-consuming. In order to make the network to train faster and cost-effective, deep learning-based transfer learning is evolved. Transfer learning allows to transfer the trained knowledge of the neural network in terms of parametric weights to the new model. Transfer learning boosts the performance of the new model even when it is trained on a small dataset. There are several pre-trained models like InceptionV3, Xception, MobileNet, MobileNetV2, VGG16, ResNet50, etc. [5, 9, 10, 15, 23, 25] that are trained with 14 million images from the ImageNet dataset. InceptionV3 is a 48 layered convolutional neural network architecture developed by Google.

In this paper, a transfer learning based approach is proposed that utilizes the InceptionV3 pre-trained model for classifying the people who are not wearing face mask. For this work, the last layer of the InceptionV3 is removed and is fine-tuned by adding 5 more layers to the network. The 5 layers that are added are an average pooling layer with a pool size equal to 5×5, a flattening layer, followed by a dense layer of 128 neurons with ReLU activation function and dropout rate of 0.5, and finally a decisive dense layer with two neurons and softmax activation function is added to classify whether a person is wearing mask. This transfer learning model is trained for 80 epochs with each epoch having 42 steps. The schematic representation of the proposed methodology is shown in Fig. 3. The architecture of the proposed model is shown in Fig. 4.

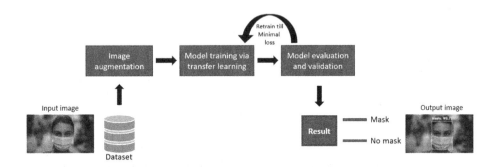

Fig. 3. Schematic representation of the proposed work.

5   Results

The experimental trials for this work are conducted using the Google Colab environment. For evaluating the performance of the transfer learning model several performance metrics are used namely Accuracy, Precision, Sensitivity,

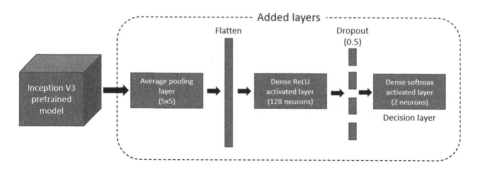

Fig. 4. Architecture of the proposed model.

Specificity, Intersection over Union (IoU), and Matthews Correlation Coefficient (MCC as represented as follows:

$$Accuracy = \frac{(TP + TN)}{(TP + TN + FP + FN)} \tag{1}$$

$$Precision = \frac{TP}{(TP + FP)} \tag{2}$$

$$Sensitivity = \frac{TP}{(TP + FN)} \tag{3}$$

$$Specificity = \frac{TN}{(TN + FP)} \tag{4}$$

$$IoU = \frac{TP}{(TP + FP + FN)} \tag{5}$$

$$MCC = \frac{(TP \times TN) - (FP \times FN)}{\sqrt{(TP + FP)(TP + FN)(TN + FP)(TN + FN)}} \tag{6}$$

$$CE = \sum_{i=1}^{n} Y_i \log_2(p_i) \tag{7}$$

In Eq. 7, Classification error is formulated in terms of Y_i and p_i where Y_i represents the one-hot encoded vector and p_i represents the predicted probability. The other performance metrics are formulated in terms of True Positive (TP), False Positive (FP), True Negative (TN), and False Negative (FN). The TP, FP, TN and FN are represented in a grid-like structure called the confusion matrix. For this work, two confusion matrices are constructed for evaluating the performance of the model during training and testing. The two confusion metrics are shown in Fig. 5 whereas Fig. 6, shows the comparison of the area under the ROC curve, precision, loss and accuracy during training and testing the model.

The performance of the model during training and testing are shown in Table 1. The accuracy of the transfer-learning model is compared with other machine learning and deep learning models namely decision tree, support vector machine [16] MobileNet, MobileNetV2, Xception, VGG16 and VGG19 when trained under the same environment, the proposed model achieved higher accuracy than the other models as represented in Fig. 7.

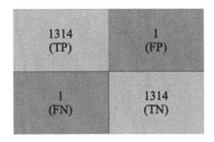 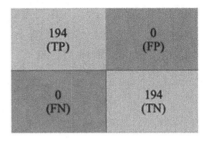

(a) Training Data   (b) Testing Data

Fig. 5. Confusion matrix.

Table 1. Performance of the proposed model.

Performance metrics	Training	Testing
Accuracy	99.92%	100%
Precision	99.9%	100%
Specificity	99.9%	100%
Intersection over Union (IoU)	99.9%	100%
Mathews Correlation Coefficient (MCC)	0.9984	1.0
Classification Loss	0.0015	3.8168E-04

5.1   Output

The main objective of this research is to develop an automated approach to help the government officials of various countries to monitor their citizens that they are wearing masks at public places. As a result, an automated system is developed that uses the transfer learning of InceptionV3 to classify people who are not wearing masks. For some random sample the output of the model shows the bounding box around the face where green and red color indicates a person is wearing mask and not wearing mask respectively along with the confidence score, as shown in Fig. 8.

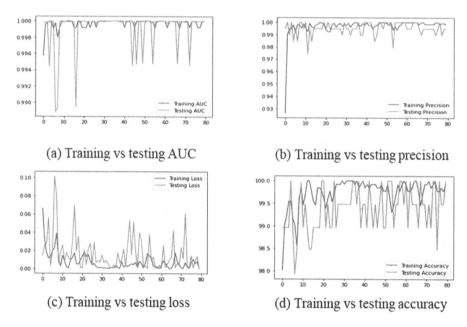

(a) Training vs testing AUC

(b) Training vs testing precision

(c) Training vs testing loss

(d) Training vs testing accuracy

Fig. 6. Comparison of performance of the proposed model during training and testing.

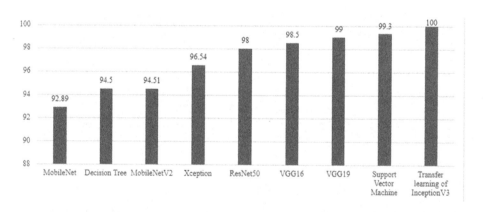

Fig. 7. Comparison of performance of the proposed model with other models.

Fig. 8. Output from the proposed model.

6   Conclusion

The world is facing a huge health crisis because of pandemic COVID-19. The governments of various countries around the world are struggling to control the transmission of the coronavirus. According to the COVID-19 statistics published by many countries, it was noted that the transmission of the virus is more in crowded areas. Many research studies have proved that wearing a mask in public places will reduce the transmission rate of the virus. Therefore, the governments of various countries have made it mandatory to wear masks in public places and crowded areas. It is very difficult to monitor crowds at these places. So in this paper, we propose a deep learning model that detects persons who are not wearing a mask. This proposed deep learning model is built using transfer learning of InceptionV3. In this work image augmentation techniques are used to enhance the performance of the model as they increase the diversity of the training data. The proposed transfer learning model achieved accuracy and specificity of 99.92%, 99.9% during training, and 100%, 100% during testing on the SMFD dataset. The same work can further be improved by employing large volumes of data and can also be extended to classify the type of mask, and implement a facial recognition system, deployed at various workplaces to support person identification while wearing the mask.

Acknowledgement. This research is supported by "ASEAN- India Science & Technology Development Fund (AISTDF)", SERB, Sanction letter no. – IMRC/AISTDF/ R&D/P-6/2017. Authors are also thankful to the authorities of "Vellore Institute of Technology", Chennai, India and "Indian Institute of Information Technology Allahabad", Prayagraj, India, for providing the infrastructure and necessary support.

References

1. Dataset. https://github.com/prajnasb/observations. Accessed 25 May 2020
2. Paris tests face-mask recognition software on metro riders. http://bloomberg.com/. Accessed 25 May 2020
3. Who coronavirus disease (COVID-19) dashboard. https://covid19.who.int/. Accessed 25 May 2020
4. Agarwal, S., Punn, N.S., Sonbhadra, S.K., Nagabhushan, P., Pandian, K., Saxena, P.: Unleashing the power of disruptive and emerging technologies amid COVID 2019: a detailed review. arXiv preprint arXiv:2005.11507 (2020)
5. Chollet, F.: Xception: deep learning with depthwise separable convolutions. CoRR abs/1610.02357 (2016). http://arxiv.org/abs/1610.02357
6. Din, N.U., Javed, K., Bae, S., Yi, J.: A novel GAN-based network for unmasking of masked face. IEEE Access **8**, 44276–44287 (2020)
7. Ejaz, M.S., Islam, M.R., Sifatullah, M., Sarker, A.: Implementation of principal component analysis on masked and non-masked face recognition. In: 2019 1st International Conference on Advances in Science, Engineering and Robotics Technology (ICASERT), pp. 1–5 (2019)
8. Feng, S., Shen, C., Xia, N., Song, W., Fan, M., Cowling, B.J.: Rational use of face masks in the COVID-19 pandemic. Lancet Respir. Med. **8**(5), 434–436 (2020)
9. He, K., Zhang, X., Ren, S., Sun, J.: Deep residual learning for image recognition. CoRR abs/1512.03385 (2015). http://arxiv.org/abs/1512.03385

10. Howard, A.G., et al.: MobileNets: efficient convolutional neural networks for mobile vision applications. CoRR abs/1704.04861 (2017). http://arxiv.org/abs/1704.04861

11. Hussain, S.A., Al Balushi, A.S.A.: A real time face emotion classification and recognition using deep learning model. J. Phys. Conf. Ser. **1432**, 012087 (2020)

12. Park, J.-S., Oh, Y.H., Ahn, S.C., Lee, S.-W.: Glasses removal from facial image using recursive error compensation. IEEE Trans. Pattern Anal. Mach. Intell. **27**(5), 805–811 (2005)

13. Khan, M.K.J., Ud Din, N., Bae, S., Yi, J.: Interactive removal of microphone object in facial images. Electronics **8**(10), 1115 (2019)

14. Li, C., Wang, R., Li, J., Fei, L.: Face detection based on YOLOv3. In: Jain, V., Patnaik, S., Popenţiu Vlădicescu, F., Sethi, I.K. (eds.) Recent Trends in Intelligent Computing, Communication and Devices. AISC, vol. 1006, pp. 277–284. Springer, Singapore (2020). https://doi.org/10.1007/978-981-13-9406-5_34

15. Liu, S., Deng, W.: Very deep convolutional neural network based image classification using small training sample size. In: 2015 3rd IAPR Asian Conference on Pattern Recognition (ACPR), pp. 730–734 (2015)

16. Loey, M., Manogaran, G., Taha, M.H.N., Khalifa, N.E.M.: A hybrid deep transfer learning model with machine learning methods for face mask detection in the era of the COVID-19 pandemic. Measurement **167**, 108288 (2020)

17. Nieto-Rodríguez, A., Mucientes, M., Brea, V.M.: System for medical mask detection in the operating room through facial attributes. In: Paredes, R., Cardoso, J.S., Pardo, X.M. (eds.) IbPRIA 2015. LNCS, vol. 9117, pp. 138–145. Springer, Cham (2015). https://doi.org/10.1007/978-3-319-19390-8_16

18. Punn, N.S., Agarwal, S.: Crowd analysis for congestion control early warning system on foot over bridge. In: 2019 Twelfth International Conference on Contemporary Computing (IC3), pp. 1–6. IEEE (2019)

19. Punn, N.S., Agarwal, S.: Automated diagnosis of COVID-19 with limited posteroanterior chest X-ray images using fine-tuned deep neural networks. Appl. Intell. 1–14 (2020). https://doi.org/10.1007/s10489-020-01900-3

20. Punn, N.S., Sonbhadra, S.K., Agarwal, S.: COVID-19 epidemic analysis using machine learning and deep learning algorithms. medRxiv (2020)

21. Punn, N.S., Sonbhadra, S.K., Agarwal, S.: Monitoring COVID-19 social distancing with person detection and tracking via fine-tuned YOLO v3 and deepsort techniques. arXiv preprint arXiv:2005.01385 (2020)

22. Qin, B., Li, D.: Identifying facemask-wearing condition using image super-resolution with classification network to prevent COVID-19 (2020)

23. Sandler, M., Howard, A.G., Zhu, M., Zhmoginov, A., Chen, L.: Inverted residuals and linear bottlenecks: mobile networks for classification, detection and segmentation. CoRR abs/1801.04381 (2018). http://arxiv.org/abs/1801.04381

24. Sonbhadra, S.K., Agarwal, S., Nagabhushan, P.: Target specific mining of COVID-19 scholarly articles using one-class approach. arXiv preprint arXiv:2004.11706 (2020)

25. Szegedy, C., Vanhoucke, V., Ioffe, S., Shlens, J., Wojna, Z.: Rethinking the inception architecture for computer vision. CoRR abs/1512.00567 (2015). http://arxiv.org/abs/1512.00567

26. Ting, D.S.W., Carin, L., Dzau, V., Wong, T.Y.: Digital technology and COVID-19. Nat. Med. **26**(4), 459–461 (2020)

27. Zhao, Z.Q., Zheng, P., Xu, S.T., Wu, X.: Object detection with deep learning: a review. IEEE Trans. Neural Netw. Learn. Syst. **30**(11), 3212–3232 (2019)

Analysis of GPS Trajectories Mapping on Shape Files Using Spatial Computing Approaches

Saravjeet Singh$^{(\boxtimes)}$ and Jaiteg Singh

Chitkara University Institute of Engineering and Technology, Chitkara University,
Rajpura 140401, Punjab, India
saravjeet.singh@chitkara.edu.in

Abstract. Locating the position of the device on the road network is a crucial component of a location-based system. The performance of location-based systems is highly affected by the mapping of user location on the digital map. Many spatial computing methods were developed by the research community to map the GPS fix on to the digital map. These mapping methods take GPS data and spatial data as input. The data used by these spatial computing algorithms is in huge amounts and is recorded using sensors and GPS receivers. While handling GPS data, these algorithms face many issues and based upon that each method has its advantages and disadvantages. This paper provides working of a few prominent methods. As per the research trends, four methods are considered and empirical evaluation and analysis are presented. To analyze the performance of these methods a standard data-set is used. 3 different routes having nodes in the range of 100 to 12598 are considered for this experiment. In this experiment mapping, results are analyzed using GPS trajectories of commutative distance 154.2 km. Performance and accuracy of considered map matching algorithms were analyzed on a total of 16804 GPS points. Results are analyzed using RMSE, accuracy ratio, and execution time. HMM-based map matching algorithm is considered as the most preferred algorithm having 94% accuracy with average execution time.

Keywords: GPS · Hidden Markov Model · Map matching · OpenStreetMap · Frechet distance

1 Introduction

As technology is advancing and smartphone become essential for many day to day activities. People are using mobile for banking, networking, entertainment, shopping, navigation, and many more. Navigation is one of the most using activities by the users using the smartphone. As per our observations, people primarily use smartphones for navigation. So there exists a huge demand for a mobile-based navigation system for route finding and route guidance activities. Spatial computing is a core component of the navigation system and is used to process spatial

© Springer Nature Switzerland AG 2020
L. Bellatreche et al. (Eds.): BDA 2020, LNCS 12581, pp. 91–100, 2020.
https://doi.org/10.1007/978-3-030-66665-1_7

data to find meaningful results. One simple example of spatial computing is map matching algorithms and is used to match the geographical information of the entity on a digital map and known as map matching algorithm. For the outdoor navigation system, map matching algorithms provide the mapping between data received from the GPS receiver with the spatial database. Main components of spatial computing are shown in Fig. 1.

Spatial Database. A spatial database is a database that provides spatial data types and spatial queries for the data storage and processing. It provides spatial indexing and spatial joins on the spatial data.

GPS Information. GPS information mainly deals with the temporal and positional information of the geographical entity. This positional information is used to find the actual location of the entity.

Spatial Computing Method. For navigation systems, map matching algorithms are main spatial computing methods. Map matching algorithm takes data from a spatial database, GPS receiver and optionally from (inertial sensors and other resources) and provides the actual location of the geographical entity.

User Output. This is actual output component which shows end result of the navigation system. It provides visual and audio output.

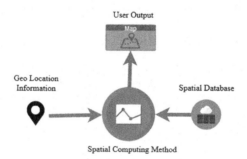

Fig. 1. Associated components in map matching process

1.1 Spatial Computing Techniques

Spatial computing techniques are main component of the navigation system. These acts as the heart of the navigation system. For this mapping application, we used map matching algorithms as a spatial computing technique. In a navigation system, these algorithms provide the actual location of the moving entity on the road network. As an output, these algorithms generate output on the GUI device. The main source of input for the map matching algorithm is shown in Fig. 2. Map matching algorithms are used for finding exact location the person

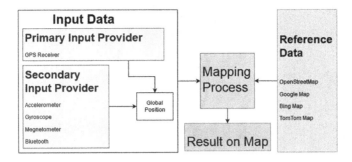

Fig. 2. Source of input for map matching algorithms

on road network and due to this, these algorithms are used in applications like route finding, geo-fencing, geo-tracking, distance between two location, shortest path problems, etc. Map matching algorithm uses two types of inputs; reference data and input data and are explained as below:

Input Data: Input data for map matching algorithm is a global position of a geographic entity (normally known as trajectories) and is fetched from GPS receiver, Bluetooth or inertial sensors. Below are the different type of input data providers:

- GPS receiver
- Accelerometer
- Gyroscope
- Megnetometer
- Bluetooth

Reference Data: Reference data is the actual data of the digital map. In map matching algorithm input data is mapped to reference data (shape file). Following are the type of reference data:

- OpenStreetMap
- Google Map
- Bing Map
- TomTom Map

Figure 3 demonstrate a typical map matching problem. A road network has n nodes. A vehicle is moving on the road, GPS receiver output (trajectory) is shown by red symbols. Location F is not on road network but it is away from the actual location. So map matching algorithm maps point F to the near by point on the road network. As in Fig. 3, point R1 and R2 are near to F and are on road network. So map matching algorithm provides best mapping of F with either R1 and R2.

Following are the few scenarios which cause poor performance of map matching algorithms [12,13]:

Fig. 3. Map matching scenario

- Errors within positioning sensors
- complex urban road networks pose
- Absence of GNSS signal in Canyon regions
- Data quality issues in reference data
- Sampling rate of received GNSS data
- Odd values in input data

Many map matching algorithms have been proposed and used by the research community. Each algorithm used different approach, models and parameters to provide the better mapping result. Research is never-ending process and to proposed new algorithm we must aware of shortcoming in available tools and techniques. So in this paper, we attempted is to identify the shortcoming of prominent map matching algorithms and then using the shortcoming for the invention of new technique. In this paper we analyzed the mapping of GPS trajectories on shape files using different spatial computing techniques. The structure of this paper is as follows: Sect. 2 provides literature study of used spatial computing methods, detail of considered algorithm for evaluation are presented in Sect. 3, Sect. 4 provides detailed result analysis and discussion, Sect. 5 provide conclusion and future scope.

2   Literature Survey

Research for vehicular navigation was started in the early 1980's using the hardware device and simple small digitized map. After the popularity of GPS, research studies were done to utilize the GPS data to find the location of moving vehicles on digital map [4]. Many studies in the early 20's proposed many methods for spatial computing techniques. These methods used the concepts like road information, vehicle information, the direction of movement, dead reckoning, Kalman filter, filtration on GNSS signal, GNSS modeling, orthogonal projection, error correction, extended Kalman filter, etc. [1,2]. To enhance the accuracy of spatial computing using map matching process many approaches based on advanced model and filters came into existence. These approaches were based on fuzzy logic, probability theory, path inference, sensor, etc. [3,7]. Probability theory played a very important role in the map matching process. Markov model (based on probability theory) provided a new direction to the map matching algorithm for online data [6,11]. Pattern and string matching techniques also used to enhance the performance of map matching algorithms [5]. Map matching using a genetic algorithm also provided good results [10]. To enhance the accuracy of the map matching algorithm, a dynamic two-dimensional algorithm was proposed [9]. Dijkstra's algorithm is further used to provide the map matching solution for the floating car data [8].

3   Considered Algorithm for the Spatial Computing

P2P: In the P2P method, a GNSS point is mapped to the closest node on the map. While implementing the P2P algorithm, only GNSS points and nodes of the road were considered. The perpendicular distance between the GNSS fix and nodes on the road were calculated. The point with minimum distance is calculated as a target matched point.

Topological Map Matching Algorithm: The topological map matching algorithm considers information of the road network, vehicle information, and GNSS fix to reconstruct the path. In this scenario, we considered road direction, road type, junction information, vehicle movement direction, and vehicle speed to fix the GNSS.

Hidden Markov Model Based Map Matching Algorithm: Hidden Markov model (HMM) based map matching algorithm uses the concept of the probability distribution to find the maximum likelihood between GPS fix and road network.

Frechet Distance Based Map Matching Algorithm: Frechet distance based map matching algorithm uses frechet distance between two-time variant curve to find the maximum likelihood between them. This algorithm uses free space diagram for finding the likelihood between two shape file and GPS trajectories.

4   Result Analysis

To analyze the performance and accuracy of map matching algorithms, an experiment is done. For the experimental study, we used P2P geometric, topological, HMM and frechet distance techniques. Three different routes were considered for the experiment and specifications of these routes are shown in Table 1. For test data (input data) collection, an android application was used to capture the GNSS receiver output after 1–5 s. 5 different users collected the data on different routes, their data collection details are shown in Table 2.

Total 5 GPS trajectories having 16804 GPS records were considered for the experiment. These trajectories were captured on the 3 routes. Reference data was selected from the OpenStreetMap dataset based on the selected routes. These

Table 1. Specification of considered route

Route no	Start address	End address	Distance
1	30.514851, 76.661137	30.300551, 76.844986	41 km
2	30.277279, 76.851552	29.946730, 76.821855	55.2 km
3	30.514925, 76.661524	30.518104, 76.658638	1.4 km

Table 2. Data collection detail

User no	Considered routes	Sampling interval	Collected GPS fix
1	1	5	3032
2	2	5	4198
3	2	2	9237
4	3	5	92
5	3	2	245

collected trajectories were mapped using the selected algorithms (output of these algorithms are shown in Fig. 4). Geometric and frechet based map matching algorithm provides incorrect results as shown in Fig. 4. Topological and HMM based map matching algorithm gives correct output as shown in Fig. 4(d) and (e).

In performed experiment map matching algorithm are analyzed by using matrices named Root Mean Square Error (RMSE), running time, and accuracy ratio. RMSE calculates the error of two datasets by comparing the mapping results as data series. Accuracy ratio calculates the mapping result by counting the correct and incorrect mapping points. According to performed experiment, HMM-based algorithm has less RMSE and high accuracy ratio whereas geometric algorithm has high RMSE and less accuracy ratio. Lower RMSE means high accuracy of the mapping. So according to RMSE value, HMM algorithm provides high accuracy. If we consider generalized result of analysis presented in Fig. 5 following is preference order of map matching algorithms:

1. HMM
2. Frechet based
3. Topological
4. Geometrical

Performance of an algorithm can be analyzed using execution time. Execution time is total time elapsed by an algorithm to provide the mapping result. Performance comparison of considered algorithms based upon the number of GPS points are shown in Fig. 6. According to performance comparison, P2P algorithm has low execution time. For less number of nodes, all algorithms have approximate same performance but at high node count, each algorithm has huge difference in execution time. If we consider generalized result of analysis presented in Fig. 6 following is preference order of map matching algorithm:

1. Geometrical
2. Topological
3. HMM
4. Frechet based

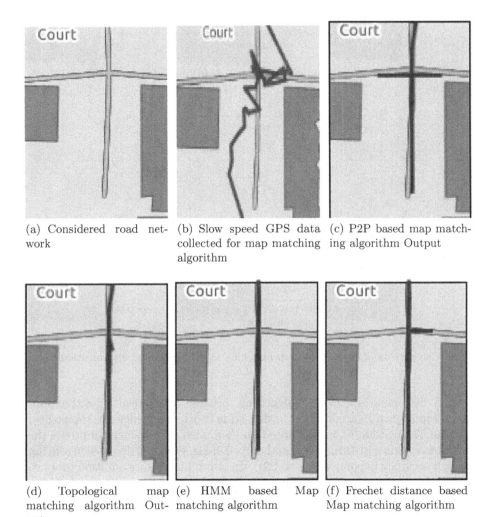

(a) Considered road network

(b) Slow speed GPS data collected for map matching algorithm

(c) P2P based map matching algorithm Output

(d) Topological map matching algorithm Output

(e) HMM based Map matching algorithm

(f) Frechet distance based Map matching algorithm

Fig. 4. Mapping output of four different selected map matching algorithms for same trajectory.

HMM based map matching algorithm have highest accuracy in comparison to other algorithm but average execution time. Topological algorithm has better

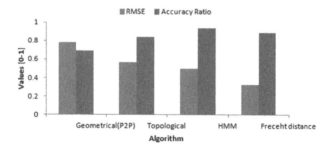

Fig. 5. Comparison of considered map matching algorithms based on RMSE and Accuracy ratio

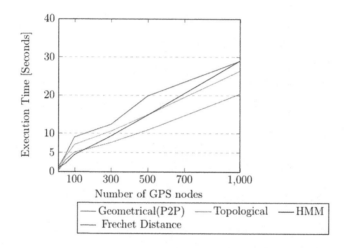

Fig. 6. Comparison of considered map matching algorithms based on execution time

accuracy in comparison to P2P algorithm, because topological algorithm considered topological information of both road network and vehicle. In topological algorithm the considered vehicle speed and direction of movement improves the algorithm accuracy at turn points and curved areas. Frenchet distance algorithm has high accuracy in comparison to P2P algorithm but require addition processing for free space calculation. Similarly P2P algorithm provides much faster result but has very low accuracy. Performance and accuracy both are important for map matching algorithm but we need to make a selection based on our requirement. Good accuracy and performance can not present in one algorithm so we need to select algorithm according to requirement. If we have sufficient processing power and space then we can select algorithm with higher accuracy. For scenario having low cost device with limited processing then preference should given to performance. According to combined analysis, if we considered average of both accuracy and performance then following is preference order to select an algorithm:

1. HMM
2. Topological
3. Frechet based
4. Geometrical

Algorithm Selection Chart

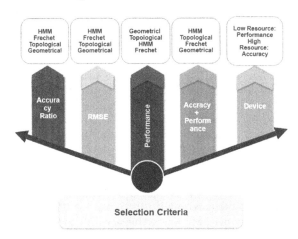

Fig. 7. Algorithm selection chart based on the user requirements and capabilities

5   Future Scope and Conclusion

In this paper, we analyzed the performance of spatial computing approaches (map matching algorithm) for mapping the GPS trajectories. Four prominent algorithms were selected and their empirical evaluation was presented. In this study, we considered point to point geometrical, topological, HMM-based, and frechet distance-based map matching algorithms. Performance and accuracy was evaluated and analyzed using the shape files of OSM data and GPS trajectories (collected using an android device). Test data was collected by five different users by traveling on three selected road network. GPS trajectories of five different users were mapped on OSM shape files. From the experimental results, it analyzed either algorithm has good performance or accuracy. For example, the HMM-based map matching algorithm has very high accuracy but poor performance and reverse is with P2P geometric map matching algorithm. As a conclusion, we presented a chart (as shown in Fig. 7) to select an algorithm based upon the user capabilities and requirement.

Accuracy and performance both are equally important but both at best range could not exist. So we need to make a selection. If the device has high processing power then accuracy should be preferred and if the device has low resources then, performance should be the preference. From this experiment, we concluded that map matching algorithm being spatial computing method still requires improvement. Research is never-ending process, so this study will act as a base for the new algorithm that have enhanced performance and accuracy. In future we will implement the changes in any of the selected algorithms so as to enhance the performance and accuracy.

References

1. Bouju, A., Stockus, A., Bertrand, F., Boursier, P.: Location-based spatial data management in navigation systems. In: Intelligent Vehicle Symposium 2002, vol. 1, pp. 172–177. IEEE (2002)
2. Greenfeld, J.S.: Matching GPS observations to locations on a digital map. In: Transportation Research Board 81st Annual Meeting (2002)
3. Hunter, T., Abbeel, P., Bayen, A.M.: The path inference filter: model-based low-latency map matching of probe vehicle data. In: Frazzoli, E., Lozano-Perez, T., Roy, N., Rus, D. (eds.) Algorithmic Foundations of Robotics X. STAR, vol. 86, pp. 591–607. Springer, Heidelberg (2013). https://doi.org/10.1007/978-3-642-36279-8_36
4. Iwaki, F., Kakihara, M., Sasaki, M.: Recognition of vehicle's location for navigation. In: Conference Record of Papers Presented at the First Vehicle Navigation and Information Systems Conference (VNIS 1989), pp. 131–138. IEEE (1989)
5. Miler, M., Todić, F., Ševrović, M.: Extracting accurate location information from a highly inaccurate traffic accident dataset: a methodology based on a string matching technique. Transp. Res. Part C Emerg. Technol. **68**, 185–193 (2016)
6. Mohamed, R., Aly, H., Youssef, M.: Accurate real-time map matching for challenging environments. IEEE Trans. Intell. Transp. Syst. **18**(4), 847–857 (2017)
7. Newson, P., Krumm, J.: Hidden Markov map matching through noise and sparseness. In: Proceedings of the 17th ACM SIGSPATIAL International Conference on Advances in Geographic Information Systems, pp. 336–343. ACM (2009)
8. Ptošek, V., Rapant, L., Martinovič, J.: Floating car data map-matching utilizing the Dijkstra's algorithm. In: Sharma, N., Chakrabarti, A., Balas, V.E. (eds.) Data Management, Analytics and Innovation. AISC, vol. 1016, pp. 115–130. Springer, Singapore (2020). https://doi.org/10.1007/978-981-13-9364-8_9
9. Sharath, M., Velaga, N.R., Quddus, M.A.: A dynamic two-dimensional (D2D) weight-based map-matching algorithm. Transp. Res. Part C Emerg. Technol. **98**, 409–432 (2019)
10. Singh, S., Singh, J., Sehra, S.S.: Genetic-inspired map matching algorithm for real-time GPS trajectories. Arab. J. Sci. Eng. **45**(4), 2587–2603 (2020)
11. Yang, C., Gidofalvi, G.: Fast map matching, an algorithm integrating hidden Markov model with precomputation. Int. J. Geogr. Inf. Sci. **32**(3), 547–570 (2018)
12. Zhao, F., Shin, J., Reich, J.: Information-driven dynamic sensor collaboration. IEEE Signal Process. Mag. **19**(2), 61–72 (2002)
13. Zhao, L., Ochieng, W.Y., Quddus, M.A., Noland, R.B.: An extended Kalman filter algorithm for integrating GPS and low cost dead reckoning system data for vehicle performance and emissions monitoring. J. Navig. **56**(2), 257–275 (2003)

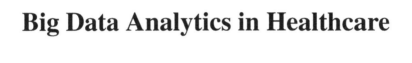

Big Data Analytics in Healthcare

A Transfer Learning Approach to Classify the Brain Age from MRI Images

Animesh Kumar[ID], Pramod Pathak, and Paul Stynes[(✉)][ID]

National College of Ireland, Dublin, Ireland
x18184731@student.ncirl.ie, {Pramod.Pathak,Paul.Stynes}@ncirl.ie

Abstract. Predicting brain age from Magnetic Resonance Imaging (MRI) can be used to identify neurological disorders at an early stage. The brain contour is a biomarker for the onset of brain-related problems. Artificial Intelligence (AI) based Convolutional Neural Networks (CNN) is used to detect brain-related problems in MRI images. However, conventional CNN is a complex architecture and the time to process the image, large data requirement and overfitting are some of its challenges. This study proposes a transfer learning approach using InceptionV3 to classify brain age from the MRI images in order to improve the brain age classification model. Models are trained on an augmented OASIS (Open Access Series of Imaging Studies) dataset which contains 411 raw and 411 masked MRI images of different people. The models are evaluated using testing accuracy, precision, recall, and F1-Scores. Results demonstrate that InceptionV3 has a testing accuracy of 85%. This result demonstrates the potential for InceptionV3 to be used by medical practitioners to detect brain age and the potential onset of neurological disorders from MRI images.

Keywords: Transfer learning · InceptionV3 · Neurological disorder · Brain age · MRI images

1 Introduction

The World Health Organization (WHO) indicates that 50 million people are suffering from neurological disorders such as Alzheimer Disease (AD). Medical practitioners are able to deduce the physiological or biological age of a person as the brain structure changes over time [8] which can assist with the early detection of neurological disorders. Age estimation can be performed based on either cortical anatomy or MRI images of grey matter, white matter, and cerebrospinal fluid present inside the brain as shown in Fig. 1.

Recently, AI-based CNN is revolutionizing the way medical data such as MRI Images are analyzed [1, 12, 16, 17]. However, conventional CNN is a complex architecture and the time to process the image, large data requirement and overfitting are some of its challenges [3, 13]. Transfer learning (TL) is a research problem in machine learning (ML) that focuses on storing knowledge gained while solving one problem and applying it to a different but related problem where training data could be partially or completely

© Springer Nature Switzerland AG 2020
L. Bellatreche et al. (Eds.): BDA 2020, LNCS 12581, pp. 103–112, 2020.
https://doi.org/10.1007/978-3-030-66665-1_8

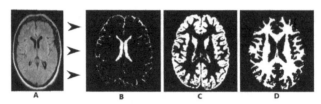

Fig. 1. Brain MRI segmentation - A. T1-weighted brain MRI, B. Cerebrospinal fluid, C. Grey matter, D. White matter

different from the testing data [14]. There are several types of Transfer Learning models such as InceptionV3 and DenseNet. Inceptionv3 is a convolutional neural network for assisting in image analysis and object detection. DenseNet is a type of convolutional neural network that utilizes dense connections between layers, through Dense Blocks, which connect each layer to every other layer in a feed-forward fashion. A DenseNet architecture is a logical extension of ResNet. Transfer learning (TL) based models such as InceptionV3 and DenseNet have been shown to decrease the processing time and model overfitting respectively [14]. It is generally used for smaller datasets for more accurate predictions.

The aim of this research is to investigate a novel transfer learning approach to classify brain age from the MRI images in order to improve the brain age classification model. The transfer learning approach uses InceptionV3. The model performance is evaluated through accuracy, precision, recall, F1-Scores, and processing time.

The contribution is a novel transfer learning approach that classifies age groups into six different categories such as 10 to 20, 21 to 30 years of age and so on where the age categories are known as bins.

This paper discusses related work in Sect. 2 with a focus on machine learning approaches to medical image classification. Section 3 discusses the research methodology used in this research. Section 4 discusses the results and discussion. Section 5 concludes the research and discusses future work.

2   Related Work

This section discusses various approaches to medical image classification such as machine learning, deep learning, and transfer learning.

Originally, analysis of the brain was mostly dependent on regression-based algorithms such as Relevance Vector Regression (RVR) [5, 6] in order to classify MRI images.

Franke et al. [5] used the RVR method to study the impact of diabetes mellitus on brain age using MRI images whilst Gaser et al. [6] used the RVR method to detect Alzheimer Disease (AD). The RVR method has a self-learning process to decide the parameters for the best model fit [6]. The RVR method had a limitation of skipping white matter lesions which is a biomarker for brain age prediction. Besteher et al. [2] included depression parameters to analyze changes in brain age. Although, the result demonstrated no major deviation in brain age. Nakano et al. [11] proposed a comparison between normal and abnormal development in a newborn baby. The model used two architecture Principal

Component Analysis Regression (PCAR) and Manifold Learning-PCAR (ML-PCAR), where ML-PCAR proved to be more accurate. However, the model failed to differentiate between 0 to 3-month-old new-borns due to the different diameters of the brain MRI.

Recently, AI has enhanced the analysis of MRI images using deep learning with techniques such as enhanced dimensionality reduction and feature extraction incorporated into the image classification models such as CNN. The CNN model has been used in the analysis of 2D [8] and 3D [12] MRI images.

Huang et al. [8] proposed a deep learning model VGG Net based on the CNN model to estimate the age of a person based on the brain MRI image. The model was applied to the IXI dataset. The results were comparable to recent research. However, the model was limited to healthy brain images and did not consider unhealthy MRI images. The research would not be able to identify if a 10-year-old child had a brain MRI of a 70-year-old person, which would indicate that the child was unhealthy and may have a neurological disorder.

Qi et al. [12] proposed an enhanced 3D CNN model with an added dense block (sub-DenseNet model) to estimate the age from MRI images. Their model showed that it helped to minimize the gradient vanishing problem and increases the fitting ability of the model. The use of a sub DenseNet demonstrates the potential to solve the overfitting problem.

Ueda et al. [16] proposed a 3D CNN model to estimate the age from brain MRI images. They used an Aoba medical center collected dataset of 1000 MRI images. The study reported improved accuracy. However, the 3D CNN architecture extracts high dimensionality features from the images at the cost of higher processing time.

Bermudez et al. [1] proposed a novel deep learning approach of using conventional CNN and volumetric feature processors to predict the brain age. They used the OASIS and IXI based datasets. The OASIS dataset images are present in GIF format with segmentation in grey matter, white matter, and cerebrospinal fluid features. The OASIS MRI images are smaller in size and consume less time to process in comparison to the NIFTY format used in the IXI dataset. The MRI images in the OASIS dataset are normalized, bias field corrected, and well documented in advance for research purposes [9] however the IXI dataset is not.

Wang et al. [17] investigated whether gray matter in brain atrophy is an established biomarker for dementia prediction in unhealthy people. This approach demonstrates the need to look at unhealthy people in order to identify Brain MRI segmentation such as grey matter.

Siar and Teshnehlab [15] proposed an AlexNet based CNN model. The age categories were divided into 5 bins ranging from 10 years to 70 years. The model is implemented with three different classification layers (SVM, Decision Tree, SoftMax). The SoftMax demonstrated the highest accuracy of 79% on 1290 images that were self-collected. The accuracy with unequal age categories bins provides an aid to medical practitioners to narrow down the patient with possible neurological disorders as older people tend to have more neurological disorders. The proposed classification of age groups in bins is of interest in this research.

CNN based models were proved to be highly efficient on large datasets using complex architectures like ResNet and AlexNet. However, it takes a longer processing time. CNN models also require a large dataset in order to minimize the overfitting problem. This would indicate a need to look at alternative models to deep learning.

Transfer learning is an approach used to address the problem of multiple domain training. The transfer learning-based models used to extract features from one domain and apply it to another similar domain for model training [14].

Ren et al. [13] proposed a transfer learning-based 3D CNN model that was trained on a UK Biobank[1] dataset that contains 9850 MRI images. The trained weights are used for model training on an NKI[2] dataset having 395 MRI images. The approach of having the transfer learning model trained on a large dataset then the transfer learning model being applied to a smaller dataset for training worked well for solving the overfitting problem.

Transfer learning-based models were also implemented for brain tumor detection [3]. The research compared 9 different transfer learning models for predicting tumor classes. The result demonstrated that the models with fewer layers performed better than models with a higher number of layers like ResNet101 with 3064 images.

Ding et al. [4] proposed a model to diagnose Alzheimer Disease from MRI images in early stages using InceptionV3. The result demonstrated that the model has a lower precision of 55% for mild cognitive impaired (MCI) on an ADNI dataset. The result also demonstrated a precision of 18% with an independent dataset that is a self-collected dataset. The result demonstrated that the model has a dependency on the clinal distribution of MRI images per class for the training dataset.

The literature review indicates that conventional CNN models face challenges such as model overfitting, and processing time with small datasets. The literature also indicates that there is a need to include MRI images from both healthy and unhealthy people in order to classify age estimation based on grey matter, white matter, and cerebrospinal fluid. Alternative approaches such as transfer learning show promise for addressing the large processing time and overfitting. For overfitting, the transfer learning models can be trained on a large dataset such as ImageNet. Then the transfer learning model can be applied to a smaller dataset for training. The literature demonstrates that transfer learning approaches like DenseNet is a useful approach for solving the overfitting problem. The review also demonstrates that InceptionV3 has better performance on a smaller dataset and reduces the overall model processing time. This review demonstrates the need for a classification model that can handle a full Brain MRI segmentation on grey matter, white matter, and cerebrospinal fluid in both healthy and unhealthy people.

3   Research Methodology

The research methodology of this study discusses the step-by-step process as shown in Fig. 2.

The first step involves the collection of data from the OASIS neuro-imaging dataset [9]. The OASIS neuro-imaging dataset consists of 436 different brain MRI images in various masking forms such as increasing the contrast of the MRI images. The data set

---

[1] https://www.ukbiobank.ac.uk/data-showcase/.

[2] https://fcon_1000.projects.nitrc.org/indi/enhanced/.

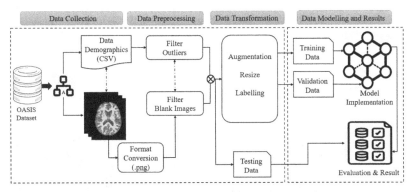

Fig. 2. Process flow diagram

is used for detecting brain age and neurological disorders such as Alzheimer disease. The mean age of the dataset is 51.35 years with age ranging from 18 years to 96 years. The changes in brain contour with ae are shown below as shown in Fig. 3.

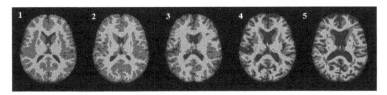

Fig. 3. Change in brain orientation with aging – 1. 20yr 2. 40yr 3. 60yr 4. 80yr 5. 96yr

The second step involves data preprocessing in which the dataset was checked for blank images. If there were blank images, then the corresponding demographic data was removed. The dataset is converted from GIPHY to (Portable Graphic Format) PNG format as the GIPHY short moving images take up 4 times the memory and increase the processing time. The PNG MRI images consisted of 436 raw and 436 masked images. The pre-processed MRI images were resized into 175X175 pixels and 3 channels using OpenCV2 library and saved with their unique ids in different class folders. Outliers were removed leaving an augmented OASIS dataset with 411 raw and 411 masked images as shown in Table 1. Images and labels were categorized into 6 different age group bins as shown in Table 1. Data was split into training and test sets. The test set was taken as 20% of the dataset. The training set was augmented, resized, and labeled as per the model requirement and split into training and validation set with a ratio of 80:20 respectively. Categorized test, train, and validation data were fed for model training and testing. Model weights and information per epochs were stored for evaluation and analysis. Model Evaluation was performed using metrics like accuracy, precision, recall, and F1-Score.

The third step involves data transformation in which the MRI images were augmented with 12 different filters for data augmentation [10] as shown in Fig. 4.

Table 1. Class distribution according to age

Age Groups	10–20	21–30	41–60	61–70	71–80	81–90	Total
Bin	1	2	3	4	5	6	–
Data	50	108	64	45	89	55	411

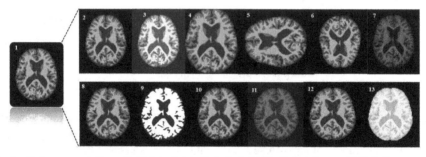

Fig. 4. Data augmentation – 1. Original data 2. Left-right flip 3. Brightness (0.2) 4. Center cropping (0.8) 5. Rotation 90 6. Upside down 7. Random contrast 8. Saturation (10) 9. Adjust contrast (8), 10. Random Hue 11. Segmented 12. Random gamma 13. Random saturation

The fourth step involves data modeling and results. Data modeling involves the implementation of InceptionV3 and DenseNet. The InceptionV3 model has a parallel processing mechanism with 11 stacked inception modules shown in different color coding in Fig. 5. Each block uses the same color inception sub-block having convolutional filters, pooling layers, and activation functions as rectified linear units. The concatenation layer is added with fully connected layers of size 1024 and 9 with SoftMax classifier [4].

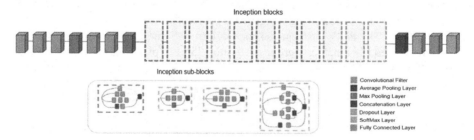

Fig. 5. InceptionV3 network architecture [4]

DenseNet is a deeper model with five dense blocks and feature reuse mechanism using a concatenation network [7]. It contains all the similar feature maps connected to each other to preserve feed-forward nature [7]. Each dense block comprises batch normalization, ReLU activation function, CNN and max-pooling layers. Pretrained models like InceptionV3 and DenseNet have a definite set of layers with varied fully connected layers as per requirement.

The training and validation dataset were passed through the image generator which fetched labels and performed real-time augmentation. The models were implemented in python language using Keras neural network library, integrated on top of TensorFlow framework. The epoch count (100) and image size 175X175 were agreed based on several model iterations and kept constant throughout the experiments for comparison. The total MRI images after augmentation for training are 7896. Model execution logging like accuracy, validation, and losses were stored in CSV file using CSV logger and improved weights were stored in google drive for future reference. An early stopping mechanism was also integrated for efficient model run. A generalized step per epoch and validation per epoch were assigned having values equals to the training or validation count upon batch size.

The InceptionV3 and DenseNet model was supported by pre-trained weights from ImageNet imported using Keras library. The hyperparameters used are batch size of 32, learning rate of 0.0001, categorical cross-entropy as loss function, and Adam as optimizer.

4   Results and Discussion

The aim of this experiment was to compare the accuracy of transfer learning models DenseNet and InceptionV3 in order to improve the brain age classification model. Model performance was evaluated using accuracy, precision, recall, F1-Scores metrics, and processing times as shown in Table 2. In order to make a useful comparison to the DenseNet and InceptionV3 experiments, this research compared the results with the deep learning model proposed by Huang et al. [8].

Table 2. InceptionV3, DenseNet and 2D CNN model comparison.

Method	Testing accuracy	Precision accuracy	Recall	F1-scores	Time (Sec)
2D-CNN	47%	11%	28%	15%	19124
DenseNet	60%	18%	17%	17%	89982
InceptionV3	85%	86%	85%	84%	7418

Huang et al. [8] proposed a 2D-CNN model based on VGG Net to estimate the age of a person based on the brain MRI image using the IXI dataset. This research replicated this experiment and extended the experiment to the augmented OASIS dataset instead of the IXI dataset. The 2D-CNN model was simulated using 5 convolutional layers. Results demonstrate that the testing accuracy with 411 raw MRI images and 411 masked MRI images. The testing accuracy was 47% as shown in Table 2. This indicates a problem of model overfitting due to unequal samples in the validation dataset. These results demonstrate that the CNN model is not suitable for analyzing MRI images on a dataset of 411 raw and 411 masked images.

The DenseNet model accuracy and model loss for the training and validation dataset is shown in Fig. 6. The model achieved a testing accuracy of 60% with precision as low as 18% for the fourth bin (61 to 70 years age) containing 45 MRI. The model accuracy and loss curves are inconsistent as shown in Fig. 6. This indicates a problem of model overfitting due to mis-representation of images using a validation dataset. These results demonstrate that the DenseNet model is not suitable for analyzing MRI images on a dataset of 411 raw and 411 masked images.

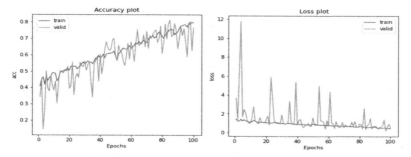

Fig. 6. DenseNet model accuracy and loss curve plots

The InceptionV3 model accuracy and model loss curves for the training and validation dataset is shown in Fig. 7. Both model accuracy and loss are consistent with a constant increase in training and validation accuracy in the accuracy plot as shown in Fig. 7.

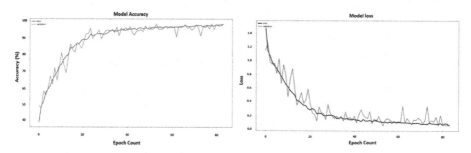

Fig. 7. InceptionV3 accuracy and loss curve

The validation accuracy and training accuracy are comparable which demonstrates that the model is not overfitting. Results show that the InceptionV3 model achieved a testing accuracy of 85% as shown in Table 2. The precision of 86% demonstrates that the model correctly classified MRI images of people with neurological disorders. The precision shows promise for using transfer learning models such as InceptionV3 for the analysis of MRI images. The result demonstrates that the InceptionV3 model's processing time is $1/12th$ the time taken by DenseNet and $2\frac{1}{2}$ times faster than 2D CNN models.

The processing time of the models has a correlation to the number of training parameters. The InceptionV3 model has a lesser number of training parameters than DenseNet. Due to which InceptionV3 processing time is $1/12th$ of DenseNet processing time on GPU. The results demonstrate that the testing accuracy for 2D-CNN is nearly half of the InceptionV3 model. It shows the inefficiency of 2D CNN on small dataset like the augmented OASIS with 411 raw and 411 masked images. The results show promise that InceptionV3 outperforms 2D-CNN and DenseNet on a smaller dataset like the augment OASIS dataset.

5   Conclusion and Future Work

This research proposed a transfer learning approach to classify brain age from MRI images in order to improve the brain age classification model. Two transfer learning models were compared namely, InceptionV3 and DenseNet. Results demonstrate that the InceptionV3 model outperformed DenseNet by 40% with a testing accuracy of 85% using an augmented OASIS dataset with 411 raw and 411 masked images after preprocessing. The results show promise for assisting medical practitioners in the early detection of neurological disorders with small datasets. Future work includes extending this research to investigate the application of transfer learning on larger datasets with 1290 to 3064 images.

References

1. Bermudez, C., et al.: Anatomical context improves deep learning on the brain age estimation task. Magn. Reson. Imaging **62**, 70–77 (2019)
2. Besteher, B., Gaser, C. and Nenadić, I.: Machine-learning based brain age estimation in major depression showing no evidence of accelerated aging. Psychiatry Res. Neuroimaging **290**, 1–4 (2019)
3. Chelghoum, R., Ikhlef, A., Hameurlaine, A., Jacquir, S.: Transfer learning using convolutional neural network architectures for brain tumor classification from MRI images. In: Maglogiannis, I., Iliadis, L., Pimenidis, E. (eds.) AIAI 2020. IAICT, vol. 583, pp. 189–200. Springer, Cham (2020). https://doi.org/10.1007/978-3-030-49161-1_17
4. Ding, Y., et al.: A deep learning model to predict a diagnosis of Alzheimer disease by using ^{18}F-FDG PET of the brain. Radiology **290**(2), 456–464 (2019)
5. Franke, K., Gaser, C., Manor, B., Novak, V.: Advanced brainAGE in older adults with type 2 diabetes mellitus. Front. Aging Neurosci. **5**, 90 (2013)
6. Gaser, C., Franke, K., Klöppel, S., Koutsouleris, N., Sauer, H.: BrainAGE in mild cognitive impaired patients: Predicting the conversion to Alzheimer's disease, PLoS ONE **8**(6), e67346 (2013)
7. Huang, G., Liu, Z., Van Der Maaten, L., Weinberger, K.Q.: Densely connected convolutional networks. In: Proceedings of the IEEE Conference on Computer Vision and Pattern Recognition, pp. 4700–4708 (2017)
8. Huang, T.-W.: Age estimation from brain MRI images using deep learning. In: 2017 IEEE 14th International Symposium on Biomedical Imaging (ISBI 2017), pp. 849–852. IEEE (2017)

9. Marcus, D.S., Fotenos, A.F., Csernansky, J.G., Morris, J.C., Buckner, R.L.: Open access series of imaging studies: longitudinal MRI data in nondemented and demented older adults. J. Cogn. Neurosci. **22**(12), 2677–2684 (2010)

10. Mikolajczyk, A., Grochowski, M.: Data augmentation for improving deep learning in image classification problem. In: 2018 International Interdisciplinary Ph.D. Workshop (IIPhDW), pp. 117–122. IEEE (2018)

11. Nakano, R., et al.: Neonatal brain age estimation using manifold learning regression analysis. In: 2015 IEEE International Conference on Systems, Man, and Cybernetics, pp. 2273–2276. IEEE (2015)

12. Qi, Q., Du, B., Zhuang, M., Huang, Y., Ding, X.: Age estimation from MR images via 3D convolutional neural network and densely connect. In: Cheng, L., Leung, A.C.S., Ozawa, S. (eds.) ICONIP 2018. LNCS, vol. 11307, pp. 410–419. Springer, Cham (2018). https://doi.org/10.1007/978-3-030-04239-4_37

13. Ren, Y., Luo, Q., Gong, W., Lu, W.: Transfer learning models on brain age prediction. In: Proceedings of the Third International Symposium on Image Computing and Digital Medicine, pp. 278–282 (2019)

14. Shao, L., Zhu, F., Li, X.: Transfer learning for visual categorization: a survey. IEEE Trans. Neural Netw. Learn. Syst. **26**(5), 1019–1034 (2014)

15. Siar, M., Teshnehlab, M.: Age detection from brain MRI images using the deep learning. In: 2019 9th International Conference on Computer and Knowledge Engineering (ICCKE), pp. 369–374. IEEE (2019)

16. Ueda, M., et al.: An age estimation method using 3D-CNN from brain MRI images. In: 2019 IEEE 16th International Symposium on Biomedical Imaging (ISBI 2019), pp. 380–383. IEEE (2019)

17. Wang, J.: Gray matter age prediction as a biomarker for risk of dementia. Proc. Natl. Acad. Sci. **116**(42), 21213–21218 (2019)

'Precision Health': Balancing Reactive Care and Proactive Care Through the Evidence Based Knowledge Graph Constructed from Real-World Electronic Health Records, Disease Trajectories, Diseasome, and Patholome

Asoke K Talukder[1](✉), Julio Bonis Sanz[2], and Jahnavi Samajpati[3]

[1] SRIT India Ltd., Bangalore, India
asoke.talukder@renaissance-it.com
[2] Medical Doctor Specialist in Family Medicine (Registered GP), Madrid, Spain
drbonis@gmail.com
[3] Srinivas Institute of Medical Science, Mangalore, India
jahnavi.samajpati@gmail.com

Abstract. Health care can be either 'Reactive Care' or 'Proactive Care'. Reactive care is self-referral where a medical help is solicited by the person or family members on suspecting illness. In Proactive care an individual seeks medical help before the appearance of symptoms in order to prevent illness, or detect and treat it early before the disease progresses or becomes chronic. There are advantages and disadvantages for both of these approaches. Reactive approach relies on healing followed by self-referrals wherein the right care is often delayed, or even neglected, resulting in accelerated disease progression. Proactive care, on the other hand, takes into account the potential risk factors in a person's health. Proactive care carries risks of overdiagnosis, overtreatment, and unnecessary interventions. In this paper we make a balance between the reactive care and the proactive care through the use of data driven algorithms, models, and knowledge graphs. We show how diseasome network constructed from tacit knowledge of Spatial Comorbidity and Temporal Comorbidity, and Patholome explicit knowledge can offer 'Precision Health'. We looked at the real-world EHR data (mostly reactive diagnosis by hundreds of doctors) to construct spatial comorbidity knowledge network. We then combined disease trajectories data (temporal comorbidity) with the spatial comorbidity. This helped us understand how diseases manifest in a target population and their interrelationships. Finally we constructed patholome disease-diagnostic-test explicit knowledge and integrated with the diseasome knowledge network to form evidence based Knowledge Graph or a Clinical Expert System. We added a Semantic Engine (Reasoning Knowledge Network) on this statistically significant knowledge graph to help a health service provider to make an accurate informed decision on balancing the reactive care and the proactive care with a focus on 'Right Care' through explainable AI (XAI). To offer the knowledge driven right care at the right time at anywhere point-of-care we used Big-Data Analytics, Statistics, Artificial Intelligence, Knowledge Discovery & Management, WebRTC, and Smartphones.

© Springer Nature Switzerland AG 2020
L. Bellatreche et al. (Eds.): BDA 2020, LNCS 12581, pp. 113–133, 2020.
https://doi.org/10.1007/978-3-030-66665-1_9

Keywords: Precision health · Comorbidity · Disease trajectory · Diseasome · Patholome · Ontologies · AI In Medicine (AIM) · Tacit knowledge · Explicit knowledge · Knowledge graph · Reasoning network · Belief network · Clinical expert system · WebRTC · Transfer knowledge · Evidence based medicine · Explainable AI (XAI) · Smartphone · Overdiagnosis and overtreatment prevention

1 Introduction

Health equity can be improved through either Reactive Care or Proactive Care. In middle and low income countries (MLC) healthcare is predominantly reactive. However, in advance economies healthcare is often proactive. In reactive care, a person who experiences some disease symptoms, or who is ill, or who believes himself to be ill, or his family members feel that he is ill, seeks the advice of a doctor. Proactive care in contrast is a type of care in which the doctor attempts to prevent future illnesses through interventions. "Proactive care has a different ethical basis from reactive care – proactive care is initiated by the doctor instead of a patient" [1].

Both reactive and proactive cares have their strengths and weaknesses. In MLC reactive health care is influenced by ignorance, socioeconomic, and other factors like complementary and alternative medicine (CAM). Reactive care or alternate medicine often trigger the healing mechanism of the body and cures the illness [2]. "Most ill states lead to recovery (even without medical attention). This is a consequence of health being an attractor state, which is in turn an inevitable consequence of evolution" [3]. Reactive care has one major weakness – often it delays the timely intervention by increasing the impact of the disease leading to complications and accelerated disease progression.

Proactive care can be considered as asymptomatic care where interventions are suggested based on the age of the person, gender of the person, preexisting diseases, and the potential risk factors in an individual's health or the environment, before the appearance of illness. Proactive care has one major downside – it tends to increase overdiagnosis, overtreatment, and unnecessary interventions [4]. Waste in health care is increasingly being recognized as a cause of patient harm and excess costs. In 2010, the Institute of Medicine (IOM) called attention to the problem, suggesting that "unnecessary services" are the largest contributor to waste in the United States (US) health care, accounting for approximately $210 billion of the estimated $750 billion in excess spending each year [5].

Comorbidity relates to one or more additional disorders co-occurring with the primary condition at the same time – it is associated with worse health outcomes, more complex clinical management, and increased health care costs. In the US, about 80% of Medicare spending is devoted to patients with 4 or more chronic conditions (comorbidities) [6]. People of any age with certain underlying medical conditions are at higher risk from COVID-19 [7]. Therefore, it is necessary to know the co-occurring, comorbid, multimorbid, or underlying conditions either in reactive or proactive care. Comorbidity can happen in both space and time. In spatial comorbidity multiple diseases co-occur at *a point time*; whereas, in temporal comorbidity the primary disease triggers many other secondary diseases in succession over *a period of time* [6]. For temporal comorbidity it is essential to diagnose disease trajectories.

Both reactive care and proactive care have one major common challenge. How to determine which condition or disease a patient or a person may have which do not have any presentation or show any symptom? This is the area of under-treatment in reactive care and overtreatment or overdiagnosis in proactive care. One of the critical examples is antibiotic resistance due to misuse of antibiotics. In the US approximately 50% of antimicrobial use in hospitals and up to 75% of antibiotic usage in long-term care facilities may be inappropriate or unnecessary [8]. In such cases a population based evidence based medicine with data driven 'Precision Health' as described in this paper will help.

In this paper we extracted knowledge from the real-world clinical and pathological data of thousands of patients diagnosed by hundreds of expert doctors. We then store the derived knowledge in a knowledgebase through concepts and their relationships. We take the best of both reactive and proactive care and attempt to model a balance between them through the knowledge graphs constructed from real-world evidence based medicine (EBM). Knowledge graphs relate structured and unstructured data to help discover hidden or unknown facts and truths. They are also necessary for creating semantic artificial intelligence (AI) applications that inherently thrive on contextual connections.

Comorbidity offers disease-disease interaction maps with disease interrelationships, which can be either spatial or temporal. For the spatial comorbidity analysis we used real-world cross-sectional EHR (Electronic Health Records) data from the public domain. We used graph theory [9] to construct 18 undirected spatial comorbidity subnetworks. We then took disease trajectory data created from 6.2 million patients' longitudinal data from the public domain. Disease trajectories describe what will usually happen between the time of diagnosis and the time of death (prognosis). Disease trajectories helped us to add the temporal component of comorbidity. We constructed a diseasome network by combining the spatial comorbidity and the temporal comorbidity networks. The diseasome network when converted into a knowledge graph will help us discover the tacit knowledge of disorders in any population. Tacit knowledge is characterized by facts and knowledge of a person that is private and acquired through personal experience, like the knowledge of an expert doctor, which is context specific and hard to formalize.

We then constructed a patholome explicit knowledge network from pathology (lab or diagnostic test) knowledge curated by medical and biological experts. In any diagnostic test there are two reference biomarkers indicating the normal or healthy range (limits). For example, the normal platelet count in the blood of a healthy individual is between 150,000 (low limit/biomarker) to 400,000 (high limit/biomarker) platelets per microliter (μL). If the diagnostic test result is lower than the low-watermark (low-reference-limit), in medical terms this state is known as Hypo state. When the value of a diagnostic test result is higher than the high-reference-limit, the state is known as Hyper state. Various diseases are associated with these Hypo/Hyper states.

We combined these three knowledge networks namely, (1) spatial comorbidity network constructed from cross-sectional EHR data, (2) temporal comorbidity network constructed from disease trajectory of 14.9 years of longitudinal data, and (3) the curated patholome network into an integrated knowledge graph for explainable artificial intelligence (XAI). To access the statistically significant (95% Confidence Interval) knowledge at the point-of-care, we stored the knowledge in the Neo4j graph database [10]. To access the knowledge at the point-of-care, we constructed a reasoning network with a user interface on a mobile smartphone that accesses this EBDPKG (Evidence Based Diseasome-Patholome Knowledge Graph) in the backend. A user can speak to or enter a disease or symptoms name or a Hyper/Hypo state of a vital in human understandable language on the smartphone. The Android App uses Google Speech recognition API and the National Library of Medicine UMLS MetaMap for NLP (Natural Language processing) [11] to convert human understandable unstructured text (2nd Generation) into a machine understandable 3rd Generation ontologies [12]. We define handwritten human readable and human understandable medical notes as *1st Generation (1G)*. *2nd Generation (2G)* medical notes are machine readable unstructured text that a machine cannot understand. *3rd Generation (3G)* notes are machine readable and machine understandable SNOMED/ICD10 ontology codes. These medical terminologies are used to fetch knowledge from the knowledge graph.

The purpose of a Knowledge Graph or an Expert System is to unleash facts that are hidden. The results presented in this paper through two 'use cases' show the power of 'Precision Health' and 'Knowledge Graph'. We used smartphone client application to access the evidence based diseasome-patholome knowledge graph (EBDPKG) via the reasoning network (cognitive engine) to empower a non-expert health worker to make an accurate informed proactive decision at the point-of-care. This could be a referral to a specialist or a referral for pathological tests to prevent a disease that might be asymptomatic today but could become an illness in the future. This will be used to predict a likely asymptomatic unknown disease from a pathological lab report. This knowledge will enable 'The Right Care at the Right Time at anywhere point-of-care'.

2   Materials and Methods

2.1   Spatial Comorbidity Network

The EHR data used in this paper is taken from public domain [13]. The data is from hospital inpatients' diagnosis taken from admissions in NSH hospitals of Madrid, Spain during 2016. Each row in the raw EHR data is an admission and the final diagnosis. All diagnoses are registered in ICD-10 codes. The ICD (International Classification of Diseases) is a globally used codification of diseases maintained by the World Health Organization (WHO) [14]. ICD10 (ICD Version 10) disease codes are grouped into 21 Chapters as following:

1. A00-B99: Certain infectious and parasitic diseases
2. C00-D49: Neoplasms
3. D50-D89: Diseases of the blood and blood-forming organs and certain disorders involving the immune mechanism
4. E00-E89: Endocrine, nutritional and metabolic diseases
5. F01-F99: Mental and behavioral disorders
6. G00-G99: Diseases of the nervous system

7. H00-H59: Diseases of the eye and adnexa
8. H60-H95: Diseases of the ear and mastoid process
9. I00-I99: Diseases of the circulatory system
10. J00-J99: Diseases of the respiratory system
11. K00-K94: Diseases of the digestive system
12. L00-L99: Diseases of the skin and subcutaneous tissue
13. M00-M99: Diseases of the musculoskeletal system and connective tissue
14. N00-N99: Diseases of the genitourinary system
15. O00-O99: Pregnancy, childbirth and the puerperium
16. P00-P96: Certain conditions originating in the perinatal period
17. Q00-Q99: Congenital malformations, deformations and chromosomal abnormalities
18. R00-R99: Symptoms, signs and abnormal clinical and laboratory findings, not elsewhere classified
19. S00-T88: Injury, poisoning and certain other consequences of external causes
20. V00-Y99: External causes of morbidity
21. Z00-Z99: Factors influencing health status and contact with health services

Table 1. Patients in different age groups

Age group (Age at onset)	Male	Female
0–9	3854	2937
10–19	1374	1327
20–29	1343	3330
30–39	2061	8372
40–49	3253	4356
50–59	4376	3978
60–69	5558	4439
70–79	6123	5320
80–120	6697	9752
Total	34639	43811

The raw EHR data comprises of 100,558 patient records with 128 attributes, and 4697 ICD10 diagnoses (diseases). Those patients that get admitted multiple times with multiple records are included only the first time. We divided the patients into 18 age groups, namely 0–9, 10–19, 20–29, 30–39, 40–49, 50–59, 60–69, 70–79, 80–120 for male and female populations. We collapsed the data and converted the data into 3 columns, namely, 'sex' (1 for male, 2 for female). 'age', and the 'diag' (diagnoses). The modified data looks like this:

```
sex...age...diag

1.....31....A15.0|Z16.39|A04.7|T36.95XA|D72.819|Z86.13|Z86.19

|R63.4||||||||||||

1.....38....A15.0|F17.200|F10.10|Z86.11|Z60.2|||||||||||||||||

1.....35....F41.9|F44.9|F14.10||||||||||||||||||
```

There are 1508 diseases that appear only in one patient – they do not co-occur. Therefore, all these 1508 diseases are discarded. The remaining diseases with at least one more co-occurring disease are potentially the nodes of the comorbidity network. Our goal is to identify which diagnoses are likely to appear together in a higher frequency than expected by chance. Table 1 shows the statistics of patients included in the comorbidity analysis. To build the relationships of the network, we defined the edges of the undirected comorbidity graph, that is, the association between coexisting diseases. We defined the "strength" or magnitude of a relationship by counting the number of times two diagnoses appear together for all the possible pairs in our data. We discard those diagnosis pairs that appear only once. If a diagnosis pair appears together many times then those diagnoses are associated; and, we will join those nodes by an edge.

We discovered patients with multiple d-separated diagnoses with marginal probabilities. In the age group of 30–40 for example, 13% of patients have "cigarette dependence" and 4% have essential hypertension. In case that cigarette dependence had no association at all with hypertension we can expect by the theory of joint probability to find $0.13 \times 0.04 = 0.0052$ (0.05%) of patients with both diagnoses just by chance. Within a total of 2369 patients in the age group 30–40 we can therefore expect to find $0.0052 \times 2369 = 12$ patients with both diagnoses just by chance. In contrast, if there was an association between cigarettes and hypertension, then we should find more than 12 patients with both diagnoses (if the association is positive) or less than 12 (if the association is negative). If the ratio of observed and expected is higher than 1 then we consider that these two diagnoses or diseases are associated statistically and included in the analysis as co-occurring diseases.

A medical doctor uses his or her training and intuition to make a medical decision. However, we are attempting to make a decision here by algorithms. Therefore, it is essential that we eliminate those fake associations that may occur just by chance and preserve only these associations that have a solid scientific foundation. For this we included an estimator of statistical hypothesis testing. The statistical test we used in our analysis is the test for proportion difference. Basically for each association between two diseases we calculate the p-value that indicates the probability of finding a difference between observed and expected that could be explained by chance and not due to a real association.

After this we build the list of edges with their p-values and the magnitude estimator of effect. The value of the estimator and its p-value allows us to select only those associations that are positive (higher than 1) and statistically significant (p-value lower than 0.05)

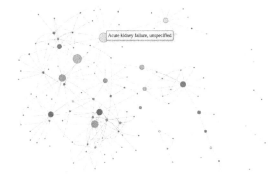

Fig. 1. The spatial comorbidity network for Male population in the age group of 60 to 69 (using interactive GraphML) – this is created for human consumption (human understandable), that displays the comorbid diseases with mouseover

at 95% Confidence Interval. The final statistically significant comorbid disease counts became 592.

Finally we built the graph that represent the comorbidity subnetworks for all 18 groups of male in the age group 0–10, 10–19, 20–29, 30–39, 40–49, 50–59, 60–69, 70–79, 80–120 years and for female in the age groups 0–10, 10–19, 20–29, 30–39, 40–49, 50–59, 60–69, 70–79, 80–120 years. We used igraph [9] to construct the comorbidity graphs with our data. For interactive visualization in the Web browser we converted these spatial comorbidity graphs into GraphML graphs. GraphML is an XML-based file format for interoperability and visualization of graphs in Web browser. Figure 1 is one such interactive graph for male in age group 60–69.

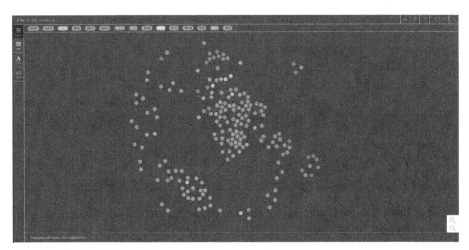

Fig. 2. The comorbidity network of Fig. 1 for Male population in the age group of 60 to 69 converted from GraphML undirected graph into a directed graph in Neo4j Graph Database – this is normalized and machine understandable

2.2 Temporal Comorbidity & Diseasome Knowledge Graph

We took the undirected 18 comorbidity interactive static graphs in GraphML described in the previous section (Fig. 1). However, we need all these disease comorbidity information accessible at anywhere point-of-care for a health care provider. Therefore, these comorbidity networks must be stored in a graph database somewhere in a server (or cloud) available over the internet for realtime access.

To construct the comorbidity knowledge graph we looked at each statistically significant relationship and made a hypothesis of causality. In each relationship, we used the high frequency node as the parent node (causative) and the low frequency node as the child node. We loaded this directed comorbidity knowledge graph into Neo4j [10] graph database as shown in Fig. 2. The fundamental difference between GraphML graph in Fig. 1 and the comorbidity knowledge graph in Fig. 2 is that Fig. 1 is a static interactive graph in GraphML for visualization on a Web browser, which can be operated and understood by human experts only. In contrast, Fig. 2 is in a knowledge database that can be accessed and interpreted by computer algorithms available at the point-of-care round the clock for access through a software application. The Neo4j knowledgebase however, can also be visualized on Web browser – as shown in Fig. 2.

We then took the disease trajectories data from public domain [15]. Disease trajectories are the pattern of disease comorbidities over time (temporal comorbidity). In other words, as the time progresses one chronic disease in a person triggers many other diseases as spatial comorbidities. We combined this trajectory data with our spatial comorbidity graphs constructed from EHR data as described in the previous section. Disease trajectories were constructed from longitudinal data from Denmark of 14.9 years of patient lives. This diseases trajectories data covers the entire spectrum of diseases of 6.2 million patients' reactive care from 1996 to 2010 [15]. EHR data being reactive, ICD10 codes used in spatial comorbidity are explicit like "*H54.1131: Blindness right eye category 3, low vision left eye category 1*", whereas in disease trajectory – because it is likely to occur in the future (predictive), the same ICD10 code in temporal comorbidity is presented as generic code like "*H54: Blindness and low vision*".

The integration of the spatial comorbidity and the temporal comorbidity gave us a true picture of disease manifestation in humans. Spatial comorbidity represents the reactive state of disease association; whereas, temporal comorbidity represents the predictive state of a chronic disease. Combination of these two graphs created a diseasome knowledge graph which quite accurately tells us the likely diseases in a population irrespective of reactive care, proactive care, and is agnostic to space or time.

2.3 Patholome Knowledge Graph

In case of patholome, we constructed the knowledge graph from curated data from literature. Medical investigations or pathological tests can be grouped into following categories:

1. Biochemical test
2. Microbiology tests
3. Radiology test

4. Genetic test
5. Immunological tests
6. Bioelectroanalysis, Electrochemical, or Biosensor based investigations

Different tests in the above categories have different criteria to measure the normal or healthy state or a disease state. There could be different disease states for Hypo (below the lower reference limit) or Hyper (above the upper reference limit). For example in pathological tests for Vitamin D – 20–80 ng/mL value is a healthy reference range. Less than 20 ng/mL will result into Vitamin D deficiency, which may cause various diseases like *"Chest pain, unspecified:R07.9"* or *"Lower abdominal pain, unspecified:R10.30"*. Whereas, Vitamin D content higher than 80 ng/mL may cause toxicity [16]. In case of electrochemical tests like Electrocardiogram (ECG or EKG) however the measurement unit is different. The height and the width of peaks or a segment in the PQRST ECG curve determines various disease states of the heart, like ST segment elevation indicates ST elevation myocardial infarction [17].

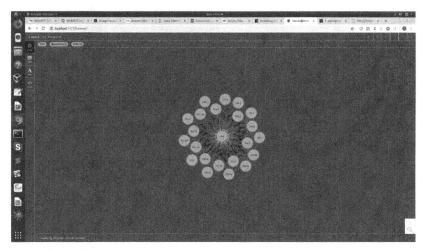

Fig. 3. The patholome network with biochemical test "Blood Count Automated Differential WBC Count" and associated 20 ICD10 codes for the Hypo state

Currently in this paper we considered only the biochemical tests related to blood. We took 9 various blood tests that are associated with 80 diseases, which are:

1. Blood Count Automated Differential WBC Count
2. Blood Count Complete Auto & Auto Differential WBC Count
3. Blood Count Complete Automated
4. Blood Count Hematocrit
5. Blood Count Hemoglobin
6. Blood Count Leukocyte WBC Automated
7. Blood Count Platelet Automated

8. Blood Count Reticulocyte Automated
9. Blood Gases Any Combination PH PCO2 PO2 CO2 HCO3

We constructed a patholome directed knowledge graph where disease is the parent and the pathology test (Hyper or Hypo) is the child. We combined the diseasome knowledge graph with patholome knowledge graph into an integrated evidence based diseasome-patholome knowledge graph (EBDPKG) or a Clinical Expert System and stored in Neo4j graph database for point-of-care access through computer algorithms.

Figure 3 shows the Knowledge graph in Neo4j for the association of ICD10 codes with single blood test namely "Blood Count Automated Differential WBC Count" with 20 ICD10 code nodes and 80 relationships. In this knowledge graph the investigative tests (WBC in this case) is the child, whereas ICD10 codes are parents.

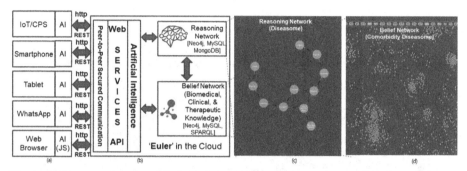

Fig. 4. Architecture diagram of AyurSuraksha with the Reasoning Network (a) User Interfaces, (b) Knowledge Graphs, (c) Reasoning Network, (d) Belief Network

2.4 Medical Decision Making Through Belief and Reasoning Networks

In this part of the work we describe how the evidence based diseasome-patholome knowledge graph (EBDPKG) is accessed and ingested (consumed) at the point-of-care through explainable artificial intelligence. For patholome network, we included only these ICD10 codes (disease concepts) that are common between diseasome and patholome networks by adding Hyper/Hypo diagnostic relationships as shown in Fig. 3.

Medical decision making is a complex process with Categorical (deterministic) reasoning at one end and the Probabilistic (evidential) reasoning at the other end of the spectrum. Here we present algorithms in our system that uses a decision tree to implement categorical reasoning for deductive decision making. We show how explicit knowledge is used in our system for medical decision making.

We constructed a deductive reasoning network using Bayesian Network Design Studio which accepts three values namely "Age", "Gender", and the "Disease" (Fig. 6(c)) for comorbidity decision. The client device may be a Smartphone, WhatsApp, or a Web Browser. A native application is developed on the smartphone that has the complete feature set that includes GPS, Camera, Microphone, Speaker, and other sensors available on the smartphone like Accelerometer, Environment sensors etc. The WhatsApp and Web interface however has limited features. Figure 4 shows the Mobile application architecture. The client interacts with the server side application through REST (Representational state transfer) API over HTTP. The client and server communicate through JSON objects as shown in Fig. 5.

3 Actionable Knowledge

Here we describe the actionable knowledge using AyurSuraksha client application for the point-of-care insights. AyurSuraksha is an Android based application that accesses the knowledge graph through the reasoning network and the belief network (Fig. 4). AyurSuraksha uses peer-to-peer protocol – either the user at the smartphone end can start a transaction; or, a transaction can be initiated by a doctor or a nurse at the hospital end like sending instructions for home care. Workflows can be sent from hospital to a person in a reactive or proactive care. In Fig. 5 we can see the multimedia workflow of CPR (cardiopulmonary resuscitation) is sent from the hospital to help a volunteer perform CPR.

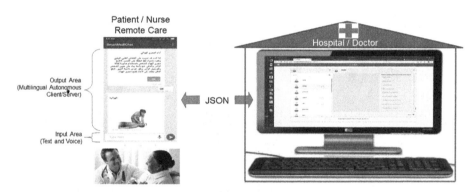

Fig. 5. External interface of the mobile application AyurSuraksha

On the smartphone – at the user end, the screen is split into two segments, upper part is reserved for the conversation history; whereas, the bottom part is used for the user (patient/person) input. Inputs can be both text and voice. For text – user enters the text in the area marked as "Type here". For voice, the user presses the microphone icon in the input area to speak into the smartphone. Figure 6 shows various major services along with the voice to text and comorbidity service. Text (direct entry) or text through 'speech to text' is treated in similar fashion. The text is human understandable unstructured data (2nd Generation). This 2nd Generation human understandable unstructured text is converted into 3rd Generation machine understandable ontology codes through NLP [11] (see Fig. 7).

Fig. 6. (a) Services in AyurSuraksha; (b) Speech recognition; and, (c) Comorbidity input for the reasoning network

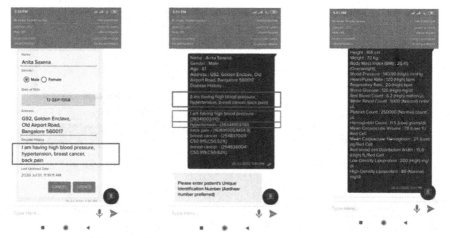

Fig. 7. Converting 2nd Generation human understandable unstructured medical text into machine understandable 3rd Generation ICD10 disease codes and SNOMED CT ontologies. Vital signs and EMR records are processed and the Hyper (High)/Hypo (Low) indications are presented to the doctor/nurse & Knowledge Graph

Figure 7 shows the "Medical Home" service. This service starts with patient interview. User enters the demographic information and the chief complaint. In the example of Fig. 7 the user enters four different diseases or symptoms or conditions:

1. "i am having high blood pressure"
2. "hypertension"

3. "breast cancer"
4. "back pain"

These are all human understandable clinical features in English texts, which a machine cannot understand. Same can be spoken to the smartphone as well. This text is converted into 3rd Generation machine understandable ontology as shown in the figure as well. It may be noted that while converting from 2nd generation to 3rd generation, the disease name is normalized as well. A doctor can understand that "i am having high blood pressure" and "hypertension" refer to the same medical condition. In computers "i am having high blood pressure" or "hypertension" are two unique unequal character strings. We process these medical text strings through a complex AI pipeline that include (1) stop-word filter, (2) spellcheck, (3) stemming & lemmatization, followed by (4) MetaMap [11] NLP and finally (5) normalize into SNOMED CT and ICD10 codes.

It can be noticed that both these strings "i am having high blood pressure" and "hypertension" have been normalized into "38341003/I10" where "38341003" is the SNOMED CT ontology code that refers to "*Hypertensive disorder, systemic arterial (disorder)*" and "I10" is the ICD10 code for "*Essential (primary) hypertension*". Though the inputs were two different text literals, the output is normalized to the same 'diagnosis' like a doctor understands that both 'high blood pressure' and 'hypertension' conditions are equivalent conditions.

Following the entry of the chief complaints or the present conditions, the user (patient for reactive care or person in proactive care) enters vitals in the input areas of application form. Same also can be fetched from the EMR (Electronic Medical Record) as well depending on the hospital requirements. The vitals are processed and the Hyper, Hypo indicators are flagged as shown in Fig. 7. The ICD code for the disease and the vitals and lab information is given to the reasoning network to obtain the comorbidity and diagnostics recommendations.

It may be argued that comorbidity may have some relationship with the population and the geography. However, the basic biology and genetic characteristics of human species are nearly constant. Unlike infectious diseases like COVID19, chronic and non-communicable diseases (NCD) are more dependent on human genes, their mutations, age, gender, and lifestyle. Data used in this study covers a wide range of doctors diagnosing a large number of patients ensuring a certain level of randomness. We believe that the results presented in this paper can be applied in other geographies and populations as well.

4   Future Work

As part of future work, we will integrate the gene-disease knowledge graph from Dis-GeNET ontology database [18]. Common genes between diseases are likely to lead temporal comorbid conditions. Figure 8 shows the gene-disease DisGeNET network in Neo4j graph browser. However, this graph is not integrated with diseasome-patholome yet.

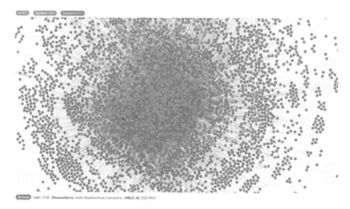

Fig. 8. DisGeNET (Disease Gene Association Network) data converted into Neo4j Knowledge Graph (will be integrated as part of future work)

The integration of DisGeNET with the diseasome-patholome knowledge graph will enhance the disease knowledge that will help understand impact of 'Precision Medicine' on disease progression. DisGeNET includes gene mutation variants as well. This will lead to 'Precision Health' where we will be able to offer personalized accurate health plans before a disease strikes. Also, in the future we will add other types of investigations (diagnostic tests). As and when new knowledge is available, these will be added as incremental knowledge by adding concepts and their connections (relationships) without any change in the graph database schema.

5   Results and Discussion

5.1   The Diseasome-Patholome Knowledge Graph

There are 13 services in AyurSuraksha application (Fig. 6(a)). In AyurSuraksha we have two types of knowledge graphs as shown in Fig. 4. These two knowledge graphs are namely (1) Reasoning network, and (2) Belief network. These 13 services are offered through 2140 nodes and 2221 relationships of reasoning. Out of this the comorbidity reasoning consists of 11 nodes and 22 relationships.

Within belief networks we have two knowledge graphs. We took the January 2020 version of the SNOMED CT and loaded it into the Neo4j graph database. The SNOMED CT knowledgebase in Neo4j is a large graph with 2,611,541 nodes and 10,267,070 edges [17]. The other belief network is the diseasome-patholome network (EBDPKG) described in this paper. This is a knowledge graph with 1989 nodes and 10,380 relationships. This knowledge graph is stored in the Neo4j graph database for realtime access at the point-of-care. The diseasome network consists of 1909 nodes and 6930 relationships. The patholome knowledge network added 80 nodes with 3450 relationships.

5.2   Use Case I

We present a use case of resource limited primary care or care for underserved population in the context of rural India. "Nearly 86 per cent of all the medical visits in India are made by ruralites with majority still travelling more than 100 km to avail health care facility of which 70–80% is borne out of pocket, landing them in poverty" [19]. The physician density in India is low compared to WHO recommendation (7.8 physicians/10,000 in the population) with 80% of physicians serving only 28% of the population living in major cities. Only 18.8% of doctors in rural India have formal degrees in modern medicine. There are many villages in India without any qualified doctor. There are Anganwadi and ASHA (Accredited Social Health Activists) health workers in India. These health workers are neither qualified doctors nor are they qualified nurses. AyurSuraksha mobile application will empower these health workers to provide acute care and improved referrals using the diseasome-patholome knowledge graph.

Our current system is designed to work in this resource limited environment. Let us assume a case where a male patient of 63 years of age becomes ill. The patient did not preserve all medical interactions of the past that includes all medical records, prescriptions, and diagnostic test reports. In India majority of the hospitals in the private or public space rely on handwritten (1st Generation) health information. This 63 years male patient therefore needs to be treated as a first time acute care patient without any reliable patient history. The patient is known to have hypertension. The ASHA health worker or the Anganwadi (mid-wife) health worker selects the age group (60–69), gender 'Male' with disease 'Hypertension' as shown in Fig. 6. AyurSuraksha client mobile application forms a JSON object with 'Age : 60–69; Sex : Male; Disease : Hypertension' and sends to the backend. At the backend server the disease name is normalized. The human understandable disease name 'hypertension' from the client smartphone is mapped onto 3rd Generation machine understandable ICD10 code of 'I10' (Essential (primary) hypertension) as shown in Fig. 7. It may be noted that in this kind of user interface there is no chances of any input errors.

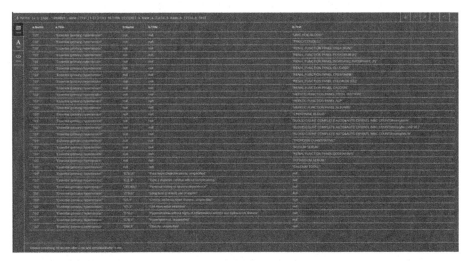

Fig. 9. Evidence Based Diseases co-existing in various age-group for ICD10 code 'I10: Essential (primary) hypertension' along with diagnostic tests in the Neo4j browser

At the server end this JSON object is processed and translated into a Neo4j Cypher command. Cypher is Neo4j's graph query language that allows users to store and retrieve data from the Neo4j graph database. The query serviced at the Neo4j graph database is:

```
MATCH  (a:I  {Age:'1060069',Name:'I10'})--(b)  RETURN  DISTINCT
a.Name,a.Title,b.Name,b.Title,b.Test
```

This Neo4j cypher command looking at the diseasome-patholome graph identified 10 diseases this person may also have at his age as per the evidence as listed below. The results from the Neo4j cypher command is shown in Neo4j browser in Fig. 9.

```
E78.00  Pure hypercholesterolemia, unspecified
E11.9 Type 2 diabetes mellitus without complications
Z87.891 Personal history of nicotine dependence
Z79.82  Long term (current) use of aspirin
I25.9 Chronic ischemic heart disease, unspecified
I25.2 Old myocardial infarction
E79.0 Hyperuricemia without signs of inflammatory arthritis and
tophaceous disease
E78.5 Hyperlipidemia, unspecified
E66.9 Obesity, unspecified
```

Within AyurSuraksha there are other services that can be used to have some leading questions to confirm whether this person has any symptom for other likely co-existing diseases like obesity, heart disease, cholesterol etc. including smoking habits. Also, the cypher command extracts following 21 possible blood tests. The health worker can now ask whether the patient has any of these reports. Depending on the current health state the diagnostic tests can also be determined.

5.3 Use Case II

In this use case we discuss a rural male of around 50+ years of age approaches an Anganwadi (midwife) health worker with some symptoms. The family members show a report that indicates high phosphorous. Above normal high phosphorous causes Hyperphosphatemia; whereas, low phosphorous causes Hypophosphatemia.

In Fig. 10 we show the result of the Neo4j cypher query that navigated the knowledge graph from a pathological test to discover diseases. Unlike in Use Case I where we navigated from a disease to diagnostic tests – here it is diagnostic test to diseases. This result is sent through a JSON object to the smartphone client.

The results from knowledge graph are two diseases as follows (Fig. 10):

```
"N18.6""End stage renal disease"
"J45.21""Mild intermittent asthma with (acute) exacerbation"
```

Fig. 10. Navigating from a pathological test (phosphorous) to the diseases for 'Inorganic Phosphorus Test'

In the Neo4j browser (Fig. 10) we see that the cypher query fetched two different diseases from the knowledge graph related to phosphorous Hyper and Hypo states. The Hyper state of phosphorous is associated with ICD10 code "N18.6", for eight age groups for "End stage renal disease". Whereas, the Hypo state of phosphorous is associated with four population groups with ICD10 code "J45.21", which is "Mild intermittent asthma with (acute) exacerbation" disease. We confirmed the accuracy of the result through literature survey as mentioned below.

Hyperphosphatemia (high phosphorus) is known to be associated with significant pathophysiology in chronic kidney disease (CKD). When kidneys are damaged, body cannot remove phosphate from blood quickly enough causing high phosphorous load in the blood. Studies have elucidated that hyperphosphatemia is a direct stimulus to vascular calcification, which is one cause of morbid cardiovascular events contributing to the excess mortality of chronic kidney disease [20]. We can see that our knowledge graph found that hyperphosphatemia is prevalent in all ages from 30 to 120 for both male and female.

"Hypophosphatemia has been recently highlighted as a reversible cause of respiratory muscle hypocontractility and reduced tissue oxygen extraction in patients with chronic obstructive lung disease and asthma" [21]. Serum phosphate concentration is generally normal in all patients with chronic obstructive lung disease and asthma. However, the phosphate concentration falls following the initiation of bronchodilator therapy. Patients "developed hypophosphatemia (serum phosphate, less than 0.8 mmol/L). Urinary phosphate level falls in parallel. A negative correlation was observed between serum phosphate and serum theophylline concentrations and a positive correlation between serum and urinary phosphate concentrations" [21]. We can see that our knowledge graph found that hypophosphatemia is prevalent in all ages from 0 to 39 for both male and female.

5.4   Decision Making in Medicine During COVID and Pre-COVID Era

The diagnosis of a disease or decision making in medicine plays the key role in medicine and patient care. Simple cases involving a single organ system with well-defined conditions result into deterministic diagnosis that follows a single flowchart. Complex cases with poorly defined conditions involving multiple organ systems demand a combination of multiple flowcharts and probabilistic decision making.

A study covering 28,570,712 consultations in 67 countries found that the average consultation time in primary care ranges from 48 s in Bangladesh to 22.5 min in Sweden.

Patients in 18 countries representing about 50% of the global population spend 5 min or less with their primary care physicians. This time includes patient interview, analyzing disease history and family history, physical examination, prescription, and care procedures [22]. Another study found that 90% of patients with poorly defined conditions remain undiagnosed in the primary care [23]. A different study discovered that 88% of patients were misdiagnosed during primary diagnosis and needed second opinion [24]. These findings justify the case of Artificial Intelligence in Medicine (AIM). AIM will reduce the Physician Burnout and increase accuracy; more importantly AIM will need machine understandable clinical/medical notes.

During COVID19 pandemic, teleconsultation has played a crucial role in health care. Teleconsultation includes fixing doctor's appointment, followed by audio or audiovisual remote consultation. This type of consultation removes the physical interaction between the patient and the doctor. However, this does not use any Artificial Intelligence (AI) or Artificial Intelligence in Medicine (AIM). Examples of such teleconsultation platforms are *"Ping An Good Doctor"* in China, *"Doctor On Demand"* in USA, or *"Practo"* in India. None of these systems to the best of our knowledge even record the diagnosis in normalized notation or codify the diagnosis in machine understandable nomenclatures like ICD10 or SNOMED CT as described in sections above. The patient data collected by these systems are unstructured and not AIM ready. Artificial Intelligence in Medicine is necessary for reducing medical errors and cognitive loads on doctors.

5.5   Contactless Care and Telemedicine in Post-COVID Era

In AyurSuraksha we used Artificial Intelligence in Medicine (AIM) and the Next Generation Web (NGW) [25] for real-time peer-to-peer audiovisual teleconsultation between the patient and the doctor to facilitate accurate and timely diagnosis. Unlike earlier generations of Web interactions which were Human to Computer, the NGW however, is real-time peer-to-peer Human to Human (browser to browser) interaction.

Unlike other telemedicine and teleconsultation, AyurSuraksha records all communications (text & data) transacted between the patient and the doctor in machine understandable formalism in document database. In addition, audiovisual communication in AyurSuraksha has the feature of recording the whole or part of the interaction.

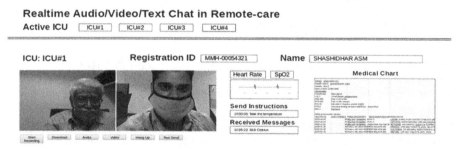

Fig. 11. Real-time private (secured) Audio-Visual and Data chat between the doctor and patient where one doctor is consulting multiple patients through realtime peer-to-peer secured Next Generation Web (picture published with consent of individuals in the picture)

AyurSuraksha supports Audio, Video, and arbitrary Data chats as part of the '*Medical services*' service. This uses realtime peer-to-peer NGW protocol WebRTC with end-to-end security (Fig. 11). The greatest advantage of Next Generation Web is that it does not require any intermediate server, or downloads, or any plugins.

Health workers like Anganwadi/ASHA personnel do the first level of screening at the point-of-care. Along with knowledge graphs AyurSuraksha offers automated triaging through its 13 services through a powerful reasoning network. The health worker can refer the patient to the doctor in charge or the District Hospital or can offer some medication as part of the acute primary care that is implemented in AyurSuraksha using Government approved guidelines. The health worker performs some routine question answer triaging to capture details. This helps performing the first level of patient interview and primary care to reduce the burden on the doctors.

6 Conclusion

The main contribution of this research work is timely accurate care using knowledge graph and Artificial Intelligence in Medicine (AIM). For reactive patients we created a model for undertreatment prevention (UP). For proactive individuals we created a model for overdiagnosis and overtreatment prevention (OOP). We stored the evidence based experts' tacit and explicit knowledge as knowledge graphs in Neo4j graph database. We transfer this experts' knowledge to non-experts at anywhere point-of-care to facilitate the "the right care at the right time" through the combined knowledge of spatiotemporal disease comorbidity and the knowledge from pathological tests.

We joined the diseasome and patholome networks constructed from spatiotemporal EHR data to form a Clinical Expert System. In this expert system the nodes or concepts are diseases, SNOMED/ICD codes, diagnostic tests. This diseasome-patholome knowledge graph will help identify an underlying condition, which may or may not show any symptom. This knowledge graph will be used by expert doctors or uncertified unskilled health workers equally. Though the patient may be a first time visitor with reactive self-referral conditions, he can be provided proactive acute care through the help of machine understandable knowledge.

Advantage with Artificial Intelligence is the transfer learning where learning from one environment (domain) can be transferred to another environment or use case. Whereas, in this work we leveraged many experts' knowledge and experiences in the form of tacit and explicit knowledge and converted into knowledge networks that helped us achieve "Transfer Knowledge". This knowledge can not only be transferred between algorithms and environments but can be transferred to unskilled individuals to function like an expert. As and when new dataset are available we can easily enhance the knowledge in the knowledge graph.

We used statistical algorithms to ensure accuracy. Because of Knowledge Transfer, we are able to show how data driven algorithms and models can help empower unskilled health workers in resource limited environment to offer health services with higher level of accuracy and efficiency. Through the use of Knowledge Graphs we implemented an Explainable Artificial Intelligence (XAI) model in health care that will also help eliminate medical errors. The current implementation enables 'Precision Health' at the population level; however, integration of DisGeNET will allow us to offer 'Personalized Precision Health' to an individual before a disease strikes.

References

1. Gillies, J.C., Baird, A.G., Gillies, E.M.: Balancing proactive and reactive care. Occas. Pap. R. Coll. Gen. Pract. **71**, 15–18 (1995)
2. Petri, R.P.: Integrative health and healing as the new health care paradigm for the military. Med. Acupunct. **27**(5), 301–308 (2015)
3. Firth, W.J.: Chaos–predicting the unpredictable. BMJ **303**(6817), 1565–1568 (1991)
4. Doctors identify 50 'unnecessary' medical interventions. https://www.mims.co.uk/doctors-identify-50-unnecessary-medical-interventions/contraception/article/1485880
5. Lyu, H., Xu, T., Brotman, D., Mayer-Blackwell, B., Cooper, M., Daniel, M., et al.: Overtreatment in the United-States. PLoS One **12**(9), e0181970 (2017)
6. Valderas, J.M., Starfield, B., Sibbald, B., Salisbury, V., Roland, M.: Defining comorbidity: implications for understanding health and health services. Ann. Family Med. **7**(4), 357–363 (2009)
7. People with Certain Medical Conditions. https://www.cdc.gov/coronavirus/2019-ncov/need-extra-precautions/people-with-medical-conditions.html
8. Morrill, H.J., Caffrey, A.R., Jump, R.L., Dosa, D., LaPlante, K.L.: Antimicrobial stewardship in long-term care facilities: a call to action. J Am. Med. Dir. Assoc. **17**(2), 183.e1–183.16. https://doi.org/10.1016/j.jamda.2015.11.013
9. Csardi, G., Nepusz, T.: The Igraph software package for complex network research. InterJ. Complex Syst. **1695**(2006). http://igraph.org
10. Neo4j Graph database. https://neo4j.com/
11. MetaMap - Tool For Recognizing UMLS Concepts in Text. https://metamap.nlm.nih.gov/
12. Yadav, S., et al.: Suśruta: artificial intelligence and bayesian knowledge network in health care – smartphone apps for diagnosis and differentiation of anemias with higher accuracy at resource constrained point-of-care settings. In: Madria, S., Fournier-Viger, P., Chaudhary, S., Reddy, P.K. (eds.) BDA 2019. LNCS, vol. 11932, pp. 159–175. Springer, Cham (2019). https://doi.org/10.1007/978-3-030-37188-3_10
13. DiseasomeCMBD2016. Github at link. https://github.com/drbonis/diseasomeCMBD2016
14. ICD-10 Version: 2019. https://icd.who.int/browse10/2019/en
15. Jensen, A.B., Moseley, P., Oprea, T.L., et al.: Temporal disease trajectories condensed from population-wide registry data covering 6.2 million patients. Nat. Commun. **5**, 4022 (2014)
16. 25-Hydroxyvitamin D2 and D3, Serum. https://www.mayocliniclabs.com/test-catalog/Clinical+and+Interpretive/83670
17. Akbar, H., Foth, C., Kahloon, R.A., Mountfort, S.: Acute ST Elevation Myocardial Infarction (STEMI) (2020). https://www.ncbi.nlm.nih.gov/books/NBK532281
18. DisGeNET - a database of gene-disease associations. https://www.disgenet.org/
19. Kurukshetra: Ministry of Rural Development, vol. 65, no. 9, 1 Jul 2017
20. Hruska, K.A., Mathew, S., Lund, R., Qiu, P., Pratt, R.: Hyperphosphatemia of chronic kidney disease. Kidney Int. **74**(2), 148–157 (2008)
21. Brady, H.R., Ryan, F., Cunningham, J., Tormey, W., Ryan, M.P., O'Neill, S.: Hypophosphatemia complicating bronchodilator therapy for acute severe asthma. Arch. Int. Med. **149**(10), 2367–2368 (1989)
22. Irving, G., et al.: International variations in primary care physician consultation time: a systematic review of 67 countries. BMJ Open **7**(10), e017902 (2017)

23. Payne, V.L., Singh, H., Meyer, A.N.D., Levy, L., Harrison, D., Graber, M.L.: Patient-initiated second opinions: systematic review of characteristics and impact on diagnosis, treatment, and satisfaction. Mayo Clinic Proc. **89**(5), 687–696 (2014)
24. Van Such, M., Lohr, R., Beckman, T., Naessens, J.M.: Extent of diagnostic agreement among medical referrals. J. Eval. Clin. Pract. **23**(4), 870–874 (2017)
25. Talukder, A.K.: Next Generation Web: Technologies and Services. In: Proceedings of BDA2020 (2020)

Prediction for the Second Wave of COVID-19 in India

Shweta Thakur, Dhaval K. Patel⬤, Brijesh Soni$^{(\boxtimes)}$⬤, Mehul Raval⬤, and Sanjay Chaudhary⬤

School of Engineering and Applied Science, Ahmedabad University, Ahmedabad, India
{shweta.t,dhaval.patel,brijesh.soni,mehul.raval, sanjay.chaudhary}@ahduni.edu.in

Abstract. COVID-19 has been declared as a global pandemic by World Health Organization on 11th March 2020. Following the subsequent stages of unlocking by the Indian Government, the active cases in India are rapidly increasing everyday. In this context, this paper carries out the comprehensive study of active COVID-19 cases in four states of India namely, Maharashtra, Kerala, Gujarat and Delhi, and predicts the arrival of second wave of COVID-19 in India. Further, since the number of cases reported varies significantly, we utilize the Multiplicative Long-short term memory (M-LSTM) architecture for predicting the second wave. In our experiment, multi-step prediction method is utilized to forecast the active cases for next six months in the four states. Since the input instances vary abruptly with time, simple Long-short term memory (LSTM) is not efficient enough to predict future instances accurately. Our results reveal that M-LSTM have outperformed simple LSTM in predicting the cases. The percentage of improvement of M-LSTM model as compared to simple LSTM is 22.3%. The error rate calculated in terms of N-RMSE (Normalized Root Mean Square Error) for M-LSTM is less than that of each state's LSTM model. A nested cross-validation method known as Day Forward Chaining improves both models' performance and avoids biased prediction errors. This technique helped in accurately predicting the active cases by degrading the error values. Our work can help the government and medical officials to better organize their policies and to prepare in advance for increase in the requirement of healthcare workers, medicines and support systems in controlling the upcoming COVID-19 situation.

Keywords: LSTM · M-LSTM · COVID-19 · Forecasting · Active cases · Second wave

1 Introduction

The first case of the novel coronavirus was in Wuhan city of China in December 2019. Then it rapidly spread across the globe [1]. The novel coronavirus SARS-COV-2 is commonly known as COVID-19 (COrona VIrus Disease of 2019) [2].

© Springer Nature Switzerland AG 2020
L. Bellatreche et al. (Eds.): BDA 2020, LNCS 12581, pp. 134–150, 2020.
https://doi.org/10.1007/978-3-030-66665-1_10

It is a pneumonia kind of virus that causes severe lung infections. This virus's most common symptoms are fever, dry cough, sore throat, headache, loss of taste and smell, chest pain, and shortness of breath. The duration of illness from the early onset of any symptom varies significantly from case to case.

Moreover, it can also be asymptomatic in many cases. A person may require hospitalization or ICU admission (with ventilators) [3]. As per the Ministry of Health and Family Welfare, the active cases on 15nd September 2020 are 995,933. The number of patients recovered is 3,942,360, which is 78.53% of the reported cases. The number of people who died with the virus is 82,066, which is 1.63% of the reported cases. However, the death rate in India as compared to other countries is low [4]. The challenge is the gap between the rise in new cases every day and beds' availability. India's Government imposed lock-downs in multiple stages across the country to reduce the community spread and minimize the effect on the existing healthcare system,. On 24th March 2020, the Government of India had announced the first lock-down of 21 days, which extended again on 14th April 2020 to 03rd May. On 01st May, an extension for two more weeks push it until 17th May, which finally extended to 30th June only for containment zones. However, to revive the country's economy, the government announced to unlock the economic activities in multiple phases, with appropriate terms and conditions of social distancing. It has indeed resulted in the rise of total active cases in the country. Since both the welfare of people and economic activities are essential, it is necessary to predict the rise in active cases at an early stage. It helps healthcare officials re-frame their policies to curb the situation.

Many statistical based prediction models like Auto-Regressive Integrated Moving Average (ARIMA) have been used in [5] to estimate and forecast the active cases [6]. ARIMA is a class of models used in forecasting that captures different temporal structures in time-series data. The author in [7] has proposed k-period performance metric to forecast time series models based on ARIMA. To predict the trend of daily confirmed cases in South Korea, Mainland China and Thailand, the author has proposed the model that uses ARIMA in [8]. Apart from the statistical models, machine learning (ML)/deep learning (DL) aided architectures have emerged as a critical player for predicting various fields. In this context, various researchers have used Recurrent Neural Network based LSTM (Long-Short Term Memory) model and its variants to predict the cases of COVID-19. For instance, the authors in [9], have applied deep LSTM, Convolutional LSTM and Bi-directional LSTM for predicting positive cases in 32 states of India. They have proposed a method that gives high accuracy for short term prediction with an error of less than 3% for daily predictions and less than 8% for weekly predictions. The authors in [10] have worked on time series forecasting of COVID-19 using deep learning LSTM networks considering Canadian data to predict the future infections. In [11], the authors have proposed shallow LSTM based neural network to predict the risk category. They have combined trend and weather data together for the prediction. A hybrid AI model namely Improved Susceptible Infected (ISI) is proposed in [12] to predict China's COVID-19 cases wherein the authors embedded LSTM network to estimate the infection rates.

The authors in [13] evaluated LSTM with Gated Recurrent Unit (GRU) model to train the datasets. The prediction results were validated on the original data using RMSE metric. The authors in [14] have predicted and estimated the mortality caused by COVID-19 with a patient information-based algorithm (PIBA) in real-time.

Moreover, the authors in [15] proposed a combination of models to capture the uncertainties in understanding the disease dynamics. They combined the forecasts for daily deaths and hospital admissions to improve the short-term forecasts' predictive accuracy. Furthermore, the works, including the analysis and data driven forecasting of COVID-19 cases in India can be found in [16–23].

Although the number of COVID-19 active cases has declined to some extent, recently, few studies concerning the second wave of the COVID-19 pandemic were found more dreadful than the first wave [24]. In this context, the authors in [25] comprehensively studied the second wave of COVID-19 in United Kingdom. The authors in [26] analyzed the second wave of COVID-19 in Canada. In contrast, the authors in [27–29] studied the arrival of the second wave of COVID-19 in North-America, Japan, and Iran, respectively. With India's massive population, it is vital to study the arrival of the second wave of COVID-19 in India. According to the authors' best knowledge, none of the work in literature predicts the second wave of COVID-19. In this context, the contribution of this paper is twofold as follows:

1. Firstly, we have performed a comprehensive study of active cases of COVID-19 in four states of India, namely, Kerala, Maharashtra, Gujarat, and Delhi. Since these are the states with maximum caseloads in the country. The Day Forward Chaining cross-validation performs unbiased estimation of prediction errors. Following that, we propose a deep learning (DL) aided prediction algorithm with forecast predictions of COVID-19 cases for next 6 months up to 01st March 2021. Most of the prior works performs one day ahead prediction (single-step). On the contrary, our work provides prediction up to six-months, which shows the insights of the upcoming second wave in India.
2. Secondly, we have proposed a model that uses the Multiplicative Long Short Memory (M-LSTM) network to predict the active cases of COVID-19 in the four states of India. The motivation to use this model is that active cases of patients in the country are highly uncertain. The LSTM model is not proficient enough to predict future instances accurately when the data varies significantly. It is overcome by the M-LSTM architecture, as will be revealed in the result section.

The flow of this paper is as under: Section 2 presents a detailed description of the data and pre-processing methods. It is to prepare the dataset in the format required to make predictions using deep learning models. Section 3 explains the description of deep learning models used to predict the active cases of COVID-19 in four states and the proposed method. Section 4 covers the experimental results to evaluate the performance of models in predicting the active cases. Section 5 concludes this work.

2   Description of Dataset

The dataset is taken from the Kaggle website [30] to study COVID-19 cases in India. The dataset contains seven files that includes age group of affected cases, the number of hospital beds in each state in India, list of ICMR testing labs, individual case level details, testing details at state level, number of COVID-19 cases in India at daily level and population of different states in India. The number of COVID-19 cases taken day-wise includes total confirmed cases, recovered cases, and total death cases. The data of active cases are segregated state-wise for particular states, and cases are from 30th January to 15th September, 2020. The four states that show high caseloads during these months, namely: Maharashtra, Gujarat, Delhi, and Kerala, are part of the study. For example, Delhi, India's capital, is located in the Northern region and surrounded by northern, western, and southern parts of Haryana and the eastern part of Uttar Pradesh. Therefore, people's movement is very high in this state, resulting in the spread of the virus fastly. The same case is valid for other selected states as well. The trend of active cases of COVID-19 in these states is in Fig. 1. Since the selected states have different weather conditions; therefore, the impact of temperature, humidity, and other external factors on the active cases is not considered in our experiment. We would like to highlight that the COVID-19 dataset varies significantly from state to state and has sufficient volume as well depending on the timestamp we consider.

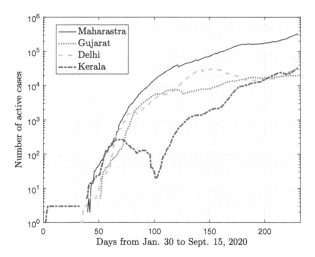

Fig. 1. Number of active COVID-19 cases for four Indian States.

3   Methods

3.1   Long Short-Term Memory Model

This model is the type of recurrent neural networks used in deep learning to solve sequence prediction problems. It is useful in speech recognition and other domains of time series classification problems. The model uses one hidden layer of LSTM units and one output layer to make the predictions. These networks take contextual information to transform the input sequence into an output sequence. The basic structure of the LSTM cell has four gates such as input gate, forget gate, control gate, and output gate, as shown in Fig. 2. In our experiment, initially, four such LSTM cells are used. During training, 8 and 16 LSTM cells add gradually to build the best model. The input gate i_t decides on the information transfer to the cell. The forget gate f_t decides which information to neglect from the previous memory. The control gate updates the cell's state from C_{t-1} to C_t. All the gates except the control gate use the sigmoid function. The equations of the input gate and forget are explained below [31].

$$f_t = \sigma(W_f).[h_{t-1}, x_t] + b_f \tag{1}$$

$$i_t = \sigma(W_i).[h_{t-1}, x_t] + b_i \tag{2}$$

The control gate that manages the update of cell from C_{t-1} to C_t follows the form equation:

$$\widetilde{C}_t = \tanh(W_c).[h_{t-1}, x_t] + b_c \tag{3}$$

$$C_t = f_t * C_{t-1} + i_t * \widetilde{C}_t \tag{4}$$

The output gate generates the output along with hidden vector h_{t-1}. The process uses the following equations:

$$O_t = \tanh(W_o).[h_{t-1}, x_t] + b_o \tag{5}$$

$$h_t = O_t * \tanh(C_t) \tag{6}$$

Here W's are the weights applied at each gate, and b's are bias values. Tanh (Tangent hyperbola) function scales the values in the range -1 to 1.

Our LSTM is trained on 148 samples and validated on 70 samples. The short term memory in simple LSTM makes it challenging to train very deep networks. Another challenge is the vanishing gradient problem due to learning by propagation through hidden layers.

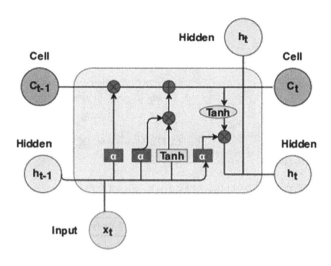

Fig. 2. The basic cell structure of LSTM [32]

3.2 Multiplicative Long Short-Term Memory Model

The use of M-LSTM can solve the above challenges. It has a multiplicative recurrent neural network and LSTM. They give outstanding performance in sequence modeling tasks. The RNN model apply factorization of hidden weights via the product rule to perform density estimation of sequences [33]. Suppose trained LSTM with fixed weights W_g encounters unexpected input x_t. In that case, cell state C_t gets disturbed, lead to agitated future h_t. By allowing hidden state to react flexibly on new input, it might able to recover from mistakes. It allows flexible input-dependent transitions that are easier to control due to the gating units of LSTM. In this model, the hidden state is transformed into an intermediate state M_t in a multiplicative way using input x_t. The element-wise multiplication allows M_t, to change its value for h_{t-1} and x_t flexibly. The internal working of M-LSTM is in Fig. 3. At timestep t, the the current observation i_t is embedded as e_{i_t} and combined with LSTM with previous cell state C_{t-1} and h_{t-1} of the last timestep t-1 which yields a new observation h_t. The inner product of h_t with the embedding of the next observation $e_{i_{t+1}}$ yields a scalar value p_{t+1} as output [33]. These models also do not adhere to vanishing gradient problems. They are proficient enough to capture long-term relationships in time series data, making them best suited for unexpected input instances. The equations of each gate used in this model follows the typical LSTM implementation. An extra intermediate state calculates the inner product of h_{t-1} with unexpected input x_t.

$$f_t = \sigma(W_f).[h_{t-1}, x_t] + b_f \tag{7}$$

$$i_t = \sigma(W_i).[h_{t-1}, x_t] + b_i \tag{8}$$

$$\widetilde{C}_t = \tanh(W_c).[h_{t-1}, x_t] + b_c \tag{9}$$

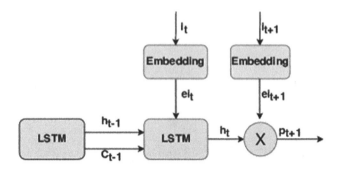

Fig. 3. Working of Multiplicative LSTM [33]

Algorithm 1: Training and building M-LSTM network

Result: The M-LSTM network

initialization;

input gates (I), control gates (C), forget gate (F), intermediate state (M) and
   output gate (O)

The M-LSTM network with hidden layers, weights, biases and activation
   functions

hyper-parameters include number of epochs, batch size, number of neurons (N)
   and number of hidden layers (H)

local variables;

look-back, Neurons

look-back ← Set to 3

foreach *epoch* **do**

    Neurons ← Set to 4, 8 and 16

    foreach *input(X_t, y) inbatch* **do**

        Reshape input instances X_t

        weight ← update using input-dim and N

        bias ← update using N

        foreach *weight and bias in network* **do**

          | *Output of each gate ← update using N and drop-out*

        end

        foreach *weight and bias in network* **do**

          *element-wise multiplication of H and X_{t+1}← update using N and
   look-back*

          weight ← update using look-back

          bias ← update using look-back

        end

    end

    return Output

end

$$C_t = f_t * C_{t-1} + i_t * \widetilde{C}_t \tag{10}$$

$$M_t = (W_m * x_t + b_m) \odot (W_m * h_{t-1} + b_m) \tag{11}$$

$$O_t = \tanh(W_o).[h_{t-1}, x_t] + b_o \qquad (12)$$

$$h_t = O_t * \tanh(C_t) \qquad (13)$$

The procedure of building and training M-LSTM network is shown in Algorithm 1. Firstly, the variables like look-back, number of epochs, number of neurons, hidden layers and activation function are initialized. For this, a constructor is used. Here, the number of neurons N are taken 4, 8, and 16 but one at a time or their combination for experiment purpose. Then, the input instances are reshaped to be feed into M-LSTM model. Next, the value of weights and bias are set by using input-dimension and number of neurons N. Using the value of N and drop-out, output of each gate is updated for each weights and biases. Lastly, the element-wise multiplication of hidden layer with the next input x_{t+1} instance is performed at intermediate state. For the next cycle, weights and biases are updated using look-back value and finally the output is generated.

3.3   Proposed Method

In our experiment, the data-set is taken from Kaggle website [30], as mentioned in Sect. 2. Open-source Python libraries like Numpy, Pandas, Tensorflow, and Keras execute the experiment on the given dataset. These are the popular libraries used to build, train, and test the deep learning models. The state-wise data of active cases are available, with cases from 30th January 2020 to 15th September 2020. It is important to note that our predictions are under the assumption that the government would not impose further lock-downs in the country, and the vaccine for COVID-19 is unavailable.

The workflow of the proposed method is in Fig. 4. Firstly, the input instances scale in the range of 0 and 1, instances are reshaped from a 2-dimensional matrix to 3-dimension. The dataset splits up into the train and test set. The train set contains two-third of the cases, while the test set contains one-third of the cases.

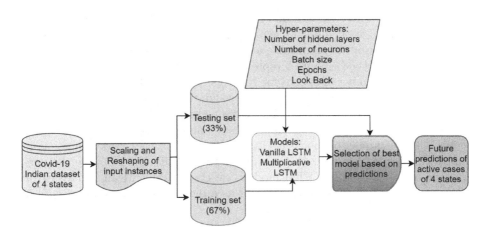

Fig. 4. Workflow of proposed method.

Table 1. List of hyper-parameters.

Hyper-parameters	Value
Number of neurons	4,8,16
Number of hidden layers	0,1,2
Batch-size	1
Look-Back	3,4 (for 1 day prediction)
Look-Back	14 (for upto 180 days prediction)
Epochs	17
Verbose	2
Optimizer	Adam
Loss function	Mean squared error

Various hyper-parameters are applied to build different versions of the model. Our experiment uses a combination of 0 to 2 hidden layers with the number of neurons taken at each layer is 4, 8, and 16. One Dense layer applies for output. The batch size is one and the number of epochs are 17. The look-back variable is taken 3 and 4 for one-day prediction, and 14 is for 180 days. Table 1 shows the list of hyper-parameters used in our experiment. Predictions are analyzed using RMSE (Root Mean Square Error) for both training and testing sets for all these models. To further improve these models' performance and avoid biased approximation of prediction errors, a nested cross-validation method is utilized, known as Day Forward Chaining. This technique accurately predicts the active cases by degrading the error values, and following that, the best model is chosen by analyzing the predictions for both LSTM and M-LSTM.

4 Results

Keras API [34] with TensorFlow backend creates and train models. We have experimented with different versions of the model by tuning the hyper-parameters, to accurately predict the active cases in various states, as discussed in the proposed method. Figure 5 represents the plot of training and validation losses versus the number of epochs for Kerala state. It can be observed from the plot that the M-LSTM model yields lower validation losses than simple LSTM. We have trained our models with eight neurons in a hidden layer, and one output layer in LSTM and M-LSTM approaches. The look-back value for the same models is 3, the number of epochs is 17, and the batch size is 1. Moreover, similar trend was observed for other states as well. To quantify and compare the performance of LSTM and M-LSTM models, we choose Root Mean Square Error and Normalised Root Mean Square Error as a metric to measure the accuracy of the prediction, expressed as:

$$\text{RMSE} = \sqrt{\frac{\sum_{i=1}^{n}(F_i - O_i)^2}{n}}, \tag{14}$$

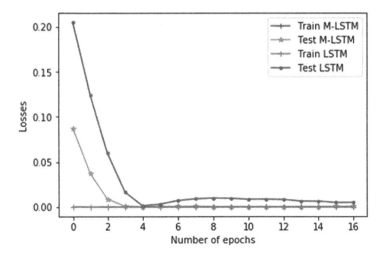

Fig. 5. Losses vs Epochs plot for Kerala state.

$$\text{N-RMSE} = \frac{\text{RMSE}}{\overline{y}}, \tag{15}$$

where 'F_i' is the expected value of active cases, 'O_i' is the observed values, 'n' is the total active cases and '$\overline{y}$' is the average active cases.

Table 2 represents the performance analysis of simple LSTM and M-LSTM models for Kerala and Maharashtra state. In this analysis we have observed the performance across different model configurations. Multiple hidden layers are also useful for analyzing both the models' behavior by considering the look-back value of 3 and 4. The number of epochs is considered 17, and the batch is 1. We have used the combination of 4, 8, and 16 neurons in each hidden layer with one output layer. Our observations proved that the N-RMSE rate of M-LSTM is less than simple LSTM in most of the cases.

Table 2. Performance analysis of Simple-LSTM and M-LSTM for Kerala and Maharashtra state.

Model Configuration					Kerala		Maharashtra	
Model No.	Hidden Layers	Look Back	Epochs	Batch Size	RMSE	N-RMSE	RMSE	N-RMSE
LSTM	4+1	3	17	1	1560.64	0.44	25763.69	0.45
M-LSTM	4+1	3	17	1	1524.11	0.43	18009.11	0.32
LSTM	8+1	3,4	17	1	1680.05	0.47	19265.48	0.34
M-LSTM	8+1	3,4	17	1	547.38	0.15	3651.54	0.06
LSTM	16+1	3	17	1	914.71	0.26	12114.06	0.21
M-LSTM	16+1	3	17	1	425.91	0.12	5084.92	0.09
LSTM	8+16+1	3	17	1	2437.51	0.68	5313.48	0.09
M-LSTM	8+16+1	3	17	1	2195.57	0.61	4797.97	0.08

Table 3. Performance analysis of Simple-LSTM and M-LSTM for Delhi and Gujarat state.

Model Configuration					Delhi		Gujarat	
Model No.	Hidden Layers	Look Back	Epochs	Batch Size	RMSE	N-RMSE	RMSE	N-RMSE
LSTM	4+1	3	17	1	2416.62	0.25	1000.60	0.15
M-LSTM	4+1	3	17	1	2037.70	0.21	734.12	0.11
LSTM	8+1	3,4	17	1	2579.63	0.27	779.18	0.12
M-LSTM	8+1	3,4	17	1	1335.67	0.14	571.18	0.09
LSTM	16+1	3	17	1	2322.19	0.24	241.47	0.04
M-LSTM	16+1	3	17	1	1757.93	0.18	298.59	0.05
LSTM	8+16+1	4,3	17	1	2548.55	0.26	1242.58	0.19
M-LSTM	8+16+1	4,3	17	1	2701.05	0.28	1124.28	0.17

Similarly, the error rates of simple LSTM and M-LSTM models has been observed for states Delhi and Gujarat in Table 3. Our primary focus is to perform one-step prediction of active cases with multiple look-backs by considering the trend of active cases in previous days. In Table 3, we have considered different model configurations and tuned our models with the same hyper-parameters as in Table 2. The observations have proved that performance of M-LSTM model in state Delhi and Gujarat is better than the simple LSTM model.

4.1 Prediction

The primary objective of this paper is to analyze the prediction of active cases in the future. Therefore, our models are trained on 67% training set and validated on 33% testing data. The number of days considered for training the model is 148, while the active case prediction is 70 days. In Fig. 6a, the predictions of active cases in Kerala state is shown in both simple LSTM and M-LSTM. The black line represents the trends of the active cases of the trained model. The number of active cases in trained models for both simple LSTM and M-LSTM overlap. The blue line indicates the predicted active cases by a simple LSTM model. In contrast, the red line represents the predicted active cases by the M-LSTM model.

In Fig. 6b, the Maharashtra state's active case prediction can be seen, which is maximum among all the other states. The trend of prediction in the Gujarat state is in Fig. 6c. One can observe that during the lock-down period, the active cases in Gujarat state decreases to some extent. This has happened due to a complete lock-down for 15 days during this period. After unlocking the non containment zones, we observed an increase in the trend of active cases. Consider the cases in Delhi state in Fig. 6d. During initial days the active cases rise at a moderate level, then suddenly it rises to its peak level. After a particular time, the cases started to fall. The current condition shows that the cases started rising again in Delhi at a moderate level. By analyzing the trends of active cases in four states, one can observe that the M-LSTM model accurately predicts the active cases while keeping the error rate minimum. On the contrary, due to the high error rate, the simple LSTM has over predicted the active cases. Now, this

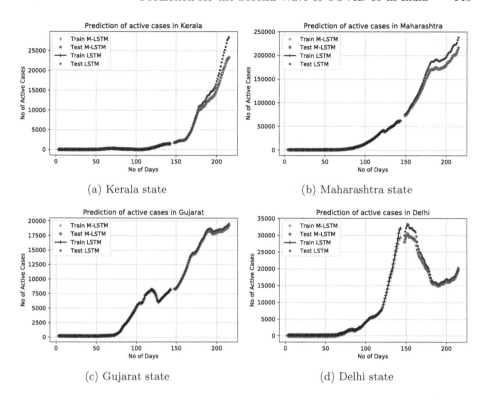

Fig. 6. Trend of predictions using LSTM and M-LSTM (One step prediction).

is a one-step prediction of active cases with multiple look-back values. Based on these observations, a model with high accuracy and less N-RMSE rate is used as the prediction model for COVID-19 data. Furthermore, to improve the performance of these models, the Day Forward Chaining cross-validation method is utilized. This method focused on the unbiased estimation of prediction errors, which accurately predicted active cases.

4.2 Day Forward Chaining

Day forward chaining is a nested cross-validation method for available time-series data. To accurately predict the time-series data using cross-validation, it is essential to withhold data structure to prevent data leakage. Therefore, instead of using k-fold cross-validation, we have utilized hold-out cross-validation to fit the models. To avoid biased estimation of prediction error on the dataset, we have applied many train/test splits and computed the average of errors over all the splits. For the training process, the Kerala state's data is taken, and a one-step prediction method is followed. One hidden layer with 8 number of neurons and one dense layer is used to build a model. The epochs, look-back, and batch-size values are 17, 3, and 1, respectively. Firstly, we have fitted the model with 5

Table 4. Day Forward chaining in Simple-LSTM and M-LSTM for Kerala state.

Day Forward Chaining	Train Size	Test Size	LSTM			M-LSTM		
			RMSE	N-RMSE	Avg. N-RMSE	RMSE	N-RMSE	Avg. N-RMSE
No. of Splits 5	57%	43%	2034.12	0.42	0.58	963.27	0.20	0.36
	67%	33%	3104.62	0.64		988.79	0.20	
	77%	23%	5200.57	1.08		3154.76	0.65	
	87%	13%	1355.16	0.28		1782.49	0.37	
	97%	3%	2317.15	0.48		1855.29	0.38	
No. of Splits 10	47%	53%	2516.01	0.52	0.45	2600.88	0.54	0.30
	52%	48%	1804.64	0.37		2004.85	0.42	
	57%	43%	2033.39	0.42		964.24	0.20	
	62%	38%	1870.50	0.39		756.33	0.16	
	67%	33%	3104.62	0.64		988.79	0.20	
	72%	28%	4042.87	0.84		1706.21	0.35	
	79%	21%	2502.46	0.52		1093.69	0.23	
	84%	16%	1304.61	0.28		1144.56	0.24	
	89%	11%	1423.19	0.29		1706.58	0.35	
	94%	6%	1332.67	0.28		1581.57	0.33	

number of splits and computed the average N-RMSE value in both the models as shown in Table 4. In simple LSTM, the average N-RMSE is 0.58, while in M-LSTM, it is 0.36.

Similarly, the model is trained with 10 number of splits. The average value of N-RMSE in simple LSTM is 0.45, and in M-LSTM, it is 0.30. We have observed that on applying 10 number of splits, the error has reduced in both the models while the M-LSTM model has outperformed the simple LSTM model. The impact of day forward chaining on the prediction of active cases can be observed in Fig. 7. It represents number of splits on the x-axis and predictions on the y-axis. We divided the dataset into 77% of the train set and 23% of the test set for one split. The hyper-parameters are tuned in the same manner as in Table 4. The error rate in the training process is calculated in terms of N-RMSE. From the figure, it is observed that the error rate degrades in both the models using cross-validation, which further has improved the performance in the prediction of active cases in this state.

4.3   Second Wave Prediction

We have estimated the predictions of active cases of COVID-19 for the second wave. In this experiment, the data trains from 30th January 2020 to 15th September 2020. The active cases forecast is for the next 180 days that is up to 01st March 2021. Figure 8a represented the predictions for Kerala state, in both the LSTM and M-LSTM based models. The hyper-parameters used to achieve these predictions are; the value of the number of neurons is 1, the value of look-back is taken 14 as the incubation period of COVID-19 is up to 14 days. The number of epochs used is 10, and the batch size is 1. Other hyper-parameters remains the same. By observing Kerala state's predictions, the active cases could reach

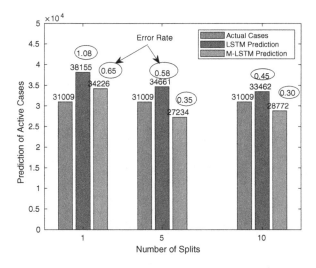

Fig. 7. Comparison of predicted cases using Day Forward Chaining in Kerala.

70,000 in the next six months. The second wave prediction for state Maharashtra and Gujarat is in Fig. 8b and 8c. The active cases in these states could extend to 8.2 lac and 32,000 respectively. In India, since the temperature, humidity, and other external factors varies from state to state, the first wave of active cases may not reach its peak in each state. Although in Kerala, Gujarat, and Delhi, the decline in the fresh cases of COVID-19 has been observed in August 2020. That may be the effect of multiple lock-downs in the country. Now, there has been seen a sharp jump again in fresh cases in September 2020. By seeing these observations, we have predicted that these states are reaching towards the second wave. There are likely chances of an increase in active cases on a large scale. Therefore, ordinary people need to take every possible precaution and maintain social distancing habits. Medical officials working in healthcare organizations are required to re-frame their policies to curb the situation. Fig. 9 shows the bar plot comparison of the predictions of active cases in both LSTM and M-LSTM

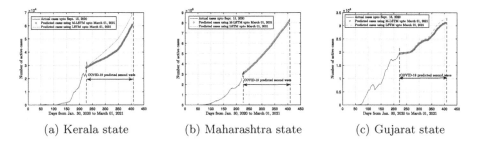

(a) Kerala state   (b) Maharashtra state   (c) Gujarat state

Fig. 8. COVID-19 second wave prediction.

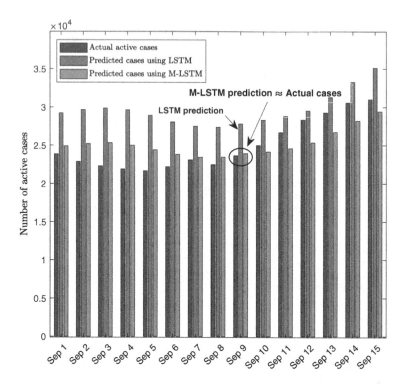

Fig. 9. Bar chart comparison of predicted cases in Kerala.

models with actual cases from September 01 to September 15, 2020. We have evaluated one day prediction in both the models and compared them with actual cases at each day. From the observations, we can analyze that the deviations of predicted cases with actual cases are comparatively less in case of M-LSTM than that of simple LSTM. We can also observe that the M-LSTM model accurately predicts the active cases while simple LSTM has over predicted the active cases due to the high error rate. The same trend of predictions can be observed from Fig. 6 and Fig. 8. The percentage gain in M-LSTM model is 22.3% as compared to simple LSTM. Hence, M-LSTM can be considered as a more accurate model in predicting the active cases.

5   Conclusion

COVID-19 is a global health emergency all over the world. An accurate model can predict the active cases in the coming future to balance the situation and advise the government and health care organizations. In this work, we propose a multiplicative-LSTM aided forecasting of active cases in India. Specifically, we consider states with maximum caseloads namely, Maharashtra, Kerala, Gujarat, and Delhi. These states have witnessed a definite increasing trend in their positive rates. Since the number of patients with active cases are uncertain and may vary abruptly with time, a simple LSTM approach is not capable of predicting future cases accurately. We found in our experiments that M-LSTM performs better than simple LSTM in predicting the active cases. The percentage of improvement of M-LSTM model as compared to simple LSTM is 22.3%. To further improve these models' performance and avoid biased estimation of prediction errors, a Day Forward Chaining cross-validation method is utilized. This technique helped in degrading the error values, hence accurately predicting the active cases.

Furthermore, using the proposed scheme we have predicted the future active cases for next 180 days i.e., up to 01st March 2021. Results suggest that the rise in active cases is heading towards the second wave of COVID-19 in India. Our experiment would help government officials organize their healthcare structure by identifying dedicated hospitals for COVID-19 patients. They could enhance their testing capacity by knowing the number of testing kits required for testing the samples of patients in each state. They can also identify the need for isolation beds, beds with oxygen support, and ICU beds for these patients.

References

1. Novel coronavirus (2019-ncov) sitiuation report-1 by world health organization. https://www.who.int/docs/default-source/coronaviruse/situation-reports/20200121-sitrep-1-2019-ncov.pdf
2. World health organization. https://www.who.int/emergencies/diseases/novel-coronavirus-2019/technical-guidance/naming-the-coronavirus-disease-(covid-2019)-and-the-virus-that-causes-it
3. Nobel, Y.R., Phipps, M., Zucker, J., et al.: Gastrointestinal symptoms and coronavirus disease 2019: a case-control study from the United States. Gastroenterology **159**(1), 373–375 (2020)
4. Ministry of health and family welfare, Government of India. https://www.mohfw.gov.in/
5. Ceylan, Z.: Estimation of covid-19 prevalence in Italy, Spain, and France. Sci. Total Environ. **729**, 138817 (2020)
6. Petropoulos, F., Makridakis, S.: Forecasting the novel coronavirus covid-19. PLOS ONE **15**(3), 1–8 (2020)
7. Barman, A.: Time series analysis and forecasting of covid-19 cases using LSTM and ARIMA models. arXiv preprint arXiv:2006.13852 (2020)
8. Dehesh, T., Mardani-Fard, H., Dehesh, P.: Forecasting of covid-19 confirmed cases in different countries with arima models. medRxiv (2020). https://doi.org/10.1101/2020.03.13.20035345
9. Arora, P., Kumar, H., Panigrahi, B.K.: Prediction and analysis of covid-19 positive cases using deep learning models: a descriptive case study of India. Chaos, Solitons Fractals **139**, 110017 (2020)
10. Chimmula, V.K.R., Zhang, L.: Time series forecasting of covid-19 transmission in Canada using LSTM networks. Chaos, Solitons Fractals **135**, 109864 (2020)
11. Pal, R., Sekh, A.A., Kar, S., Prasad, D.K.: Neural network based country wise risk prediction of covid-19. arXiv preprint arXiv:2004.00959 (2020)
12. Zheng, N., et al.: Predicting covid-19 in China using hybrid AI model. IEEE Trans. Cybern. **50**(7), 2891–2904 (2020)

13. Bandyopadhyay, S.K., Dutta, S.: Machine learning approach for confirmation of covid-19 cases: positive, negative, death and release. medRxiv (2020)

14. Wang, L., et al.: Real-time estimation and prediction of mortality caused by covid-19 with patient information based algorithm. Sci. Total Environ. **727**, 138394 (2020)

15. Bowman, V.E., Silk, D.S., Dalrymple, U., Woods, D.C.: Uncertainty quantification for epidemiological forecasts of covid-19 through combinations of model predictions. arXiv preprint arXiv:2006.10714 (2020)

16. Antia, H.M.: COVID-19: data analysis and modelling. PhysicsNews **50**(2) (2020)

17. Nishant, N., et al.: Investigation on the covid-19 outbreak in India: lockdown impact and vulnerability analysis. J. Geograph. Inform. Syst. **12**, 334–347 (2020)

18. Tiwari, S., Kumar, S., Guleria, K.: Outbreak trends of coronavirus disease-2019 in India: a prediction. Disaster Med. Pub. Health Preparedness **115**, 1–6 (2020)

19. Sarkar, K., Khajanchi, S., Nieto, J.J.: Modeling and forecasting the covid-19 pandemic in India. Chaos, Solitons Fractals **139**, 110049 (2020)

20. Gupta, R., Pandey, G., Chaudhary, P., Pal, S.K.: SEIR and regression model based covid-19 outbreak predictions in India. arXiv preprint arXiv:2004.00958 (2020)

21. Tomar, A., Gupta, N.: Prediction for the spread of covid-19 in India and effectiveness of preventive measures. Sci. Total Environ. **728**, 138762 (2020)

22. Chakraborty, T., Ghosh, I.: Real-time forecasts and risk assessment of novel coronavirus (covid-19) cases: a data-driven analysis. Chaos, Solitons Fractals **135**, 109850 (2020)

23. Rafiq, D., Suhail, S.A., Bazaz, M.A.: Evaluation and prediction of covid-19 in India: a case study of worst hit states. Chaos, Solitons Fractals **139**, 110014 (2020)

24. Xu, S., Li, Y.: Beware of the second wave of covid-19. Lancet **395**(10233), 1321–1322 (2020)

25. Adebowale, V., et al.: Covid-19: call for a rapid forward looking review of the UK's preparedness for a second wave–an open letter to the leaders of all UK political parties. BMJ 369 (2020). https://www.bmj.com/content/369/bmj.m2514

26. Vogel, L.: Is Canada ready for the second wave of covid-19? CMAJ **192**(24), E664–E665 (2020)

27. Vaid, S., McAdie, A., Kremer, R., Khanduja, V., Bhandari, M.: Risk of a second wave of covid-19 infections: using artificial intelligence to investigate stringency of physical distancing policies in North America. Int. Orthop. **44**(8), 1581–1589 (2020)

28. Suda, G., et al.: Time-dependentchanges in the seroprevalence of covid-19 in asymptomatic liver diseaseoutpatients in an area in Japan undergoing a second wave ofcovid-19. Hepatol. Res. **50**(10), 1196–1200 (2020)

29. Ghanbari, B.: On forecasting the spread of the covid-19 in Iran: the second wave. Chaos, Solitons Fractals **140**, 110176 (2020)

30. Kaggle website. https://www.kaggle.com/sudalairajkumar/covid19-in-India

31. Sun, Q., Jankovic, M.V., Bally, L., Mougiakakou, S.G.: Predicting blood glucose with an lstm and bi-lstm based deep neural network. In: Proceeding of IEEE Symposium on Neural Networks and Applications, pp. 1–5 (2018)

32. Hochreiter, S., Schmidhuber, J.: Long short-term memory. Neural Comput. **9**(9), 1735–1780 (1997)

33. Krause, B., Lu, L., Murray, I., Renals, S.: Multiplicative LSTM for sequence modelling. arXiv preprint arXiv:1609.07959 (2017)

34. Chollet, F., et al.: Keras (2015). https://github.com/fchollet/keras

Texture Feature Extraction: Impact of Variants on Performance of Machine Learning Classifiers: Study on Chest X-Ray – Pneumonia Images

Anamika Gupta[1], Anshuman Gupta[1](✉), Vaishnavi Verma[1], Aayush Khattar[1], and Devansh Sharma[2]

[1] S.S. College of Business Studies, University of Delhi, New Delhi, India
anamikargupta@sscbsdu.ac.in, {anshuman.18514,vaishnavi.18537, aayush.18576}@sscbs.du.ac.in
[2] Department of Computer Science, University of Delhi, New Delhi, India
devanshsharma80@gmail.com

Abstract. Image textures are a set of image characteristics used for identifying regions of interests (ROIs) in images. These numerical features can thus be used to classify images in various classifiers. This paper introduces the task of classifying Chest X-ray images with Machine Learning Classifiers and to see the impact of variations on the result of classification. For this purpose, second-order statistical features (GLCM texture features) are extracted from all the images with preprocessing and classification is performed using these features. Various variants are applied for image processing. First-order features are included, the image is divided into multiple regions, different values of distance for GLCM are used. Several evaluation metrics are used to judge the performance of the classifiers. Results on Chest X-ray (Pneumonia) dataset shows remarkable improvements in the accuracy, F1-Score, and the AUC of the classifier.

Keywords: Texture extraction · GLCM · First order statistics · Pre-processing

1 Introduction and Related Work

Pneumonia is a form of acute respiratory disease which can be caused by an infection due to a virus, bacteria, and sometimes by other microorganisms. Other causes of pneumonia include allergic reactions from medications or conditions such as autoimmune diseases. Pneumonia is a fatal illness in which the air sacs get filled with pus and other liquid. A person can get infected with pneumonia but conditions like cystic fibrosis, asthma, diabetes, heart failure, a history of smoking, a poor ability to cough, and a weak immune system can be risk factors making someone more prone to pneumonia [15]. People infected with pneumonia feel difficulty in breathing. Every year, around 15% of children under the age of 5 die due to pneumonia [19]. With the advancement in medical

© Springer Nature Switzerland AG 2020
L. Bellatreche et al. (Eds.): BDA 2020, LNCS 12581, pp. 151–163, 2020.
https://doi.org/10.1007/978-3-030-66665-1_11

technology, the treatment of most of the diseases has become easier, including pneumonia. Diagnosis of pneumonia can be done by observing the chest X-ray, CT scan of the lungs, ultrasound of the chest, needle biopsy of the lung, and MRI of the chest [15]. X-ray images are preferred over CT scan images, as the technology for X-ray imaging is easily available and widespread. Therefore, the use of Chest X-ray images, along with computer-aided technology, is becoming popular today as this approach is more cost-efficient and can benefit a larger audience. Figure 1 shows the chest X-ray scans of a normal person and pneumonia infected person.

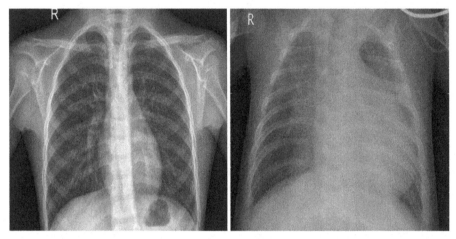

Fig. 1. Chest X-ray images of a normal person (left) and pneumonia infected patient (right) (Source: Kaggle [11]).

Image processing is the manipulation of an image to extract some meaningful information from the image. The first part of image processing is extracting some useful features from the image known as feature extraction. Feature extraction involves the careful calculation of textural features in images so as to get sufficient information for further classification and analysis while cutting back on any redundant information that might be misleading for the same. Texture feature extraction can thus be used to save memory, time, and computation costs. In this paper, we use GLCM (Grey level co-occurrence matrix) for feature extraction. It is a traditional approach to texture analysis with a number of applications, especially in medical image analysis. Haralick's fourteen GLCM features are calculated and used for the classification of Pneumonia chest X-ray images [10].

The pre-processing is necessary to resolve several types of problems including noisy data, redundancy in data, missing data values, etc. as a lot of irrelevant and redundant information can heavily affect the performance of Machine Learning models. To produce more understandable and accurate results, several inadequate information is removed [6]. To detect extreme points in the dataset also known as outliers, DBSCAN is used as the outlier detection technique [4]. Redundant and irrelevant features are removed using RFE [13]. The normal class has relatively fewer samples as compared to the

Pneumonia class in the dataset. Hence, SMOTE is used for oversampling the minority class. Standardisation is used to scale the features in the dataset [7].

In several earlier studies, GLCM technique has been used for diagnosing various medical conditions. Pugalenthi et al. [16] used the technique for the detection of tumors using brain MRIs, where features were extracted using GLCM, followed by machine learning classifier to classify the MRI images using the textural features extracted by GLCM [16]. It has also been used to classify the bone X-ray images into fractured and non-fractured categories, where the second-order statistical information of gray levels between neighboring pixels extracted by GLCM are fed to different classifiers such as Logistic Regression and Decision trees for abnormality detection [14]. Ankita et al. used the technique to classify lung cancer datasets. Features were extracted using GLCM and classification was done on the images using SVM classifier [3]. Although GLCM has been prominently used in many pieces of research for feature extraction, there are several instances where various other methods have been used for feature extraction from radiographs like chest X-ray images and brain MRIs. The Wavelet transform and Curvelet transform methods have been used for the classification of early-stage lung cancer diagnosis of chest X-ray images [9]. Chandra et al. also proposed a Hierarchical feature extraction method, which has been used to extract features from chest X-ray images for detection of tuberculosis related abnormalities in chest X-ray images [5].

2   Background

Feature extraction can be performed using First Order Statistical features and Second-order Statistical features. Kurtosis, skewness, mean, variance, etc. are some of the first-order features which only consider the pixel under observance. GLCM is one of the most popular approaches for computing second-order statistical features. GLCM considers the spatial relationship of adjacent pixels in the texture analysis. GLCM provides a matrix that shows how often pairs of pixels with a specific value and in a specified spatial relationship occur [2]. Haralick et al. [10] described fourteen textural features - Angular Second Moment, Contrast, Correlation, Variance, Inverse Difference Moment, Sum Average, Sum Variance, Sum Entropy, Entropy, Difference Variance, Difference Entropy, Two Information measures of Correlation, and Maximal Correlation Coefficient - that can be extracted using GLCM for texture analysis of images.

Various classification techniques have been popular in literature for predicting the behaviour of unknown tuples. Some of the most popular ones are KNN, Neural networks, Support vector machine, Bayes classifier, and Decision tree classifier. KNN classifiers use data and classify new data points based on a majority vote of its neighbours [12]. SVM can be utilized to perform image classification and to optimize the classification of images [12]. Naive Bayes classifiers are probability driven ML models based on the Bayes theorem [17]. Decision Tree Classifiers classify the data samples by learning some conditional rules of the input features and are also used to train the image understanding system to accomplish supervised machine learning [1]. The neural network model is made up of neurons, which are small computational units that learn some function of the input features. Many such neurons are combined in a neural network for learning some complex functions directly from the inputs. A neural network has the ability to learn complex features from simpler ones and thus reduce the need for hand-engineered features for classification tasks of raw data, like images in various image recognition and classification tasks, and get the output directly [8].

Various evaluation metrics have been proposed in the literature for comparing the performance of the classifiers such as Confusion Matrix, Precision, Classification Accuracy, Recall, Specificity, F1 Score, and AUC ROC Curve. Confusion Matrix is a matrix that shows a visualized representation of the performance of a machine learning model [18]. It describes 4 metrics as its entries as shown in Table 1. True Positive (TP) are those samples in which both the actual and predicted class were true, that is, the object classified was true in reality and the model also classified it as true; True Negatives (TN) are those samples which the model classified as false and the target was also false; False Positives (FP) are when the actual class of a sample is false but the predicted class is true, that is, the target of the sample was false but the model predicted it as true; False Negatives (FN) are those samples for which the actual class is true but the predicted class is false, that is, the target of the sample was true but the model predicted it as false [18]. Table 2 describes other classification metrics used in the study.

Table 1. Confusion matrix and its entries for a binary classification problem.

	Predicted Negative (0)	Predicted Positive (1)
Actual Negative (0)	TN	FP
Actual Positive (1)	FN	TP

Various pre-processing techniques such as Standardisation, Oversampling, Outlier Detection, and Feature Ranking have been used in order to improve the performance of various Machine Learning algorithms such as KNN, SVM, Naive Bayes, Decision tree, and Neural network. The process of scaling one or more features/attributes of a dataset such that their mean is equal to zero and the standard deviation is equal to one is called standardisation. If x_i is the value of a feature for the ith sample in the dataset consisting of m samples, $\bar{x}$ is the mean value of that feature, σ is the standard deviation, and z_i is the standardised feature value for that sample then for each feature in the dataset standardisation can be performed using Eq. (1).

$$z_i = \frac{x_i - \bar{x}}{\sigma}, \ where \ i = 1, 2, \dots, m \tag{1}$$

The process of increasing the number of samples for the minority class by creating new samples that are similar to the existing data for that class is called oversampling. Synthetic Minority Oversampling Technique (SMOTE) is a technique that draws a line between existing samples that are close in the feature space and then randomly picks a new sample along the line [7]. Outlier detection is the process of identifying samples that are very different from other samples and differ from the overall pattern in the dataset.

Table 2. Description of classification metrics along with their formulas.

Metric	Definition	Formula
Precision	It is defined as the ratio of the number of correct positive predictions to the total number of samples that were predicted positive	$\frac{TP}{TP+FP}$
Classification accuracy	It is defined as the proportion of the total samples correctly predicted out of all the samples	$\frac{TP+TN}{TP+FP+FN+TN}$
Recall	It is defined as the proportion of positives correctly predicted by our model out of all the samples that were actually positive	$\frac{TP}{TP+FN}$
Specificity	It is defined as the proportion of negatives correctly predicted by our model out of all the samples that were actually negative	$\frac{TN}{TN+FP}$
F1 score	The harmonic mean of recall and precision is known as F1 Score	$2 * \frac{Precision*Recall}{Precision+Recall}$
AUC-ROC curve	It is defined as the area under the Receiver Operator Characteristic (ROC) curve. It tells about how capable our model is in distinguishing the classes	

Density-Based Spatial Clustering of Applications with Noise (DBSCAN) is an outlier detection technique based on the density-based clustering algorithm. It classifies various samples into three categories, namely core points, border points, and outliers using the clustering method and uses the concept of reachability for categorizing various points [4]. Machine Learning models are prone to overfitting when the number of features becomes very large. To mitigate this, features are ranked according to their importance and the least important features are discarded. In Recursive Feature Elimination (RFE), the model is fitted on the entire set of features and the importance of each feature is calculated, following which the weakest features are eliminated. This process is repeated recursively until the required number of features are selected [6].

3   Methodology and Experiments

The study in this paper focuses on analyzing the impact of applying various techniques that affect the performance of the classification on various evaluation metrics. The basic model of extracting features using the GLCM method has been developed. Standardisation, outlier handling, feature selection, and imbalance handling have been experimented with on the Chest X-Ray dataset. Experiments regarding the division of the image into multiple parts, incorporating first-order statistical features, changing the distance of pixels have been performed to check the suitable combinations. Observations have been reported in all the experiments.

3.1 Experimental Setup

Experiments are performed on the Chest X-ray images of Pneumonia patients obtained from the Kaggle website. The dataset consists of a total of 5,856 Chest X-Ray images in 3 different folders dividing the data into 3 separate sets for training, testing, and validation. The training set consisted of 5216 samples, test set consisted of 624 samples and validation set consisted of 16 samples. Since, the validation set was very small, it was not used for evaluation. All the folders are further subdivided into 2 more additional folders containing different categories of images: Normal and Pneumonia. The dataset consists of X-Rays of patients of age 1-5 as part of their routine clinical checkup from the Guangzhou Women and Children's Medical Center, Guangzhou [11].

The size of the images given in the dataset was not uniform. Since comparison cannot be performed on different sizes, the images have been resized to 1024 × 1024 pixels. Fourteen Haralick's GLCM Features are extracted from the images with four different orientations (0, 45, 90, 135) and a distance of two pixels. The mean of four orientations is computed and used as one feature. Five classifiers namely, K-Nearest Neighbours (KNN), Support Vector Machine (SVM), Naive Bayes, Decision Tree, and Neural Network are used to classify the obtained features.

Seven evaluation metrics namely, train accuracy, test accuracy, precision, recall, specificity, F1 score, and AUC-ROC are computed to judge the performance of the classifiers. Running the above experiment, we got the results as shown in Table 3.

Table 3. Classification results for the base experiment.

	Neural Network	KNN	SVM	Naïve bayes	Decision Tree
Train accuracy	0.5805	0.8288	0.9988	0.7977	1.0
Test accuracy	0.5544	0.6555	0.6266	0.7676	0.7484
Precision	0.7478	0.6866	0.6260	0.7980	0.7484
Recall	0.4333	0.8256	1	0.8410	0.9
Specificity	0.7564	0.3718	0.0043	0.6453	0.4957
F1 score	0.5487	0.7497	0.77	0.8190	0.8172
AUC-ROC	0.6185	0.6787	0.6375	0.8248	0.6979

The results have been shown in the form of a bar graph (Fig. 2) for better understanding. Here, we can observe that the highest test accuracy is 76.76% for Naïve- Bayes classifier, highest precision is 79.8% for Naïve-Bayes classifier, highest recall is 100% for SVM, highest specificity is 75.6% for neural network, the highest F1 score is 81.89% for Naïve-Bayes and highest AUC-ROC is 82.48% for Naïve-Bayes classifiers.

3.2 Variants

We experimented with various variants of the above experiment and compared the results. Variants used are listed below:

Base Experiment Results

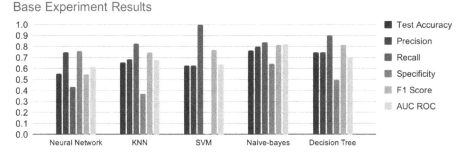

Fig. 2. Base experiment results (with 14 GLCM features).

Variant 1. Study the impact of standardization, Outlier removal, oversampling, and feature ranking. The sub-variants are as follows:

1.0. Fourteen GLCM features, distance = 2 (Base Experiment).
1.1. Standardise the features: StandardScaler is used to scale the feature vectors.
1.2. Standardise and remove the outliers: The outliers are removed using the DBSCAN technique.
1.3. Standardise and remove the outliers, oversample the minority class.
1.4. Standardise and remove the outliers, oversample the minority class, feature ranking: Selected the best features using the RFE feature ranking method.

The results of the above experiments are shown in Table 4. We observe here that standardisation of the features increases the test accuracy, F1 score, and AUC but preci- ↙ sion, recall, and specificity go down. Removing the outliers improves the test accuracy and F1 score further, but precision, recall, and specificity are going down. However, oversampling the minority class or feature ranking is not improving any of the evaluation metrics. It remains almost the same as variant 1.3. Since the difference between normal and abnormal cases is not much, the dataset can be considered as a more or less balanced dataset. Hence, oversampling doesn't seem to improve the results. Further, the number of features is very less, hence feature ranking doesn't improve the result.

Table 4. Classification results for Variant 1.0 to Variant 1.4

	Variant 1.0 (Base)	Variant 1.1	Variant 1.2	Variant 1.3	Variant 1.4
Train accuracy	100%	100%	100%	100%	100%
Test accuracy	76.76%	77.56%	**78.52%**	78.52%	78.52%
Precision	**79.8%**	78.55%	77.23%	78.57%	78.57%
Recall	**100%**	94.6%	96.1%	94.1%	94.1%
Specificity	**75.64%**	61.9%	54.27%	58.9%	58.9
F1 score	81.89%	83.76%	**84.4**	84.0	84.0
AUC-ROC	82.4%	**85.89%**	82.9%	82.4%	81.2%

Variant 2. Study the impact of dividing into multiple regions. The sub-variants are as follows:

2.0. Fourteen GLCM features, normalised, outliers removed, d = 2 (Base Experiment).
2.1. Base experiment and Divide the image into 4 regions.
2.2. Base experiment and Divide the image into 16 regions.
2.3. Base experiment and divide the image into 64 regions.

The results for the above experiments are shown in Table 5.

Table 5. Classification results of Variants 2.0 to 2.3.

	Variant 2.0 (Base)	Variant 2.1	Variant 2.2	Variant 2.3
Train accuracy	100%	100	100	100
Test accuracy	78.52%	79.4	**87.6**	86.6
Precision	78.57%	79.6	**87.3**	86.5
Recall	94.1%	94.3	**97.1**	95.8
Specificity	58.9	61.9	**77.3**	75.9
F1 score	84.0	85.1	**90.4**	89.7
AUC	81.2%	83.9	**92.0**	90.4

From Table 5 we observe that dividing the image into 4 regions improves all the evaluation metrics. Dividing the image into 16 regions improves all the evaluation metrics drastically. Dividing into 64 regions yields better than dividing into 4 quadrants but not better than 16 regions. So, we conclude that dividing the image into smaller parts for feature extraction certainly improves the performance but after a certain limit, the performance starts deteriorating.

Variant 3. Study the impact of including first-order features. The sub-variants are as follows:

3.0. Only GLCM features.
3.1. GLCM features and first-order features.

The results of the above experiments are as illustrated in Table 6.

We observe from Table 6 that most of the evaluation metrics yield better results when first-order features are also extracted along with GLCM features (second-order features).

Table 6. Classification results of Variants 3.0 and 3.1

	Variant 3.0	Variant 3.1
Train accuracy	100%	**100**
Test accuracy	78.52%	**79.6**
Precision	78.57%	**79.9**
Recall	**94.1%**	91.7
Specificity	58.9	**63.6**
F1 score	84.0	**84.9**
AUC	81.2%	**84.3**

Variant 4. Study the impact of including first-order features and dividing the image into sub-images. The sub-variants are as follows:

4.0. GLCM and first-order features (Base Experiment).
4.1. Base experiment and divide the image into 4 regions.
4.2. Base experiment and divide the image into 16 regions.
4.3. Base experiment and divide the image into 64 regions.

The results of the above experiments are shown in Table 7.

Table 7. Classification results of Variants 4.0 to 4.3

	Variant 4.0	Variant 4.1	Variant 4.2	Variant 4.3
Train accuracy	100	100	**100**	100
Test accuracy	79.6	81.4	**88.4**	85.7
Precision	79.9	82.7	**89.7**	85.5
Recall	91.7	95.8	**96.1**	96.1
Specificity	63.6	69.2	**83.3**	73.9
F1 score	84.9	86.5	**90.0**	89.0
AUC	84.3	91.1	**92.4**	91.0

Dividing the image into 4 regions improves the performance of the classifiers on all evaluation metrics as shown in Table 7. Dividing into 16 regions improves the results further. However, dividing into 64 regions doesn't yield results better than 16 regions.

Variant 5. Study the effect of changing the pixel distance for the GLCM. The sub-variants are as follows:

5.0. 14 GLCM features, 5 first-order features and dividing the image into 16 regions with $d = 2$ (Base Experiment).

5.1. Change the distance d = 4.
5.2. Change the distance d = 8.
5.3. Change the distance d = 16.

The results of the above experiments are as shown in Table 8.

Table 8. Classification results of Variants 5.0 to 5.3

	Variant 5.0	Variant 5.1	Variant 5.2	Variant 5.3
Train accuracy	100	100	100	100
Test accuracy	88.4	**89.5**	88.4	87.2
Precision	**89.7**	89.7	87.8	87.7
Recall	**96.1**	94.3	94.61	94.35
Specificity	**83.3**	82.0	78.2	78.6
F1 score	90.0	**91.8**	91.1	90.1
AUC	92.4	**93.2**	93.2	92.6

It can be observed from Table 8, that considering more distance between the pixels to compute GLCM features doesn't improve the performance of the classifiers. Some metrics improved while others' performance degraded.

3.3 Inter-variant Comparative Analysis

Comparison of Variant 1 and Variant 2. Variant 1 consisted of the application of several preprocessing techniques after the extraction of GLCM features. In Variant 2, the x-ray images were divided into subregions and GLCM features were extracted for each region followed by the application of preprocessing techniques as used in Variant 1.

On studying the results obtained in both variants (Table 4 and Table 5) we observe that results majorly improved from Variant 1 to Variant 2 as shown in Fig. 3. The test accuracy changed from 78.52% to 87.6% (11.56% increase), precision from 79.8% to 87.3% (9.4% increase), recall from 100% to 97.1% (2.9% decrease), specificity from 75.64% to 77.3% (2.19% increase), F1-score from 84.4% to 90.4% (7.1% increase) and AUC-ROC from 85.89% to 92% (7.11% increase).

Comparison of Variant 1, 3, 4 and 5. Variant 3 uses first-order features along with the fourteen GLCM features extracted for the image. Variant 4 combines Variant 2 (dividing image into subregions) with Variant 3. In Variant 5, the best experiment settings from Variant 4 are selected and the distance parameter for the GLCM is varied. We compared the best results from all the four variants with respect to test accuracy, F1-score, and AUC-ROC metrics as shown in Fig. 4. From Variant 1 to Variant 3, the test accuracy slightly changed from 78.52% to 79.6% (1.37% increase), F1-score from 84.4% to 84.9% (0.6% increase) and AUC-ROC from 85.89% to 84.3% (1.85% decrease). From Variant 3 to Variant 4, the test accuracy increased from 79.6% to 88.4% (11.05% increase), F1-score from 84.9% to 90% (5.67% increase) and AUC-ROC from 84.3% to 92.4% (9.6%

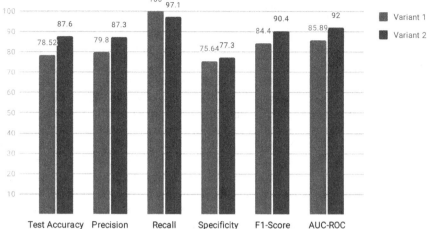

Fig. 3. Comparison of Variant 1 and Variant 2 to study the individual impact of first-order features and division of the image into regions.

increase). From Variant 4 to Variant 5, the test accuracy improved from 88.4% to 89.5% (1.24% increase), F1-score from 90% to 91.8% (2% increase) and AUC-ROC from 92.4% to 93.2% (0.86% increase). Overall, there was a 14% increase in test accuracy, 9.24% increase in F1-score and 8.5% increase in AUC-ROC from Variant 1 to Variant 5, highlighting the effectiveness of the experiments performed which can also be observed from the upward trend shown in Fig. 4.

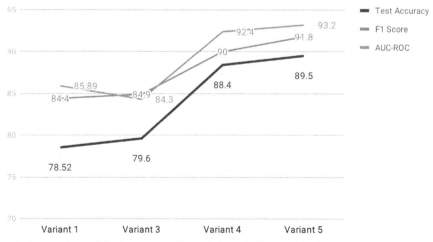

Fig. 4. Comparison of the best results from Variants 1, 3, 4, and 5 with respect to test accuracy, F1-score, and AUC-ROC.

4 Conclusion and Future Scope

The study was mainly focused on the use of the GLCM technique for extracting second-order statistical features (GLCM texture features) and the impact of variants on the performance of Machine Learning classifiers for the purpose of identifying Pneumonia patients from the Chest X-ray.

The base experiment showed the highest test accuracy of 76.76%. From variant 1 we concluded that the pre-processing techniques such as standardisation, outlier removal, oversampling, and feature ranking slightly improved the accuracy. The pre-processed images were then divided into smaller regions in variant 2 and then the GLCM features were extracted for each region which increased the accuracy. We then combined the 14 GLCM features with four first-order features in variant 3 and found that even without division of the image the inclusion of the first-order features improved our performance. Thus, we combined both the approaches of variant 2 and variant 3 in variant 4 and found that the combined results were even better. In order to increase the accuracy further, we observed the effect of changing the pixel distance in variant 5 and it can be concluded that increasing the pixel distance increased the test accuracy to some extent but after a certain limit the accuracy started deteriorating. The maximum accuracy achieved after applying all these variants was 89.5%.

Hence, GLCM based analysis of Chest X-ray images for the extraction of textural features provides good accuracy and can be used for the detection of Pneumonia disease.

References

1. Agarwal, C., Sharma, A.: Image understanding using decision tree based machine learning. In: ICIMU 2011: Proceedings of the 5th International Conference on Information Technology & Multimedia, pp. 1–8. IEEE, Kuala Lumpur (2011). https://doi.org/10.1109/icimu.2011.612 2757
2. Aggarwal, N., Agrawal, R.K.: First and second order statistics features for classification of magnetic resonance brain images. J. Sig. Inf. Process. **3**(2), 146–153 (2012). https://doi.org/10.4236/jsip.2012.32019
3. Ankita, R., Kumari, C.U., Mehdi, M.J., Tejashwini, N., Pavani, T.: Lung cancer image-feature extraction and classification using GLCM and SVM classifier. Int. J. Innov. Technol. Explor. Eng. **8**(11), 2211–2215 (2019). https://doi.org/10.35940/ijitee.K2044.0981119
4. Çelik, M., Dadaşer-Çelik, F., Dokuz, A.Ş.: Anomaly detection in temperature data using DBSCAN algorithm. In: 2011 International Symposium on Innovations in Intelligent Systems and Applications, pp. 91–95. IEEE, Istanbul (2011). https://doi.org/10.1109/inista.2011.594 6052
5. Chandra, T.B., Verma, K., Singh, B.K., Jain, D., Netam, S.S.: Automatic detection of tuberculosis related abnormalities in chest x-ray images using hierarchical feature extraction scheme. Expert Syst. Appl. **158**, 113514 (2020). https://doi.org/10.1016/j.eswa.2020.113514
6. Chatterjee, S., Dey, D., Munshi, S.: Integration of morphological preprocessing and fractal-based feature extraction with recursive feature elimination for skin lesion types classification. Comput. Methods Programs Biomed. **178**, 201–218 (2019). https://doi.org/10.1016/j.cmpb.2019.06.018
7. Chawla, N.V., Bowyer, K.W., Hall, L.O., Kegelmeyer, W.P.: SMOTE: synthetic minority over-sampling technique. J. Artif. Intell. Res. **16**, 321–357 (2002). https://doi.org/10.1613/jair.953

8. Giacinto, G., Roli, F.: Design of effective neural network ensembles for image classification purposes. Image Vis. Comput. **19**(9–10), 699–707 (2001). https://doi.org/10.1016/S0262-885 6(01)00045-2

9. Gindi, A., Attiatalla, T.A., Sami, M.M.: A comparative study for comparing two feature extraction methods and two classifiers in classification of early stage lung cancer diagnosis of chest x-ray images. J. Am. Sci. **10**(6), 13–22 (2014). https://doi.org/10.7537/marsjas10061 4.03

10. Haralick, R.M., Shanmugam, K., Dinstein, I.H.: Textural features for image classification. IEEE Trans. Syst. Man Cybern. SMC **3**(6), 610–621 (1973). https://doi.org/10.1109/TSMC. 1973.4309314

11. Kaggle, Chest X-Ray Images (Pneumonia) Dataset Page. https://www.kaggle.com/paultimot hymooney/chest-xray-pneumonia. Accessed 13 Sept 2020

12. Kim, J.I.N.H.O., Kim, B.S., Savarese, S.: Comparing image classification methods: K-nearest-neighbor and support-vector-machines. In: Proceedings of the 6th WSEAS International Conference on Computer Engineering and Applications, and Proceedings of the 2012 American conference on Applied Mathematics, vol. 1001, pp. 133–138. WSEAS, Wisconsin (2012)

13. Kotsiantis, S.B., Kanellopoulos, D., Pintelas, P.E.: Data preprocessing for supervised learning. Int. J. Comput. Sci. **1**(2), 111–117 (2006)

14. Mall, P.K., Singh, P.K., Yadav, D.: GLCM based feature extraction and medical X-RAY image classification using machine learning techniques. In: 2019 IEEE Conference on Information and Communication Technology, pp. 1–6. IEEE, Allahabad (2019). https://doi.org/10.1109/ cict48419.2019.9066263

15. NHLBI Website, Pneumonia. https://www.nhlbi.nih.gov/health-topics/pneumonia. Accessed 13 Sept 2020

16. Pugalenthi, R., Rajakumar, M.P., Ramya, J., Rajinikanth, V.: Evaluation and classification of the brain tumor MRI using machine learning technique. J. Control Eng. Appl. Inform. **21**(4), 12–21 (2019)

17. Rish, I.: An empirical study of the Naive Bayes classifier. In: IJCAI 2001 Workshop on Empirical Methods in Artificial Intelligence, pp. 41–46 (2001)

18. Sunasra, M.: Performance Metrics for Classification problems in Machine Learning. https://medium.com/@MohammedS/performance-metrics-for-classification-problems-in-machine-learning-part-i-b085d432082b. Accessed 15 Sept 2020

19. World Health Organization (WHO), Pneumonia facts. https://www.who.int/news-room/fact-sheets/detail/pneumonia. Accessed 13 Sept 2020

Computer-Aided Diagnosis of Thyroid Dysfunction: A Survey

Bhavisha S. Parmar$^{(\boxtimes)}$ and Mayuri A. Mehta

Department of Computer Engineering, Sarvajanik College of Engineering and Technology, Surat, India
bhavishaparmar02@gmail.com, mayuri.mehta@scet.ac.in

Abstract. Thyroid is a hormone secreting gland that is crucial for the regulation of all the metabolic activities in our body. When the thyroid gland over-functions or under-functions, thyroid dysfunction occurs. In recent years, several computer-aided diagnosis techniques have been proposed in literature to assist doctors to diagnose thyroid dysfunction more accurately. Different techniques use different input modalities to diagnose thyroid dysfunction. Only one of the techniques uses the concept of deep learning to diagnose thyroid dysfunction and the rest of the techniques are machine learning based techniques. In this paper, we present a broad review of computer-aided thyroid dysfunction detection techniques. The paper is presented in five folds. Firstly, we discuss the various types of thyroid dysfunction. Second, we briefly illustrate clinical methods used to diagnose thyroid dysfunctions, along with the shortcomings of the clinical diagnostic process. Third, we discuss computer-aided thyroid dysfunction detection techniques and propose their classification based on the input modality used by them. Fourth, we present a summary of computer-aided techniques which reveals strengths, open research problems and scope of the improvement in this research area. Fifth, we identify a set of parameters to compare computer-aided techniques and present their comparison based on the identified parameters.

Keywords: Thyroid · Thyroid gland · Thyroid dysfunction · Thyroid disorder · Thyroid disease · Computer-aided diagnosis · Feature extraction · Classification · Medical imaging · Machine learning · Deep learning · Image processing

1 Introduction

Thyroid is a butterfly-shaped gland that is located in front of the neck and lies below Adam's apple [1]. The thyroid gland produces hormones that help the body to function normally. These hormones regulate metabolism, blood pressure, heart rate and body temperature [2, 3]. An abnormality in the functioning of thyroid gland is referred as thyroid dysfunction. The two main causes of thyroid dysfunction are autoimmune disease and nutrient deficiency. The various types of thyroid dysfunction are hyperthyroidism, hypothyroidism, goiter, thyroiditis, thyroid nodules and thyroid cancer [4–8].

Thyroid dysfunction is clinically diagnosed by different diagnostic tests such as blood tests, thyroid scan, RadioActive Iodine Uptake (RAIU) test and ultrasound imaging test

© Springer Nature Switzerland AG 2020
L. Bellatreche et al. (Eds.): BDA 2020, LNCS 12581, pp. 164–189, 2020.
https://doi.org/10.1007/978-3-030-66665-1_12

[9–17]. However, clinical diagnosis has several drawbacks as follows: (1) The clinical diagnosis tests are subjective in nature. 2) Diagnostic tests vary according to various laboratory settings. 3) Clinical diagnosis is a lengthy process to diagnose thyroid dysfunction because laboratory tests take a long time. 4) Early medication cannot be prescribed due to dissimilarity in diagnosis outcome. To overcome these drawbacks of clinical diagnosis, some of the clinical tests have been computerized. Diagnosis using computerized technique helps doctors to perform diagnosis accurately, to bring objectivity in diagnosis and to improve the evaluation of thyroid dysfunction. Besides, computerization makes diagnosis faster; therefore, early medication can be provided during the initial phase itself.

Several computer-aided thyroid dysfunction detection techniques are available in literature. In this paper, we present their comprehensive details, summarization and comparative evaluation. In addition, based on the input data used by the techniques to detect the thyroid dysfunction, we categorize them into three major categories: (1) thyroid dysfunction detection using blood tests (2) thyroid dysfunction detection using thyroid image and (3) thyroid dysfunction detection using infant's crying signal. Techniques that use thyroid image are further categorized into five sub-categories: dysfunction detection using SPECT image, dysfunction detection using ultrasound image, dysfunction detection using histopathology image, dysfunction detection using hyperspectral image and dysfunction detection using thermal image. In order to present an analytical evaluation of these techniques, we have identified a set of parameters and subsequently, have presented a parametric evaluation. To the best of our knowledge, this is the first attempt to survey computer-aided thyroid dysfunction detection techniques available in literature. The objective of this survey is to provide a complete study and comparative analysis of computerized techniques which will help researchers and the doctor community to understand technological advancements in this field. Moreover, this survey will serve as a catalogue of existing thyroid dysfunction detection techniques to help researchers to know the scope of the research in this field and to come up with a new efficient technique.

The rest of the paper is structured as follows: Sect. 2 gives a brief overview of thyroid and various types of thyroid dysfunction. In addition, clinical diagnostic tests that have been computerized for automated detection of dysfunction are also discussed in this section. In Sect. 3, we first introduce the proposed taxonomy of computer-aided thyroid dysfunction detection techniques. Then, we illustrate the various techniques available under each category. Subsequently, we present the summary of these techniques in the form of key characteristics and open research problems. An analytical parametric evaluation of computerized thyroid dysfunction detection techniques is presented in Sect. 4. Section 5 specifies conclusion and future scope.

2   Theoretical Background

This section gives an overview of relevant theory of thyroid dysfunction for clear and better understanding of computer-aided thyroid dysfunction detection. Before further discussion, we first describe the abbreviations used throughout the paper in Table 1.

Table 1. List of abbreviations

Abbreviation	Description
ANN	Artificial Neural Network
BPSO	Binary Particle Swarm Optimization
CART	Classification And Regression Trees
CFNN	Cascade Forward Neural Network
CNN	Convolutional Neural Network
DT	Decision Tree
FTDNN	Focused Time Delay Neural Network
GLCM	Gray Level Co-occurrence Matrix
GRNN	General Regression Neural Network
KNN	K-Nearest Neighbors
MFC	Mel Frequency Cepstral
MLP	MultiLayer Perceptron
MLPNN	MultiLayer Perceptron Neural Network
MNN	Mobile Neural Network
PCA	Principle Component Analysis
PNN	Probabilistic Neural Network
RBF	Radial Basis Function
ROI	Region Of Interest
SVM	Support Vector Machine
T3	Triiodothyronine
T4	Thyroxine

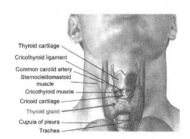

Fig. 1. Gross anatomy of thyroid gland [1]

2.1 Thyroid Dysfunction

The thyroid gland is located at the base of the neck. As shown in Fig. 1, it surrounds the trachea below the cricoid cartilage [1, 18]. The function of the thyroid gland is to convert the iodine found in foods into thyroid hormones such as T3 and T4 [2, 3, 19].

Thyroid gland controls heart, muscle, brain development, digestive function and bone maintenance. Its proper functioning depends on the intake of iodine in the diet [20, 21]. The two major causes of thyroid problems are autoimmune disease and nutrient deficiency. As shown in Fig. 2, various types of thyroid dysfunction are hyperthyroidism, hypothyroidism, goiter, thyroiditis, thyroid nodules and thyroid cancer.

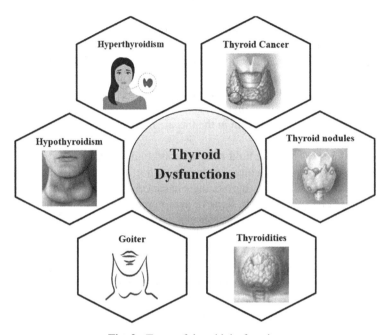

Fig. 2. Types of thyroid dysfunction

Hyperthyroidism [22–24] occurs when the thyroid gland produces an excessive amount of T4 hormone and hence, it is also called an overactive thyroid. The causes of hyperthyroidism include graves' disease, thyroid nodules and thyroiditis. Hyperthyroidism causes a variety of signs and symptoms such as tachycardia, unintentional weight loss, arrhythmia, palpitations, increased appetite, anxiety and irritability, nervousness, tremor, sweating, changes in menstrual patterns, goiter, fatigue, increased sensitivity to heat, muscle weakness, difficulty in sleeping, skin thinning and brittle hair. If hyperthyroidism is left untreated, then it leads to serious complications such as heart problems, brittle bones, eye problems, graves' dermopathy and thyrotoxic crisis.

Hypothyroidism [25–27] is an underactive thyroid that produces an inadequate amount of thyroid hormones. The several causes of hypothyroidism are autoimmune disease, over-response to hyperthyroidism treatment, thyroid surgery, radiation therapy and medications. The signs and symptoms of hypothyroidism include fatigue, increased sensitivity to cold, weight gain, puffy face, dry skin, constipation, muscle weakness, hoarseness, muscle aches, stiffness or swelling in joints, heavier or irregular menstrual periods, thinning hair, slowed heart rate, depression, impaired memory and enlarged thyroid gland. During its initial stage, it does not cause noticeable or prominent symptoms. Hypothyroidism affects women more frequently than men. If hypothyroidism is left untreated, then it leads to severe consequences such as heart problems, mental health issues, peripheral neuropathy and myxedema.

Goiter [28–30], also called enlarged thyroid, enlarges the thyroid gland abnormally. It may affect anyone; however, it is common in women. The types of goiter are colloid goiter, nontoxic goiter and toxic nodular. The causes of goiter include iodine deficiency, graves' disease, hashimoto's thyroiditis, inflammation, nodules, thyroid cancer and pregnancy. The symptoms of goiter are swelling at the base of the neck, a tight feeling in the throat, coughing, hoarseness, difficulty in swallowing and breathing.

Thyroiditis [31–33] is an inflammation of the thyroid gland that initially causes a leak of hormones. The types of thyroiditis are hashimoto's thyroiditis, subacute thyroiditis, post-partum thyroiditis and silent thyroiditis. The common symptoms of thyroiditis are fatigue, swelling at the neck and pain in the throat. The symptoms may vary depending on the stage of inflammation.

Thyroid nodules [34–36] are fluid-filled or solid lumps that form within the thyroid gland. The causes of thyroid nodules are overgrowth of normal thyroid tissue, thyroid cyst, chronic inflammation of the thyroid, thyroid cancer, multinodular goiter and iodine deficiency. There are no noticeable signs or symptoms of thyroid nodules. As thyroid nodules become larger, the problems arise such as an enlarged thyroid gland, difficulty in swallowing and breathing, pain at the base of the neck and hoarse voice. The major complications due to thyroid nodules are hyperthyroidism and a problem in breathing.

Thyroid cancer [35–38] is a malignant disease form in the cells of the thyroid gland. The causes of thyroid cancer are inherited genetic syndromes, iodine deficiency and radiation exposure. The types of thyroid cancer are papillary thyroid cancer, follicular thyroid cancer, medullary cancer and anaplastic thyroid cancer. Generally, symptoms of thyroid cancer are not visible during its early stage. The symptoms noticeable over the time are throat pain, a lump in the neck, vocal changes, hoarseness, difficulty in swallowing, cough and swollen lymph nodes.

2.2 Clinical Tests for Diagnosis of Thyroid Dysfunction

As shown in Fig. 3, the clinical tests performed to diagnose the presence of thyroid dysfunction are mainly categorized into four categories [9–17]: blood tests, thyroid scan, RAIU test and ultrasound imaging test. They differ from each other with respect to the type of data used to diagnose thyroid dysfunction.

Blood Tests. Different types of blood tests are available to diagnose thyroid dysfunction [9–12]. Doctors prescribe one or more blood tests in certain cases based on the preliminary examination. The types of blood tests are TSH, T4, T3 and thyroid antibody. In these tests, a healthcare professional takes the blood samples and sends them to the laboratory for examination.

TSH Test. In this test, a healthcare professional checks the amount of TSH in blood [9]. TSH is a hormone produced in the pituitary gland that specifies how much T3 and T4 should be released from thyroid gland. The high level of TSH in blood indicates the

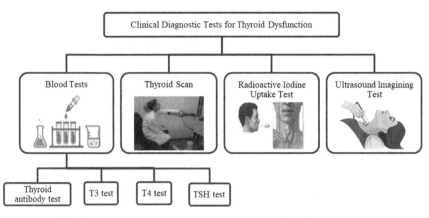

Fig. 3. Clinical diagnostic tests for thyroid dysfunction detection

presence of hypothyroidism in patients and the low level of TSH in blood indicates the presence of hyperthyroidism.

T4 Test. This test measures the amount of the T4 hormone in the blood [9]. The high level of the T4 hormone specifies the existence of hyperthyroidism and the low level of T4 hormone specifies the existence of hypothyroidism [10]. In particular cases such as when a patient is pregnant or taking oral contraceptives, when a patient is suffering from illness, or when a patient is on medications for asthma, skin or other health problems, the high or the low level of the T4 hormone may not be considered as the thyroid problem.

T3 Test. Typically, T3 and T4 hormones work together to regulate the metabolic activities of our body. Therefore, to identify the thyroid dysfunction, both T3 and T4 tests are mandatory [9, 10]. The high level of the T3 hormone specifies the existence of hyperthyroidism and the low level of T3 hormone specifies the existence of hypothyroidism. In some cases, the amount of T4 hormone is found normal; however, the amount of T3 hormone may be high. Hence, measuring both T3 and T4 levels are useful in proper diagnosis of thyroid dysfunction.

Thyroid Antibody Test. In this test, a patient is checked for various autoimmune diseases by measuring his/her level of thyroid antibodies [11, 12]. The autoimmune thyroid disorders are graves' disease which is caused by hyperthyroidism and hashimoto's disease which is caused by hypothyroidism. Health care professionals may prescribe for thyroid antibody test when other blood tests result in thyroid disease.

Thyroid Scan. Thyroid scan test is performed to know the thyroid gland's size, shape and position [13]. In this test, the radioactive iodine is given to the patient to know the cause of hyperthyroidism and thyroid nodules. The healthcare professional firstly injects a small amount of radioactive iodine or similar substance into the patient's vein. Subsequently, the scanning is performed which normally takes 30 min. In case if a patient has swallowed the radioactive iodine or substance, the scan is performed after 24 h. The result of thyroid scan is a picture of thyroid gland that depicts clear vision of thyroid nodules and other autoimmune diseases. This test is not applicable to pregnant or breastfeeding women.

Radioactive Iodine Uptake Test. RAIU test, also called thyroid uptake test, is performed to check the proper functioning of thyroid gland and to know the cause of hyperthyroidism [14]. In this test, firstly a patient is asked by a healthcare professional to avoid the food containing high levels of iodine for a week. Then, a patient is asked to swallow a small amount of radioactive iodine in the form of capsule or liquid form. Subsequently, the Gamma probe device is used and placed in front of the neck. This device measures the radioactive iodine of thyroid gland which helps doctors to decide upon the presence/absence of thyroid dysfunction. Generally, thyroid scan and RAIU tests are performed at the same time.

Ultrasound Imaging Test. Ultrasound is the most sensitive imaging modality for the examination of thyroid gland [15]. Ultrasound imaging is non-invasive, less expensive test which does not use any ionizing radiation [16]. A device called a transducer is used to perform this test. It is a safe and painless device. A technician puts this device on the patient's neck to perform this test. The test usually takes 30 min. This test is mainly used to confirm the presence of a thyroid nodule, to measure the dimensions of thyroid gland accurately, and to differentiate between benign and malignant thyroid [15, 17]. Unlike above tests, this test is not applicable to know whether thyroid gland is underactive or overactive.

3   Computer-Aided Thyroid Dysfunction Detection Techniques

In this section, we propose the categorization of thyroid dysfunction detection techniques. Next, we describe the various techniques under different categories. Then, we present a comprehensive summary of these techniques in the form of key characteristics and open research issues for the ready reference of the research community.

Several thyroid dysfunction detection techniques have been proposed in literature. As shown in Fig. 4, we categorize them into three categories based on the input data used by them: thyroid dysfunction detection using blood test records, thyroid dysfunction detection using image and thyroid dysfunction detection using infant's crying signal. These techniques aim to automate diagnosis of thyroid disorders and diseases.

3.1   Thyroid Dysfunction Detection Using Blood Tests

The majority of the techniques [39–48] available in literature detect thyroid dysfunction using blood test records. Figure 5 shows the major steps involved in these techniques: preprocessing of blood test records, feature extraction from preprocessed records and classification. These techniques are machine learning based thyroid dysfunction detection techniques that use thyroid blood test dataset from the UCI repository or diagnostic lab. The preprocessing step removes missing data-values, noisy data and redundant data as well as checks the number constraint using the masking method. Then, the features are extracted from the preprocessed records. After selecting appropriate feature subset from

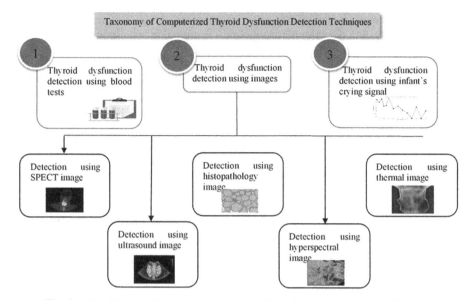

Fig. 4. Classification of computer-aided thyroid dysfunction detection techniques

the extracted features, classification is performed using classifiers such as naïve bayes, RBF network, DT, CART, SVM, KNN and ANN. The performance of these techniques is evaluated using metrics such as accuracy, precision, recall, true positive rate, false positive rate, F-measure and ROC (Receiver Operating Characteristic) area.

3.2 Thyroid Dysfunction Detection Using Image

Different image modalities are used to examine the presence of thyroid dysfunction. The image modalities investigated in literature are SPECT image, ultrasound image,

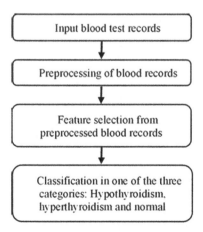

Fig. 5. Steps of thyroid dysfunction detection using blood test records

histopathology image, hyperspectral image and thermal image. Below we discuss the various thyroid dysfunction detection techniques that use one of these image modalities.

Thyroid Dysfunction Detection using SPECT Image. The automatic thyroid dysfunction detection technique has been proposed in [49]. The flow of this technique is depicted in Fig. 6. As shown in Fig. 6, there are mainly two stages: training stage and diagnosis stage. Input to this technique is thyroid SPECT images. The SPECT images collected by authors were not in sufficient quantity to train the model; therefore, authors augmented data using Mixup augmentation method and generated 2000 images. Pre-trained CNN model is used to improve feature extraction and to reduce overfitting. Specifically, authors have fine-tuned DenseNet architecture as modified network architecture by adding trainable weight parameters to each skip connection. The optimized flower pollination algorithm is used to optimize the learning rate. After completion of the training stage, the trained CNN model is used to diagnose thyroid dysfunction. This technique classifies thyroid dysfunction into one of the four categories: grave's disease,

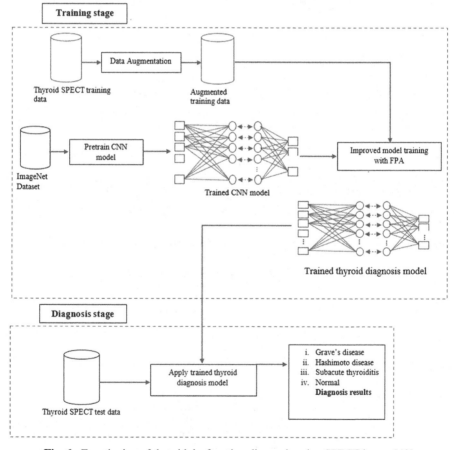

Fig. 6. Functioning of thyroid dysfunction diagnosis using SPECT image [49]

hashimoto's disease, subacute thyroiditis or normal. The performance of these techniques is evaluated using recall, precision, accuracy, specificity and F1-score parameters.

Thyroid Dysfunction Detection using Ultrasound Image. A technique named thyroid disorder detection using image segmentation in medical images has been proposed in [50]. Input to this technique is an ultrasound image of the thyroid gland. The major steps involved in this technique are described in Fig. 7(a). Figure 7(b) shows the result of steps mentioned in Fig. 7(a). Firstly, the input image is converted into a grayscale image to change the ultrasound image to gray shades without apparent color. Subsequently, the median filter is applied to grayscale images to remove noise for later processing and to enhance the quality of the ultrasound image. After successfully enhancing the image, segmentation is performed to simplify the representation of an image and to get the meaningful information for further analysis. Segmentation mainly locates objects and boundaries such as lines, curves and edges in the image. The region growing method is applied to find the relation between the pixels and neighborhood pixels.

Begin
1. Input ultrasound image
2. Convert input image into grayscale image
3. Apply median filter to grayscale image to enhance the quality of image
4. Compute image segmentation
5. Locate suspicious thyroid region
End

(a)

1. Input image   2. Grayscale image   3. De-noised image

4. Segmented image   5. Output image with identified suspicious

(b)

Fig. 7. (a): Steps of thyroid disorder detection using ultrasound image, (b): Pictorial representation of steps of Fig. 7(a)

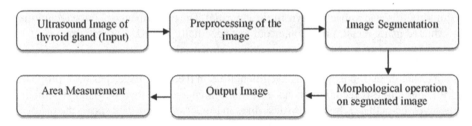

Fig. 8. Functioning of thyroid gland's area estimation technique

In [51], authors have proposed an automated system for area estimation of the thyroid gland. The working process of this technique is shown in Fig. 8. Input to this technique is ultrasound images. PCA based preprocessing is applied to reduce the large dimensionality of data. After dimensionality reduction, the preprocessed image is segmented using the ROI concept to detect the abnormal region based on centroids and clusters. The morphological operation is then performed for smoothening the detected region. Next, the masking operation is applied to filter an image. Normalization is performed on the masked image to change value ranges of pixel intensity. Finally, various areas in the image are estimated to know the factors such as thyroid size, shape and area.

Thyroid Dysfunction Detection using Histopathology Image. A technique that detects thyroid follicular lesions based on nuclear structure has been presented in [52]. As shown in Fig. 9, input to this technique is histopathology images. In this technique, first preprocessing of nuclei (set of nucleus) is performed by applying segmentation, gray level extraction and intensity normalization. Then, features are extracted to classify thyroid follicular lesions into one of the three classes viz. normal

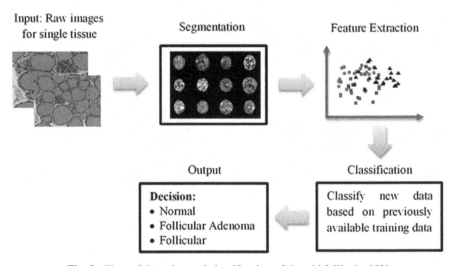

Fig. 9. Flow of detection and classification of thyroid follicular [52]

thyroid nuclei, follicular adenoma or follicular carcinoma. In this technique, a numerical based feature space approach is used to automatically classify nuclei.

The general steps for automatic classification of nuclei are as follows: 1) choose a portion from the available data for training. 2) Train a classifier to detect the class of nucleus. 3) Assign a single class nucleus by testing individual nucleus and then choose the majority vote. The unique chromatin patterns are identified to separate patterns from other classes. For the classification of thyroid follicular, several classification algorithms such as MNN, SVM-Quadratic and SVM-RBF have been used. Among them, the SVM-RBF algorithm yields higher classification accuracy.

Thyroid Dysfunction Detection using Hyperspectral Image. In [53], authors have proposed a technique to detect thyroid disorder combining tongue color spectrum and thyroid hormones. In this technique, a correlation between tongue color spectra and thyroid hormones is calculated. Firstly, the acquired data is normalized to spectral reflectance for spectral distribution to diffuse reflection standard. Then, spectral absorbance is calculated based on Beer-Lambert law for negative logarithm. Next, a median filter is applied to remove noise and to represent the tongue area. Subsequently, all the filtered data are analyzed using the principal component analysis method to find a vector of high correlation that specifies correlation between blood examination and tongue color spectra.

Thyroid Dysfunction using Thermal Image. A technique named diagnosis of hypo and hyperthyroid using MLPN network has been proposed in literature [54]. The main steps of this technique are shown in Fig. 10. Input to this technique is thermal images. This technique uses a median filter to remove noise from the image. After noise removal, the image is enhanced using histogram equalization. Next, Otsu's thresholding method is applied to segment the image. Subsequently, features are extracted using a Gabor filter

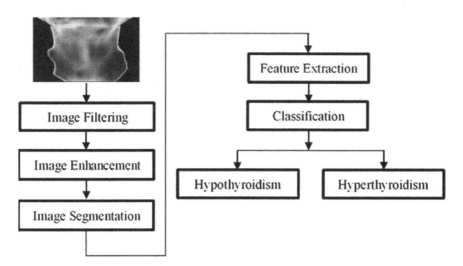

Fig. 10. Working flow of thyroid disorder diagnosis using thermal image [54]

and GLCM. The extracted features are then classified into one of the two classes viz. hyperthyroidism and hypothyroidism using MLPNN.

In [55], authors have proposed a technique to detect various thyroid disorders. The first few steps of this technique are similar to the technique presented in [54]. Specifically, input to this technique is also thermal images. Like previous technique, the input thermal image is filtered using the median filter. The filtered image is then enhanced by histogram equalization for further analysis of an image. Image segmentation is also done by the same Otsu's thresholding method. The differing point is that in this technique, five textural features viz. haar wavelet features, coefficient of local variation feature, BDIP feature, normalized multi-scale intensity difference and area are considered for classification. Classification is carried out using Bayesian classifier.

Thyroid Dysfunction Detection using Infants Crying Signals. A   technique that detects hypothyroidism in infants has been proposed in [56]. The functioning of this technique is depicted in Fig. 11. It was observed that the crying signal of infants suffering from hypothyroidism differs from the crying signal of healthy infants. Based on this observation, this technique was proposed to detect the hypothyroidism in infants from their crying signal. The very first step of this technique is the extraction of the features (MFC coefficient) from the crying signal. The input crying signal is first segmented and then segmented signals are broken down into overlapping frame-blocks. Then, Fast Fourier Transform is applied on each frame to generate a frequency spectrum. Next, the frequency spectrum is filtered using a Mel-scale filter. Subsequently, Mel-spaced filter banks are designed to detect frequencies tones in linear scale and logarithmic scale

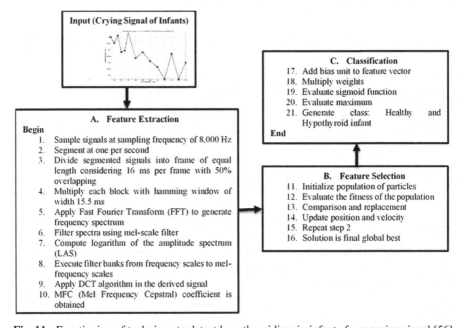

Fig. 11. Functioning of technique to detect hypothyroidism in infants from crying signal [56]

lower than 1 kHz and higher than 1 kHz respectively. Finally, Discrete Fourier Transform is applied to the derived signal and MFC coefficient is calculated. After calculating coefficients, BPSO is used for selecting appropriate coefficients. The next step is the classification of hypothyroid infant cry. For classification, a three-layer MLP is used with a sigmoid function. The MLP architecture is comprised of an input feature vector, 5 hidden nodes and 2 output neurons. The classification accuracy is calculated as a mean.

At this juncture, we present a compendium of the above discussed techniques mainly to highlight the scope of the research in this area. Table 2 describes the key characteristics of the existing thyroid dysfunction detection techniques along with the open research issues in computer-aided thyroid dysfunction detection. Among the techniques described in Table 2, the techniques in [39–48] detect the thyroid dysfunction from blood test records and the remaining techniques [49–55] detect the thyroid dysfunction from thyroid image. There is only one technique presented in [56] that detects thyroid dysfunction from infant crying signals.

Table 2. Summary of the thyroid dysfunction detection techniques

Technique	Key characteristics	Scope of the improvement
Women thyroid prediction using data mining techniques [39]	• Thyroid prediction only for women • Prediction from blood test records • Uses Ensemble I and Ensemble II models for prediction	• Can be extended to predict thyroid in men too • Ensemble model can be integrated with different classifiers to enhance prediction accuracy • Can be modified to further classify the thyroid disease
Thyroid prediction using machine learning techniques [40]	• Prediction from blood test records • Less number of attributes are taken into consideration for prediction • Classification is carried out using four classifiers: ANN, KNN, SVM and DT • Amongst these classifiers, ANN provides better accuracy	• Accuracy can be further enhanced using advanced machine learning techniques • Appropriate feature extraction technique should be used to extract relevant features
Diagnosing thyroid by neural networks [41]	• Diagnosis from blood test records • Diagnosis is assessed using five different neural networks: MLP, PNN, GRNN, FTDNN and CFNN	• Appropriate feature extraction techniques should be used to extract the useful features

(*continued*)

Table 2. (*continued*)

Technique	Key characteristics	Scope of the improvement
Thyroid classification using differential evolution with SVM [42]	• Thyroid classification from blood test records • Feature selection is done using wrapper model • Classification is carried out by using radial basis kernel with SVM	• Less number of attributes can be considered to reduce complexity and processing time of the algorithm • Embedded methods of machine learning can be used to select more relevant features • Classification accuracy can be further increased using advanced classification approach
Thyroid prediction using data mining techniques [43]	• Prediction from blood test records • Data mining techniques used are DT, backpropagation NN, SVM and density based clustering • Prediction is performed by finding correlation between thyroid functions and thyroid disorders	• Dimensionality reduction can be performed to remove redundant data • Less number of attributes can be considered for dysfunction detection to minimize prediction time
Diagnosis of thyroid ailments using data mining classification techniques [44]	• Diagnosis from blood test records • Classification is carried out using four data mining classification techniques: KNN, SVM, DT and NB • Amongst classification techniques, DT provides higher accuracy	• Less number of attributes can be considered to reduce complexity and processing time of the algorithm • Diagnosis accuracy can be further increased considering advanced machine learning techniques

(*continued*)

Table 2. (*continued*)

Technique	Key characteristics	Scope of the improvement
Predict thyroid from classification algorithms [45]	• Prediction from blood test records • Feature selection using four techniques: mutual information, random subset feature selection, sequential forward selection and statistical dependency • Among all, sequential forward generates more number of features • Three classifiers are used for thyroid dysfunction classification: KNN, ANN and FANN • FANN provides higher accuracy compared to other classifiers	• Less number of attributes can be considered for thyroid dysfunction detection to reduce complexity of the algorithm • Dimensionality reduction can be applied to remove repeated data and thereby to reduce prediction time • Accuracy can be further enhanced using advanced classification techniques
Optimized classification model for prediction of hypothyroid [46]	• Prediction from blood test records • Optimization is performed using ranker search technique • Classification is carried out using naïve bayes classifier	• Less numbers of attributes can be considered for prediction to reduce processing complexity • Dimensionality reduction can be applied to remove irrelevant data • Accuracy can be further increased using better efficient classifiers
Diagnosis and classification of hypothyroid disease using data mining [47]	• Diagnosis using blood test records • Classification is performed using data mining techniques: decision stump, Bayes net, multilayer perceptron, C4.5 and REP tree • Amongst various data mining techniques, C4.5 provides higher accuracy	• Dimensionality reduction can be performed to eliminate repeated data • Less number of attributes can be considered to reduce diagnosis time

(*continued*)

Table 2. (*continued*)

Technique	Key characteristics	Scope of the improvement
Thyroid disease diagnosis using PNN and SVM [48]	• Diagnosis from blood test records • Feature selection is performed using genetic algorithm • Classification is done using two classification techniques: PNN and SVM • SVM gives better results than PNN	• Higher number of data should be considered to increase diagnosis accuracy • Feature selection can be performed using better methods to enhance accuracy
Thyroid diagnosis from SPECT images using convolutional neural network [49]	• Diagnosis from SPECT image • Feature extraction using pre-trained CNN model • Flower pollination algorithm is used to optimize learning rate	• Model can be trained with larger dataset for improved diagnosis • Thyroid disorder can be further classified for right treatment • Deep learning models can be used to generate better accuracy
Thyroid detection using image segmentation [50]	• Detection from thyroid gland image • Wiener filter is applied to enhance the image quality • Thresholding method is used to segment the image • Region growing method is applied to extract the regions	• Thyroid growth can be calculated from the features such as area, volume and centroid for better diagnosis of thyroid disorder • Thyroid disorder can be further classified for better treatment
Automated segmentation of thyroid gland using PCA and area estimation [51]	• Detection from ultrasound image • Data reduction is performed using PCA • ROI concept is used to segment the abnormal region based on cluster and centroids • Morphological process is applied to smooth the detected region • Estimates thyroid size, shape and area	• Diagnosis accuracy can be further increased considering deep learning methods • Thyroid disorder can be further classified to prescribe right medication

(*continued*)

Table 2. (*continued*)

Technique	Key characteristics	Scope of the improvement
Detection and classification of thyroid follicular lesions based on nuclear structure [52]	• Detection from histopathology image • Thyroid cancer is detected from follicular lesions • Discriminate features of nuclei are extracted from follicular carcinoma and follicular adenoma • Classification is performed using multiclass SVM	• Higher number of patient cases should be considered to get enhanced results • More number of features can be extracted using deep learning to get improved accuracy • Work can also be further extended by generating result of follicular lesions stages
Detecting hyper/hypothyroidism using principal component analysis [53]	• Detection from hyper spectral image • Image is filtered using median filter • Principle component analysis method is used to calculate correlation between blood examination and tongue colour spectra • Detection is based on correlation between tongue colour spectra and thyroid hormones	• Angle should be measured using enhanced technique to obtain accurate measurement • Other symptoms can also be considered for improved detection
Diagnosis of hypo and hyperthyroid using MLPN network [54]	• Detection from thermal thyroid gland • Image is filtered using median filter • Image is enhanced using histogram equalization method • Image segmentation is performed using Otsu's thresholding method • MLPN is applied to classify the thyroid gland images	• Image segmentation can be improved using fuzzy logic • Thyroid disorder can be further classified for better diagnosis and treatment

(*continued*)

Table 2. (*continued*)

Technique	Key characteristics	Scope of the improvement
Hypo and hyperthyroid disorder detection using Bayesian classifier [55]	• Disorder detection from thermal images • Image is enhanced using histogram equalization method • Segmentation is performed using Otsu thresholding method • Classification is carried out using Bayesian classifier	• Diagnosis accuracy can be further increased by using advanced classifiers • Less number of attributes can be considered to reduce the complexity of the algorithm and thereby to reduce the detection time
Detection of hypothyroidism in infants using binary particle swarm optimization [56]	• Detection from baby's crying signal • Feature extraction based on Mel frequency cepstral coefficient • Feature selection is performed using binary particle swarm optimization • Classification is performed using three-layer MLP classifier	• Enhanced feature extraction method can be used to obtain better coefficients • Crying signals can be preprocessed to produce right signals • Better fitness function can be used to obtain improved results

4   Analysis and Discussion

From our detailed study of thyroid dysfunction detection techniques, we have identified three sets of parameters to evaluate them exhaustively. The first set of parameters gives the general information about the thyroid dysfunction detection techniques and it includes two parameters that are gender/age-group and the types of thyroid dysfunction detected. The second set provides the information related to the dataset used by these techniques and it includes two parameters that are data type and dataset details. The third set gives the technical information and it includes parameters such as preprocessing, feature extraction technique and classification approach.

Table 3 illustrates the general information about these techniques. Gender/age-group specifies the gender and age-group to which thyroid technique is applicable. Types of thyroid dysfunction parameter represents the various types of thyroid dysfunctions which are detected by the technique.

Table 3. General details of thyroid dysfunction detection techniques

Technique	Gender/age group	Types of thyroid dysfunction diagnosed
Women thyroid prediction using data mining techniques [39]	Women (age: 0 years onwards)	Hypothyroidism, hyperthyroidism, euthyroid
Thyroid prediction using machine learning techniques [40]	Men & Women (age: 0 years onwards)	Thyroid
Diagnosing thyroid by neural networks [41]		Hyperthyroidism, hypothyroidism, resistant thyroid
Thyroid classification using differential evolution with SVM [42]		Hyperthyroid, hypothyroid
Thyroid prediction using data mining techniques [43]		Hypothyroidism, hyperthyroidism
Diagnosis of thyroid ailments using data mining classification techniques [44]		Hypothyroidism, hyperthyroidism
Predict thyroid from classification algorithms [45]		Thyroid, under function
Optimized classification model for prediction of hypothyroid [46]		Primary hypothyroid, secondary hypothyroid
Diagnosis and classification of hypothyroid disease using data mining [47]		Hypothyroidism
Thyroid disease diagnosis using PNN and SVM [48]		Thyroid
Thyroid diagnosis from SPECT images using convolutional neural network [49]		Graves' disease, hashimoto, sub-acute thyroiditis
Thyroid detection using image segmentation [50]		Thyroid
Automated segmentation of thyroid gland using PCA and area estimation [51]		Thyroid Cancer
Classification of thyroid follicular lesions based on nuclear structure [52]		Follicular adenoma, follicular carcinoma

(continued)

Table 3. (*continued*)

Technique	Gender/age group	Types of thyroid dysfunction diagnosed
Detecting hyper/hypothyroidism using principal component analysis [53]		Hyperthyroidism, hypothyroidism
Diagnosis of hypo and hyperthyroid using MLPN network [54]		Hyperthyroidism, hypothyroidism
Hypo and hyperthyroid disorder detection using Bayesian classifier [55]		Hypothyroidism, hyperthyroidism
Detection of hypothyroidism in infants using binary particle swarm optimization [56]	Infants – men & women both (age: 0-2 years)	Hypothyroidism

Table 4 represents the dataset information of these techniques. Data type specifies the input modality used for the detection of thyroid dysfunction. Dataset details gives the information regarding dataset source, tools used to collect the data and/or specification of the data. The two major sources of thyroid data are private or public hospitals and UCI repository.

Table 4. Dataset details of thyroid dysfunction detection techniques

Technique	Data type	Dataset details
[39], [40], [41], [42], [43], [45], [46], [47], [48]	Blood test records	UCI repository
[44]	Blood test records	Diagnostic lab, Kashmir
[49]	SPECT images	Heilongjiang provincial hospital
[50]	Ultrasonography images	Not specified
[51]	Ultrasound images	Not specified
[52]	Tissues blocks	University of Pittsburgh Medical Center, Device used to collect data: Olympus BX51 microscope, 24bit RGB channels, 0.074 microns/pixels
[53]	Hyperspectral images	Hyperspectral camera, Spectral range: 400-800 nm, Resolution: 5-nm

(*continued*)

Table 4. (*continued*)

Technique	Data type	Dataset details
[54]	Thermography images	Digital infrared camera
[55]	Thermal images	Thermal camera FLIR E-30, Temperature range: −20 to 250 °C, Thermal sensitivity: 0.10 °C, Pixel: 160 x 120
[56]	Infant crying signal	Instituto Technologic Superior de Atlixco of Mexico

Table 5 gives the technical information of these techniques. Preprocessing specifies the prior steps carried out during the preprocessing of data and/or the method used to carry out those steps. Feature extraction technique specifies the technique used to extract and select the features from the input data. Classification approach specifies the classifier used to detect the types of thyroid dysfunction.

Table 5. Technical details of the thyroid dysfunction detection techniques

Technique	Preprocessing	Feature extraction technique	Classification approach
[39]	Not specified	Bagging, Boosting	DT, Neural network
[40]	Not specified	Mutual Information	ANN, SVM, DT, KNN
[41]	Not specified	Not specified	MLP, PNN, GRNN, FTDNN, CFNN
[42]	Not a number (NaN) constraint is checked using masking method	Hybrid Differential Evolution	SVM
[43]	Not specified	Gabor, Wavelet	DT, Backpropagation Neural Network, SVM, Density kernel based clustering
[44]	Not specified	Not specified	KNN, SVM, DT, NB
[45]	Not specified	Mutual information, Random subset, Sequential forward, Statistical dependency	ANN, KNN
[46]	Not specified	Ranker search	NB
[47]	Not specified	Not specified	MLP, Bayes Net, RBF, C 4.5, REP tree, CART, Decision stump

(*continued*)

Table 5. (*continued*)

Technique	Preprocessing	Feature extraction technique	Classification approach
[48]	Not specified	Genetic Algorithm	PNN, SVM
[49]	Not specified	CNN	Softmax
[50]	Noise removal using wiener filter algorithm	Thresholding, Otsu's method, K – means Clustering	Not specified
[51]	Data reduction using PCA	ROI, Masing, Normalization	Not specified
[52]	Gray level extraction, Intensity normalization	Not specified	SVM
[53]	Not specified	PCA	Not specified
[54]	Noise removal using median filter, image enhancement using histogram equalization	Gabor filter, GLCM	MLPNN
[55]	Noise removal using median filter, image enhancement using histogram equalization	ROI	Bayesian classifier
[56]	Not specified	MFC, BPSO	MLP

It has been observed that the majority of the techniques use thyroid hormone test records (blood test reports) to detect thyroid dysfunction and provide good accuracy. Less amount of work has been done to detect the thyroid dysfunction from image data. Technique based on SPECT images identifies thyroid autoimmune disease with 100% accuracy [49]. The aforementioned techniques have brought several benefits to the healthcare industry. However, still there are following significant challenges which must be addressed for successful deployment of computer-aided thyroid dysfunction detection techniques in healthcare.

- Diagnosis is carried out considering limited number of symptoms which may lead to inaccurate results
- Automated models are trained with records of a limited number of patients. Such models may provide inaccurate diagnosis
- Crying signal of baby used to diagnose thyroid dysfunction may or may not be accurate
- Existing techniques give limited diagnosis accuracy
- In several techniques, limited number of features are extracted which may lead to misclassification
- Some techniques use more number of attributes for detection which increases processing and diagnosis time
- Several techniques do not detect the type of thyroid disorder

- Some techniques take higher time to diagnose thyroid dysfunction
- Incorrect temperature range for thermal imagining gives misclassification or wrong diagnosis

5 Conclusion

Thyroid dysfunction is the most common disease in the endocrine field and results in serious consequences. Hence, precise and early diagnosis of thyroid dysfunction is crucial for early medication as well as to prevent the patient from severe complications. In this paper, firstly we have presented an overview on thyroid dysfunction, its types and various clinical diagnostic tests. Secondly, we have presented a systematic comprehensive review of computer-aided thyroid dysfunction detection techniques. Thirdly, a parametric comparison of thyroid dysfunction detection techniques based on identified parameters has been presented. It is observed from parametric comparison that the majority of the techniques available in literature diagnose thyroid dysfunction using thyroid hormone test records and provide satisfactory accuracy. Techniques have also been proposed to diagnose thyroid dysfunction from medical images. Only one technique has been proposed that identifies thyroid dysfunction from infant's crying signals. In addition, it has also been observed that the majority of the available techniques are traditional machine learning based techniques. Hence, there is a wide research scope in this domain using deep learning techniques. Our exhaustive survey will help the research community to get knowledge of both clinical and computerized thyroid dysfunction detection techniques. Moreover, our survey will assist doctors to comprehend knowledge of computerized thyroid detection techniques, so that they can take advantage of technological advancements for automated computerized diagnosis.

References

1. Mitra, S.: Thyroid anatomy and physiology of thyroid hormone secretion. Manage. Thyroid Disord. Made Easy. 1 (2009)
2. Dev, N., Sankar, J., Vinay, M.V.: Functions of thyroid hormones. Thyroid Disord. 11–25 (2016)
3. Khanorkar, S.: Functions of thyroid hormones and diseases of thyroid gland. Insights Physiol. 451 (2012)
4. Jolobe, O.M.P.: Thyroid disorders—an update. Postgrad. Med. J. **77**(904), 144 (2001)
5. Monaco, F.: Classification of thyroid diseases: suggestions for a revision. J. Clin. Endocrinol. Metab. **88**(4), 1428–1432 (2003)
6. Galofré, J.C., Díez, J.J., Cooper, D.S.: Thyroid dysfunction in the era of precision medicine. Endocrinología y Nutrición **63**(7), 354–363 (2016)
7. Unnikrishnan, A., Menon, U.: Thyroid disorders in India: an epidemiological perspective. Indian J. Endocrinol. Metab. **15**(6), 78 (2011)
8. Mohamedali, M., Maddika, S.R., Vyas, A., Iyer, V., Cheriyath, P.: Thyroid disorders and chronic kidney disease. Int. J. Nephrol. **2014**, 1–6 (2014)
9. Sheehan, M.T.: Biochemical testing of the thyroid: TSH is the best and, oftentimes, only test needed – a review for primary care. Clin. Med. Res. **14**(2), 83–92 (2016)

10. Koulouri, O., Moran, C., Halsall, D., Chatterjee, K., Gurnell, M.: Pitfalls in the measurement and interpretation of thyroid function tests. Best Pract. Res. Clin. Endocrinol. Metab. **27**(6), 745–762 (2013)

11. Fröhlich, E., Wahl, R.: Thyroid autoimmunity: role of anti-thyroid antibodies in thyroid and extra-thyroidal diseases. Front. Immunol. **8**, 521 (2017)

12. Ceccarini, G., Santini, F., Vitti, P.: Tests of thyroid function. Endocrinol. Thyroid Dis. 1–23 (2017)

13. Amdur, R.J., Mazzaferri, E.L.: Definitions: thyroid uptake measurement, thyroid scan, and wholebody scan. In: Amdur, R.J., Mazzaferri, E.L. (eds.) Essentials of Thyroid Cancer Management, pp. 49–54. Springer, Boston (2005). https://doi.org/10.1007/0-387-25714-4_6

14. Caplan, R.H.: Thyroid uptake of radioactive iodine. JAMA **215**(6), 916 (1971)

15. Hegedü, L.: Thyroid Ultrasonography as a Screening Tool for Thyroid Disease. Thyroid. **14**(11), 879–880 (2004)

16. Chaudhary, V., Bano, S.: Thyroid ultrasound. Indian J. Endocrinol. Metab. **17**(2), 219 (2013)

17. Sholosh, B., Borhani, A.A.: Thyroid ultrasound part I: technique and diffuse disease. Radiol. Clin. North Am. **49**(3), 391–416 (2011)

18. Goodman, H.M.: Basic Medical Endocrinology. Elsevier, Oxford (2009)

19. Mense, M.G., Boorman, G.A.: Thyroid gland. Boorman's Pathol. Rat. 669–686 (2018)

20. Beynon, M.E., Pinneri, K.: An overview of the thyroid gland and thyroid-related deaths for the forensic pathologist. Acad. Forensic Pathol. **6**(2), 217–236 (2016)

21. Galofré, J.C., Díez, J.J., Cooper, D.S.: Thyroid dysfunction in the era of precision medicine. Endocrinología y Nutrición **63**(7), 354–363 (2016)

22. Little, J.W.: Thyroid disorders. part i: hyperthyroidism. Oral Surg. Oral Med. Oral Pathol. Oral Radiol. Endodontology **101**(3), 276–284 (2006)

23. Leo, S.D., Lee, S.Y., Braverman, L.E.: Hyperthyroidism. The Lancet **388**(10047), 906–918 (2016)

24. Ross, D.S., et al.: 2016 American thyroid association guidelines for diagnosis and management of hyperthyroidism and other causes of thyrotoxicosis. Thyroid **26**(10), 1343–1421 (2016)

25. Gaitonde, D.Y., Rowley, K.D., Sweeney, L.B.: Hypothyroidism: an update. Am. Fam. Phys. 244–251 (2012)

26. Vanderpump, M.P.: Epidemiology of thyroid disease. In: Encyclopedia of Endocrine Diseases, pp. 486–495 (2018)

27. Taylor, P.N., et al.: Global epidemiology of hyperthyroidism and hypothyroidism. In: Yearbook of Paediatric Endocrinology (2018)

28. Medeiros-Neto, G., Camargo, R.Y., Tomimori, E.K.: Approach to and treatment of goiters. Med. Clin. N. Am. **96**(2), 351–368 (2012)

29. Dauksiene, D., et al.: Factors associated with the prevalence of thyroid nodules and goiter in middle-aged euthyroid subjects. Int. J. Endocrinol. **2017**, 1–8 (2017)

30. Mesele, M., Degu, G., Gebrehiwot, H.: Prevalence and associated factors of goiter among rural children aged 6–12 years old in Northwest Ethiopia, cross -sectional study. BMC Pub. Health. **14**(1), 130 (2014)

31. Little, J.W.: Thyroid disorders. Part II: hypothyroidism and thyroiditis. Oral Surg. Oral Med. Oral Pathol. Oral Radiol. Endodontology **102**(2), 148–153 (2006)

32. Caturegli, P., Remigis, A.D., Rose, N.: Hashimoto thyroiditis: clinical and diagnostic criteria. Autoimmun. Rev. **13**(4–5), 391–397 (2014)

33. Pearce, E.N., Farwell, A.P., Braverman, L.E.: Thyroiditis. N. Engl. J. Med. **348**(26), 2646–2655 (2003)

34. Polyzos, S., Kita, M., Avramidis, A.: Thyroid nodules - stepwise diagnosis and management. Hormones **6**(2), 101–119 (2007)

35. Jibawi, A., Cade, D.: Thyroid nodules and cancer. In: Current Surgical Guidelines. pp. 389–398. (2009)

36. Durante, C., Grani, G., Lamartina, L., Filetti, S., Mandel, S.J., Cooper, D.S.: The diagnosis and management of thyroid nodules. JAMA **319**(9), 914 (2018)
37. Gimm, O.: Thyroid cancer. Cancer Lett. **163**(2), 143–156 (2001)
38. Sessions, R.B., Davidson, B.J.: Thyroid cancer. Med. Clin. N. Am. **77**(3), 517–538 (1993)
39. Yadav, D.C., Pal, S.: To generate an ensemble model for women thyroid prediction using data mining techniques. Asian Pac. J. Cancer Prevent. **20**(4), 1275–1281 (2019)
40. Begum, A., Parkavi, A.: Prediction of thyroid disease using data mining techniques. In: 5th International Conference on Advanced Computing & Communication Systems (2019)
41. Obeidavi, M.R., Rafiee, A., Mahdiyar, O.: Diagnosing thyroid disease by neural networks. Biomed. Pharmacol. J. **10**(02), 509–524 (2017)
42. Geetha, K., Santhosh Baboo, S.: Efficient thyroid disease classification using differential evolution with SVM. Indian J. Sci. Develop. Res. **88**(3), 110 (2016)
43. Shankar, K., Lakshmanaprabu, S.K., Gupta, Deepak., Maseleno, Andino, de Albuquerque, Victor Hugo C.: Optimal feature-based multi-kernel SVM approach for thyroid disease classification. J. Supercomput. **76**(2), 1128–1143 (2018). https://doi.org/10.1007/s11227-018-2469-4
44. Sidiq, U., Aaqib, S.M., Khan, R.A.: Diagnosis of various thyroid ailments using data mining classification techniques. Int. J. Sci. Res. Comput. Sci. Eng. Inf. Technol. **5**, 131–136 (2019)
45. Tyagi, A., Mehra, R., Saxena, A.: Interactive thyroid disease prediction system using machine learning technique. In: 5th International Conference on Parallel, Distributed and Grid Computing (2018)
46. Dash, S., Das, M.N., Mishra, B.K.: Implementation of an optimized classification model for prediction of hypothyroid disease risks. In: International Conference on Inventive Computation Technologies (2016)
47. Pandey, S., Miri, R., Tandan, S.R.: Diagnosis and classification of hypothyroid disease using data mining techniques. Indian J. Eng. Res. Technol. **2** (2013)
48. Saiti, F., Naini, A.A., Shoorehdeli, M.A., Teshnehlab, M.: Thyroid disease diagnosis based on genetic algorithms using PNN and SVM. In: 3rd International Conference on Bioinformatics and Biomedical Engineering (2009)
49. Ma, L., Ma, C., Liu, Y., Wang, X.: Thyroid diagnosis from SPECT images using convolutional neural network with optimization. Comput. Intell. Neurosci. **2019**, 1–11 (2019)
50. Razia, S., Rao, M.R.N.: Thyroid disorder detection using image segmentation in medical images. Indian J. Sci. Develop. Res. (2016)
51. Gomathy, V., Snekhalatha, U.: Automated segmentation using PCA and area estimation of thyroid gland using ultrasound images. In: 2015 International Conference on Innovations in Information, Embedded and Communication Systems (2015)
52. Wang, W., Ozolek, J.A., Rohde, G.K.: Detection and classification of thyroid follicular lesions based on nuclear structure from histopathology images. Cytometry Part A 9999A. (2010)
53. Yamamoto, S., Ogawa-Ochiai, K., Nakaguchi, T., Tsumura, N., Namiki, T., Miyake, Y.: Detecting hyper-/hypothyroidism from tongue color spectrum. In: 10th International Workshop on Biomedical Engineering (2011)
54. Vaz, V.A.S.: Diagnosis of hypo and hyperthyroid using MLPN network. Indian J. Innov. Res. Sci. Eng. Technol. **3**(7), 14314–14323 (2014)
55. Mahajan, P., Madhe, S.: Hypo and hyperthyroid disorder detection from thermal images using Bayesian Classifier. In: 2014 International Conference on Advances in Communication and Computing Technologies (2014)
56. Zabidi, A., Khuan, L.Y., Mansor,W., Yassin, I.M., Sahak, R.: Binary particle swarm optimization for feature selection in detection of infants with hypothyroidism. In: 2011 Annual International Conference of the IEEE Engineering in Medicine and Biology Society (2011)

Information Interchange of Web Data Resources

Generic Key Value Extractions from Emails

Rajeev Gupta$^{(\boxtimes)}$

Microsoft STCI, Hyderabad, India
Rajeev.gupta@microsoft.com

Abstract. Web information extraction systems are widely used to find and understand relevant parts of a text, combine multiple such parts, and produce a structured representation of the information. There are various scenarios where HTML formatted emails are generated by filling a *template* with user and transaction-specific values from databases. These emails are sent for human consumption among other things. Examples are such emails include *flight confirmation emails, restaurant reservation emails, bills, hospital records*, etc. In majority of these B2C emails information is presented in the form of *key-value* pairs i.e., user or transactions specific values are presented in an HTML format with their associated keys. In this paper, we describe a *generic* method to extract these key-value pairs which can be used for various applications. We analyze the pairs for a number of applications including identifying *semantically similar keywords* and creating clusters of keywords which then can be used for building information extraction wrappers. We show that just using *word-embeddings* is a poor substitute for finding *similar* keys. We use a number of features—types of values, cooccurrence graph of keys, etc., and combine them to present a keyword similarity algorithm which gives more than 50% improvement in *homogeneity* of the clusters, in comparison to just using word embeddings, using various real-world data.

Keywords: Information extraction · Key-value pairs · Semantic similarity · Graph similarity

1 Introduction

Email is the most frequently used web application. More than 60% of the email traffic constitutes business to consumer (B2C) emails [1]. Most B2C (business to consumer) emails are generated by filling a *template* (form) with user or transaction-specific values from databases. Many other machine-generated contents in B2C environment share similar characteristics. In these scenarios, usually, users only look for *key-value* pairs, while the other static or visual content is used for locating them. For example, in a *hotel confirmation email*, the user is interested in the values of the *check-in date, check-out date, confirmation number, hotel name, hotel address, number of guests*, etc. Various approaches exist to extract information from these HTML formatted emails and present the information in a structured format [16–18]. The information can be presented to the user in a more structured format or can be used in various other applications, e.g., generating alerts, identifying intents by personal assistants like Alexa, Cortana, etc. In

L. Bellatreche et al. (Eds.): BDA 2020, LNCS 12581, pp. 193–208, 2020.
https://doi.org/10.1007/978-3-030-66665-1_13

this paper, we consider extracting these key-value pairs from B2C emails using a *generic* method– across different scenarios and templates (i.e., DOM structures). Specifically, unlike the traditional information extraction systems [16–18] which work for a particular website, template, or scenario, we present a system that can be used to extract key-value pairs from B2C emails irrespective of their domain or usage. In this paper, we present several applications using the key-value pairs thus extracted. We mainly focus on algorithms to identify *semantically equivalent keys* across these emails. Here is a motivating example for the same:

Example 1: A user wants to know the *confirmation number* from a hotel confirmation email while searching her mailbox. The user may not know the exact keyword string to do the search. The mail may use the *hotel record id* as the key to indicate the confirmation number. Thus, we need to augment the keyword search using semantically equivalent keys.

Here are some of the other use cases of the extracted key-value pairs:

U1: After extracting key-value pairs from our techniques, a rule-based system can be used to represent the key-value pairs into a structured format. For example, key-values from a *flight confirmation email* can be used to create different legs of the flight using other visual signals from the mail; and presented to the user.

U2: A B2C email contains lots of text which may be common across emails of similar types, or which doesn't convey much useful information to the user. Key-value pairs are the most *information-heavy* text. Thus, using key-values for clustering and classification is likely to remove lots of *noise* from these emails.

U3: Extracted keys from emails belonging to the same scenarios (e.g., hotel confirmation, bills, airline reservations, etc.) can be clustered together to create keyword dictionaries based on their semantic similarity. These keyword dictionaries can be used to extract information from B2C emails. For example, we can get text following keyword *confirmation number* or its semantically similar keys to get the confirmation numbers from the emails.

One obvious solution of finding semantic equivalent keys is to get *language embeddings* like Word2Vec [2], GloVe [3], etc., for various keys and use some similarity measure to find the nearest keywords. In this paper, we show that using just language embedding has lots of precision and recall issues. Here is an example of semantically similar keywords we extracted:

Example 2: For the keyword search, let *"Airline confirmation code"* be the query keyword. Depending on the input emails, we expect a majority of these as semantic similar keywords (or phrases) to be used for the search: *Portugalia airlines confirmation code, Israir airlines confirmation code, American airlines confirmation code, Numéro de reservation, Sabre record locator #, Agency reference number,* etc.

Here is the outline of the paper: Literature related to key-value extraction and finding similar keys is presented in Sect. 2. In Sect. 3 we describe different template and scenario agnostic *features* we use to generically extract *key-value* pairs from HTML formatted emails. We also present performance results of the same in this section. In Sect. 4, we describe algorithms to find semantically similar keys. Several key and value characteristics are used for these algorithms. One of them is a *graph similarity method* as explained in Sect. 5. The performance of the various similarity algorithms is presented in Sect. 6. We present the performance of classification and semantically similar keys clustering, using the extracted key-values. The paper concludes in Sect. 7.

2 Related Works

There are numerous works in literature describing algorithms to extract information from HTML web pages. Some of the techniques involves humans writing rules after going through a set of example documents. Among other techniques some are supervised whereas some others are semi-supervised and unsupervised [16]. All these techniques work on only a single template or HTML DOM structure, i.e., one needs to generate rules for information extraction for each template separately. The key-value extraction technique we describe in this paper is generic, i.e., the same algorithm can be used to extract key-value pairs from all varieties of templates.

Authors of [19] describe a rule induction approach which automatically defines *programs* to extract information from a given set of *examples.* While this approach works for a given set of examples or the HTML pages following the similar patterns of layouts, it does not work for the diverse kinds of templates. Authors of [9] show that information extraction from emails does not work satisfactorily across templates. They use a combination of machine learning and template-based rules to extract a specific attribute from a particular template, e.g., extracting *check-out date* from an email from, say, *info@hilton.com* following a particular HTML template. Unlike these machine learning and rule-based techniques, our aim to develop a single system which can extract key-value pairs from all the emails from any sender domain and any scenario. Further, unlike these techniques, our key-value extraction algorithms are completely unsupervised.

Authors of [20] propose a method to extract information from HTML tables. Our algorithms compliment their work—while they extract information from tables based on provided examples with high F1 score, our work is more generic, possibly with lower F1 score. Our work can be extended by using their techniques to increase the F1 score for a given corpus of emails.

For identifying semantically similar words, phrases, and sentences various word embedding algorithms are proposed. Word2Vec [2] and GloVE [3] are extensively used to find contextual similarity between words. These are also used to get semantic similarity. But, these embeddings, in general, do not indicate semantic similarity in the sense of synonym, antonym, hypernym, hyponyms, etc. One can use lexical databases like WordNet [15] to get better semantic similarity but, unfortunately, it is not extensible to get similarity of derived words, phrases, and sentences. Authors of [14] describe a machine learning-based model trained on common-crawl data with multi-task optimizations to represent each keyword phrase using a 500-dimension vector of real numbers. Cosine similarity between these phrase vectors can be used to measure the semantic similarity between keywords. In this paper, we show that merely using word embedding doesn't

Fig. 1. Different layouts for important information

give us good solution for semantic similarity of keys extracted from emails. First, these embeddings are trained on common crawl which doesn't adequately reflect the B2C emails. Second, training embeddings over emails has its own limitations: B2C emails convey more information using HTML formatting and tables compared to a typical document used for common crawl. Thus, the context for various phrases used in B2C emails is very different than what we get in common crawl documents (e.g., *Israir airlines confirmation code* is not like *Sabre record locator*). In this paper we combine the language embeddings provided by [14] with other methods to find the semantically similar keys.

3   Extracting Key Value Pairs

A typical information extraction system over an HTML document is designed for a particular template and scenario [16], e.g., different wrappers are designed to extract *books* from *library catalogue*, *product shipping information* from *logistic emails*, *hotel address* from *hotel confirmation emails*, etc. These specific scenarios help in designing output schema and extraction by providing the kind of information we need to look for. Further, even the slight change in the template leads to failure in extractions. Figure 1 shows various example layouts for key-value pairs in B2C emails such as *flight* and *hotel confirmations.* Each of these emails can be viewed as a combination of fixed and variable parts. The fixed part is same across all the emails created from a single template. These templates are filled by user and transaction specific values forming the variable part. For the purpose of this paper, we typically consider text from this variable part as *values* from the B2C emails. Specific text used to identify the semantic of these values is called as *keys*. For the email shown in Fig. 1(c), keys include Origen, Destino, Salida, Llegada, etc., with corresponding values being Gerona-Costa Brava, Pisa G Galliei, 17/04/2016 08:50, 17/04/2016 10:15, etc., respectively.

In this paper, we are interested in an extractor which can extract key-value pairs irrespective of the email template or scenario. Thus, the extractor working for the mail shown in Fig. 1(a) should also work for that in Fig. 1(c). To extract key-value pairs, we first extract all the *text-nodes* of the HTML using in-order DFS parsing. These are the leaf level nodes containing all the text which is displayed to the user. All the keys and values are part of these text nodes. We use a number of features of a text node to identify whether a text node contains a key-value pair. In Sect. 3.1 we describe the extraction of key-value pairs using textual features whereas in Sect. 3.2 we describe extracting those from HTML tables.

3.1   Key-Value Extractions from Text Separators

We use the text in the leaf nodes to identify the key-value pairs. Specifically, we consider the cases where the keys and corresponding values are separated by separators like a colon (:), hash (#), etc. It should be noted that the keys and the corresponding value can be there in the same text node or different text nodes while being separated by the specified separator. We use different regular expressions along with the HTML DOM structures to identify such key-value pairs. These expressions use the presence of specific HTML tags (e.g., *<span>*, *<p>*, etc.) and text to identify key value pairs from one or multiple consecutive text nodes. But the mere presence of the separator doesn't indicate key-value pairs. For example, as shown in Fig. 1(c) the separators can occur in various other situations, e.g., date and time entities, any long text, etc. We use entity detectors [4] and filters to remove these cases.

3.2   Key-Value Extractions from Tables

HTML Table is a popular format to arrange keys and values in web pages. In various HTML pages, *<table>* tag is used in a complicated and nested manner. Hence, we write a number of *conditions* to identify the tables containing key-value pairs. We only take leaf-level *<table>* tags for key-value extraction. Further, since such tags are also used for visual presentation of individual string, only the tables with more than one row and more than one column are considered.

Identifying Table Orientation. The table orientation can be vertical or horizontal. In a horizontal table keys and values are horizontally aligned (Fig. 1(b)) whereas, in a vertical table, they are vertically aligned (Fig. 1(c)). The vertical table is more popular when a single key has more than one value. For example, details of various legs in a flight confirmation, or different guests in a hotel reservation are usually represented in a vertical table. We use a triplet of *<color, emphasis, font>* of the text (called *format*) to determine the table orientation. If all the cells of a row or a column have the same format, we call that as the format of the row or the column. Further, if the first column of a table has a different format compared to the other columns, we take that as the horizontal table; whereas, if the first row has a different format compared to other rows, we take that as the vertical table.

There are situations where we can identify the presence for a table but not its orientation as the format of all the table cells is the same. In those cases, we use textual features to detect whether the table is horizontal or vertical. These features include the presence of entities, the number of characters in a *<div>* node, number of words, etc.

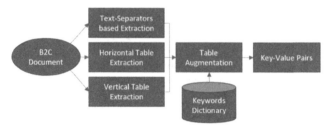

Fig. 2. Key-value extraction workflow

Specifically, we get text features of the first two rows and the first two columns. Depending on the differences in these textual features we declare the table as horizontal, vertical or *undecided*. If we are still undecided, we use table augmentation, as described later, to identify keys. Figure 2 shows the summary of the key-value extraction workflow.

Extracting Key Values from Tables. After identifying HTML tables and categorizing them as horizontal or vertical tables, we get the keys from the text nodes in the first column or the first row, respectively. Specifically, all the text in a text node in the corresponding first row or first column forms a *key*. To get *values* for these *keys*, we compare X-Y coordinates of the text nodes in the same table with those of the text nodes containing keys, looking for horizontal or vertical alignment of keys and values in horizontal and vertical tables, respectively. To get coordinates of the text nodes we use headless browser PhantomJS [22] which simulates the layout of the page and gives the X-Y coordinates of each text node.

Table Augmentation. This step is performed to extract key-value pairs from tables whose orientation is undecided. After the previous steps of key-value extractions (using separator and table orientation methods), we store the extracted keys in a keyword dictionary. We perform another pass to process the undecided tables. We take text from the text nodes of the first row and first column of these tables. We look for these texts in the keyword dictionary. If we get more than two dictionary keys in a row/column, we declare that as a vertical/horizontal table and all the texts from other cells of the same row/column are also designated as keys. For example, in Fig. 2(b), format of all the rows and columns is same. But, if from some other email (from the same scenario) we know that *Addresse email* and *date d'arrivee* are keys, then the fact that *Code d'invitation mobile* is appearing in the same column we declare that as a key. After getting keys we use the XY-coordinates to extract values from other rows and columns, as explained in the previous para.

Improving Precision. As we extract key-value pairs from a large number of emails belonging to the same scenario, the same keys are likely to be repeated in a number of emails (as they are likely to be part of the *fixed* portion of one or more templates). Thus, even in case of extracting some spurious keys (false positives) from an email, we use frequency of emails in which a key appears to filter out the spurious keys, thus, improving the precision. We remove all the keys which are occurring less than a *threshold* times from the keyword dictionary. Next, we present the precision and recall results for the key-value extractions.

3.3   Performance of Key-Value Extractions

For measuring the performance of various algorithms presented in this paper, we use 1-day's representative emails in a popular email service provider for *logistics, flight* and *hotel reservations*. As emails are private data, we use *k*-anonymization [11] to scrub PII data from the emails. For anonymization, we first get a number of emails with the same template and identify their fixed and variable parts. We need to scrub only the variable parts. We identify a set of popular entities [4] such as dates, phone numbers, person names, addresses, etc., from the variable parts. Then, we replace the identified PII entities with the entities of the same types. If a text is not part of any such entity, we replace that text with a random text. The anonymization process ensures that the characteristics of *keys* and *values* are not changed while scrubbing the user-specific information. We used two methods to generate the ground truth for measuring the performance of generic key-value extraction: human annotations and template specific wrappers.

Performance Using Human Annotations. In the first approach, we got a sample of anonymized logistics emails (e.g., from *DTDC, DHL*, B*lue-dart*, etc.) annotated by human annotators identifying all the key-value pairs from them. We matched these with the extracted key-value pairs. A pair was considered a match only if both the key and its corresponding value matched exactly. We got a precision of 65% and recall of 60%. It should be noted that as B2C emails are diverse with many different ways of presenting information, in general, we do not expect to get a very high precision and recall for the extracted key-value pairs. In fact, some of these numbers are comparable to the precision and recall for unseen templates of *flight* emails in a supervised machine learning approach [21]. We can write more rules using the textual and DOM features outlined above to improve precision and recall of the key-value extraction system [20]. Still, if we have a corpus of emails from the same scenario (e.g., hotel confirmation, logistics, etc.) we are able to identify all the important keys for the analytical applications for which we are going to use the key-value pairs.

Performance Using Golden Dataset.

Getting human annotation is costly– it took 3 annotators more than a week to annotate 2000 emails. Instead, we use a supervised information extraction pipeline to get a golden set. We use a number of template specific wrappers [9] to extract information from emails with very high precision and recall. For example, from a hotel reservation email, we extract attributes like *check-in date, check-out date, hotel address, guest names*, etc. These attributes are used to create a hotel reservation JSON objects [10]. To measure the performance of key-value extraction, we

Table 1. Performance of key-value extraction

Attribute name	% recall
reservationId	78
checkinTime	67
checkoutTime	50
reservationFor.telephone	59
aggregatorReservationId	32
guestName	10

compare these extracted objects with the extracted key-value pairs. If any of the extracted values are the same as the values in the JSON object corresponding to that email, we consider that we have correctly extracted that pair. Table 1 shows the recall numbers for different attributes from hotel confirmation emails. We have high recall value for *reservationId*, reasonably good recall values for *check-in* and *check-out* times and very low recall for the *guest name*. There are various reasons for that. We found that in many of the mails, the *guest name* doesn't have a key. To handle such situations, we can use entity detectors [4] to get values of certain types without looking for keys. Here are the other precision and recall issues we faced: The ground truth was inconsistent. Keys like *conditions*, *privacy policy*, etc., were not marked by human annotators or the HTML wrappers (as they were not part of the output JSON schema), while they were extracted by the generic techniques. Such precision issues can be handled by black-listing certain keywords. Recall issues were there as in various cases HTML tables were there without using any *<table>* tag. Similar issues were observed where keys and values are present in different formats without using any textual separators. Some of these issues can be resolved by writing additional rules for key-value extraction [12].

4   Finding Similar Keys

In this section, we describe various methods that can be used for identifying similar keys. The first measure we consider is based on natural language processing. In Sect. 4.2 and Sect. 4.3 we describe two more methods. We show that just using natural language similarity does not give us the desired results.

4.1   Natural Language Similarity

To know whether two keys are similar, we can use various natural language processing concepts. First, we find whether two keywords have the same stemming output. To get a stemmed representation of a keyword consisting of multiple words (e.g., checking-in) we stem each individual word, ignoring all digits and special characters, and concatenate them. Using this method, we find the similarity between *Checking-in* and *Check-in*, whereas "1st *Night*" is similar to "4th *Night*". Another NLP construct we can use is word embedding. GloVE [3] and Word2Vec [2] are extensively used to find contextual similarity between words. These are also used to get semantic similarity. But as we mentioned in Sect. 2, these do not indicate semantic similarity in the sense of synonym, antonym, hypernym, hyponyms, etc. According to these embeddings, similarity between *check-out date* and *departure date* is lower compared to, say, *guest name* and *hotel name*. Using common crawl embedding [14] ensures that the similarity between *check-in date* and *check-out date* is lower compared to that between *check-in date* and *arrival date*. Still, there are few problems in this way of getting word embeddings. These embeddings are trained on text from various webpages which doesn't adequately reflect the B2C emails. Further, embeddings do not capture lexical similarity, so we get similar embeddings for *Window* and *Room*, *Check-in* and *Check-out*, etc. These deficiencies led us to explore other additional similarity methods.

4.2   Value Similarity

This measure is based on the intuition that semantically similar keys have similar or the same types of values. For example, keywords like *Check-in date*, *Arrival date*, and

Reporting date have all the values as *dates*. Similarly, keywords like *Confirmation #* and *Booking confirmation* have digits and capital letters in their values. Thus, to get *value features*, we get all the values for a keyword and use their aggregate features to characterize the keyword. The value features include:

1. Average number of words in the value
2. Average number of characters
3. Presence and number of digits
4. Presence of capital letters
5. Presence of alphanumeric values, and
6. Presence of various entities like dates, telephone numbers, people names, etc.

The weighted similarity between these features vectors gives the value similarity between the keywords. We learned weights of these features by a logistic regression model created using ground truth of values corresponding to similar keys.

Learning Value Feature Weights. As explained in Sect. 3.3, we use a number of template specific wrappers to extract information from the emails with high precision and recall. We collect values of individual attributes (i.e., all values of *confirmation numbers*, *check-in dates*, etc.). These act as ground truth, i.e., weights of different features should be such that all the values of the same attributes should have high similarity whereas values of different attributes should lead to low similarity. We preprocess the output object attributes so as to keep a single bucket for attributes like *check-in date* and *check-out date* (as they have high *value similarity*).

Issues with Value Similarity. Even if a pair of keywords have high language and value similarities, they may not have the same or similar semantic meaning. For example, *Check-in date* and *Check-out date* have the language similarity of 0.7 and value similarity of 0.95 (both are dates), but for *hotel reservation* emails, they are different keywords with different semantic meanings. We try resolving these issues in the method described next.

4.3 Cooccurrence of Keys

In this section, we describe the intuition behind the similarity calculation based on cooccurrence of keywords, i.e., keywords which occur in the same email:

Assertion 1. *Two distinct keys with the same semantic meaning are less likely to be part of the same email.*

This assertion is based on the assumption that an email will not use different keys with the same semantic meaning. For example, we don't expect *check-in date* and *arrival-date* to appear in the same email.

Assertion 2. *Similar semantic keys will be there in the emails from the same scenario.*

For example, we are more likely to get similar keys for "*Airline confirmation number*" from the emails from the airline providers. Specifically, if email pairs have several key-pairs with the same or similar semantic meaning, the emails are more likely to be from the same scenario. We use these two assertions to model the semantically similar keys as *node similarity* in a graph, as described in the next section.

5   Keywords Graph

In a keywords graph, a node is created for each keyword extracted from a corpus of emails D. An edge between the keywords K_i and K_j is assigned a weight W_{ij} using set of emails D_i and D_j in which the keyword K_i and K_j appear respectively:

$$W_{ij} = \frac{|D_i \cap D_j|}{|D_i \cup D_j|}$$

The edge weights indicate the degree of the cooccurrence between two keywords, i.e., more the edge weight between two keywords more is the chance of the keywords being in the same email. As stated by *Assertion* 1, keywords corresponding to nodes having edges with higher values of weights are less likely to be similar. Similarly, using *Assertion* 2, we can say that if two nodes have a similar structure with respect to several other nodes (i.e., have edges with the same set of nodes) then they are likely to have a higher chance of being similar. How should we calculate the similarity between nodes, i.e., associated keywords? Can we use popular graph embeddings [2, 6] and use them to get cooccurrence similarity between keywords? We explore different options of node similarity next.

5.1   Node Similarity

In Fig. 3 we consider sub-structures of a keyword graph where we show edges only if the edge weights are higher than a certain threshold (i.e., they occur in the same emails frequently). As explained in *Assertion* 1, two nodes K_1 and K_2, which have high edge weight between them are less likely to be having high similarity (Fig. 3(a)). In Fig. 3(b) two nodes, K_3 and K_5, are collocated with the same node K_4, while not being collocated themselves, they are more likely to be similar (*Assertion* 2). Similarly, we can have another keyword K_6 having an edge

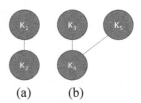

(a)   (b)

Fig. 3. Sub-structures for node-similarity

of a high weight with K_5. Since K_3 and K_5 are more likely to be similar and since K_5 and K_6 are less likely to be similar; K_3 and K_6 will have a lower possibility of being similar. Thus, we can see that two nodes have higher chance of being similar if their path length in the keyword graph is *even* whereas the *odd* path lengths indicate the dissimilarity. We need to define *keyword similarity* based on these intuitions over subgraphs.

5.2 Graph Embeddings

Graph embeddings are latent vector representations of the graph which gives a mapping of the individual nodes of the graph to a vector of real numbers so that the mapping preserves proximity measure defined over the graph. The proximity measure can be defined using *structural similarity* (i.e., similarly connected nodes are closer) and *community similarity* (i.e., more connected nodes are closer). *Node2vec* [5] is a popular method of generating graph embedding. In this technique, node-paths are generated using random walks starting each node. Edge weights are used to calculate transition probabilities for the random walk. The paths are used in a manner similar to *word2vec* [2] method of word embeddings to generate graph embeddings of the individual nodes. Another popular work *struc2vec* [6] calculates the structural similarity between each node pair in the graph for different neighborhood sizes. These structural similarities are used to create a weighted multi-layer graph such that each layer corresponds to different levels of structural similarity. This multi-layer graph is used to generate a context for each node using the random walk. In *struc2vec* two nodes are termed similar if the degree sequence of two nodes at each level (k-neighborhood) is similar.

Both *node2vec* and *struc2vec* don't satisfy the intuition explained in the previous section. In *node2vec*, all the nodes having a similar context (say, keywords appearing in lots of emails together) are assigned a similar embedding. For semantic similarity between keywords, we don't need the structural similarity. Further, we found that if a graph contains a clique, all the nodes in the clique have similar *struc2vec* embedding. *Struc2vec* embedding considers the length of the path in the neighborhood similarly irrespective of whether the path length is odd or even. As we explained in Sect. 5.1, path length plays an important role in graph similarity. Next, we show how we calculate graph similarity between two nodes using path lengths.

5.3 Path Length Algorithm

As we explained in Sect. 5.1, we should get a higher similarity measure between keywords if the path-lengths between corresponding nodes is even, and less similarity if the path-lengths is odd. But there can be multiple paths between two nodes, with different path lengths. It should be noted that we are interested in the similarity between two keywords only if they are from the same scenario (e.g., *flights*, *medical*, *hotel*, etc.). Thus, we should get a similarity between two nodes only if one or more paths exist between them (i.e., have common keywords cooccurrence between them). Here is how we calculate the *graph similarity measure* among keywords:

- Create a keyword graph while having the edges which have *edge weight* more than the *cooccurrence_threshold* (Say, 0.1). This removes the spurious edges.
- Get all the paths between two keywords with length less than a *path_length_threshold* (τ, say 5).
- For each path, we get *path weight* as the product of all its ($-1 \times$ *edge weights*). If the path length is odd, its weight will be negative, otherwise, it will be positive.
- Get the average weights of all paths between two nodes as the graph similarity measure between the nodes.

Using performance over real data we show that this algorithm performs better compared to other popular graph embeddings-based similarity algorithms. It should be noted that we depend on available emails to detect the cooccurrence similarity. If the input data doesn't contain *sufficient combinations*, it may lead to false positives. For example, if we have two keywords-sets collocating frequently in emails $\{K_1, K_2, K_3, K_4\}$ and $\{K_1, K_2, K_5, K_6\}$. In the absence of any other evidence, K_3 and K_4 will be similar to both K_5 and K_6. But we also know that K_3 and K_4 are not similar. We use the other similarity methods to break this dichotomy, as explained next.

5.4 Aggregate Similarity Algorithm

We combine the three similarities: *natural language similarity*, *value similarity*, and *cooccurrence similarity*, to get a single similarity measure between keywords. As explained in the previous section, we use *cooccurrence similarity* for filtering keyword-pairs, i.e., we consider a pair of keywords similar only if their cooccurrence similarity is above a positive threshold, but vice-versa is not true. We normalized *natural language similarity* (S_e) and *value similarity* (S_v) such that their values are between 0 and 1; and used Euclidian distance to combine them. Specifically, we associate weights W_e and W_v with S_e and S_v, respectively, to calculate the aggregate similarity measure as:

$$S = \sqrt{\left(\frac{W_e^2 S_e^2 + W_v^2 S_v^2}{W_e^2 + W_v^2}\right)}$$

We use this measure to get the similarity between the keywords. This is used for clustering the keywords as we show in the next section.

6 Classification and Clustering Using Key-Values

In this section, we explain how we use the key-value similarity for the classification of emails and the clustering of keys. We use 1-day representative emails from a popular email server for several scenarios to measure their performance.

6.1 Classification Using Key-Values

In this section, we show how the extracted key-value pairs can be used to classify emails into different classes (i.e., scenarios). We developed a binary classifier [13] where keys were represented using one-hot vectors whereas value features were added by replacing each '1' in the one-hot vector by the corresponding value feature vector (Sect. 4.2). We used emails from two scenarios (*hotel confirmation* and *flight reservation*) to measure the performance of classification. We performed experiments with two sets of data mix.

Classification Across Scenarios. In the first set, we took emails that had either hotel confirmation or flight reservation, i.e., we removed the emails which were tagged as both in the ground truth (human annotations). Just keys-based classifier resulted in precision/recall (PR) of 92/90. By adding value features, we didn't get any addition to the PR values.

Granular Classification. In the second experiment, we took emails that were only marked as *hotel confirmation* while all other emails were taken as the negative set (these emails also had *hotel cancellations*). With email keys as features, we got the PR of 80/77 whereas by adding value features the PR increased to 81/80. Possibly, by adding value features, we can distinguish keys which may be there in both positive and negative classes but have different semantic meaning. Further, we can't claim to gain much by adding value features as keys with the same semantic meaning may have different value features (as we are getting features for individual values in an email rather than an aggregate across emails). For example, a *date* can be represented as 07-05-2019 as well as July 5th, 2019; and they will have different value features. This shows that, in general, we can use key-value pairs for scenario identification, but it doesn't work very well for granular sub-scenario level classes.

6.2   Keywords Clustering

We use the similarity measure described in Sect. 5.4 to cluster semantically similar keys together. We specifically used DBSCAN [6] algorithm to cluster the keywords. For each keyword, we put the other keywords in the same cluster if their similarity is higher than a given threshold ϕ_s. We keep on expanding the cluster until we get more keywords with the similarity higher than ϕ_s. We remove the clusters from the dictionary if the number of keywords in the cluster is less than a threshold ϕ_c. Typical values of ϕ_s and ϕ_c are kept at 0.2 and 2 respectively. We used 9800 emails for *hotel confirmation* and 27345 emails for *flight reservation* to measure the performance of clustering. Table 2 shows some clusters of keywords for *flight reservations* obtained using the aggregate keyword similarity measure. Our algorithm clusters *Ecredits number* and *American airlines confirmation code* together although they don't share any word.

Table 2. Example clusters for *flight reservation*

Key phrase clusters
Confirmation number (pnr), Your trip confirmation number, Porter confirmation number, United confirmation number, Booking reference, Locator, Ecredits number, World travel record locator, American airlines confirmation code
e-mail, your e-mail address, email address, we will send your ticket to the following e-mail address, mailing address
total fare, fare total, fare equivalent, airport price per adult, 2 adult fare, 2 child fare, fare price
company name, customer, contact, billing name, gol cust service
taxes, fees and charges, taxes & fees

Effect of Cooccurrence Similarity. Table 3 shows the importance of the cooccurrence similarity for *flight reservations* email. The table shows example keywords (K_1, K_2, K_3) and their different similarity values S_e (natural language similarity), S_v (value similarity), and S_c (cooccurrence similarity). The low value of S_c ensures that `Hand baggage allowance` and `Check-in baggage allowance` are in different clusters despite having high language and value similarity.

Table 3. Effects of cooccurrence similarity

Clusters with similarity values
K_1: Hand baggage allowance, K_2: Check-in baggage allowance, S_e: 0.92, S_v: 0.99, S_c: 0.07
K_1: Arrival terminal, K_2: Departure terminal, S_e: 0.92, S_v: 0.99, S_c: 0.16
K_1: First name, K_2: Last name, S_e: 0.92, S_v: 0.99, S_c: 0.25
K_1: Expedia itinerary number, K_2: Travelocity itinerary number, K_3: itinerary number, $S_e(1,2)$: 0.92, $S_v(1,2)$: 0.99, $S_c(1,2)$: 0.98, $S_e(1,2)$: 0.98, $S_v(1,2)$: 0.99, $S_c(1,2)$: 0.55

Effect of Path Length Algorithm. Rather than using *node2vec* or *struc2vec* we present the effect of using the *path length algorithm* to filter key phrases. The average *cosine distance* between *node2vec* vectors of keywords of the same cluster was 0.54. A similar measure for *struc2vec* was 0.32. Using the path length algorithm, more than 70% of the keyword pairs within the same cluster had a higher similarity. That leads to better clustering homogeneity as explained next.

Homogeneity and Completeness. Homogeneity and completeness are popular measures for clustering algorithms. Homogeneity (H) measures the percentage of keywords in clusters we discover which are also in the same clusters of the ground truth. Completeness (C) measures the percentage of keywords from a ground truth cluster which are part of the same discovered clusters. To get the ground truth of clusters, we used the output of the template specific wrappers (as described in Sect. 3.3) and got the keys that designate the same attribute in the output. E.g., all the keys which give the value of confirmation number were put into the same *ground truth* cluster. We experimented with number of weight pairs for language and value similarities. We got the best results when both the weights were same, proving the importance of the value similarity. If we only use the language embedding (i.e., not using the *cooccurrence similarity* and keeping W_v = 0), we get H/C of 58/48. By using the *aggregate similarity measure* we were able to increase H/C to 83/54 ($W_e = 1$ and $W_v = 1$).

7   Conclusions

In this paper, we described a method of generic key-value extraction from B2C emails. Extracted key-values can be used for a number of extraction and analytical purposes. We used the key-values for the classification of emails and the clustering of keywords. We specifically showed that keys clustering can be used for identifying semantically similar keys from a corpus of emails. Our similarity measure led to good improvement in homogeneity and completeness measures of the clustering. Currently, we are working on method to generate a common embedding for each node which can combine the graph embedding with language embeddings and value features. Using the extracted key-values and keyword dictionaries (clusters) for creating semantic event cards [9] is another area of our future work.

References

1. Ailon, N., Karnin, Z.S., Liberty, E., Maarek, Y.: Threading machine generated email. In: Proceedings of the Sixth ACM International Conference on Web Search and Data Mining, WSDM 2013 (2013)
2. Mikolov, T., Sutskever, I., Chen, K., Corrado, G., Dean, J.: Distributed Representations of Words and Phrases and their Compositionality. CoRR. abs/1310.4546 (2013)
3. Pennington, J., Socher, R., Manning, C.D.: GloVe: global vectors for word representation. In: Proceedings of the 2014 Conference on Empirical Methods in Natural Language Processing (EMNLP), 25–29 October 2014, Doha, Qatar (2014)
4. Kumaran, G., Allan, J.: Text classification and named entities for new event detection. In: Proceedings of the 27th Annual International ACM SIGIR Conference on Research and Development in Information Retrieval, pp. 297–304 (2004)
5. Grover, A., Leskovec, J.: node2vec: scalable feature learning for networks. CoRR, vol, abs/1607.00653 (2016)
6. Ester, M., Kriegel, H.-P., Sander, J., Xu, X:. A density-based algorithm for discovering clusters in large spatial databases with noise. In: KDD 1996 (1996)
7. Leonardo, F.R., Ribeiro, P.H.P.S., Figueiredo, D.R.: Struc2vec: learning node representations from structural identity. In: KDD 2017 (2017)
8. Hammer, J., Garcia-Molina, H., Cho, J., Aranha, R., Crespo, A.: Extracting semi-structured Information from the Web. Technical report. 1997–38, Stanford Info Lab. http://ilpubs.stanford.edu:8090/250/
9. Sheng, Y., Tata, S., Wendt, J.B., Xie, J., Zhao, Q., Najork, M.: Anatomy of a privacy-safe large-scale information extraction system over email. In: Proceedings of the 24th ACM SIGKDD International Conference on Knowledge Discovery & Data Mining. KDD 2018 (2018)
10. Guha, R.V., Brickley, D., Macbeth, S.: Schema.org: evolution of structured data on the web. Commun. ACM, February 2016
11. Bayardo, R.J., Agrawal, R.: Data privacy through optimal k-anonymization. In: Proceedings of the International Conference on Data Engineering, ICDE 2005 (2005)
12. Cohen, W.W., Hurst, M., Jensen, L.S.: A flexible learning system for wrapping tables and lists in HTML documents. In: Proceedings of the 11th international conference on World Wide Web (WWW 2002) (2002)
13. sklearn.linear_model.LogisticRegression. https://scikit-learn.org/stable/modules/generated/sklearn.linear_model.LogisticRegression.html
14. Cer, D., Yang, Y., Kong, S., Hua, N., Limtiaco, N., et al.: Universal Sentence Encoder. arXiv: 1803.11175v2 [cs.CL] 12 Apr 2018
15. Pedersen, T., Patwardhan, S., Michelizzi, J.: WordNet similarity: measuring the relatedness of concepts. In: Demonstration Papers at HLT-NAACL 2004 (2004)
16. Chang, C.-H., Kayed, M., Girgis, M.R., Shaalan, K.F.: A survey of web information extraction systems. IEEE Trans. Knowl. Data Eng. **18**(10), 1411–1428 (2006)
17. Zheng, S., Song, R., Wen, J.-R., Lee Giles, C.: Efficient record-level wrapper induction. In: Proceedings of the 18th ACM Conference on Information and Knowledge Management, CIKM 2009 (2009)

18. Penna, G.D., Magazzeni, D., Orefice, S.: Visual extraction of information from web pages. J. Vis. Lang. Comput. **21**(1), 23–32 (2010)
19. Gulwani, S., Jain, P.: Programming by examples: PL meets ML. In: Asian Symposium on Programming Languages and Systems, November 2017
20. Tengli, A., Yang, Y., Ma, N.L.: Learning table extraction from examples. In: Proceedings of the 20th international conference on Computational Linguistics (COLING 2004) (2004)
21. Iyer, A., Jonnalagedda, M., Parthasarathy, S., Radhakrishna, A., Rajamani, S.K.: Synthesis and machine learning for heterogeneous extraction. In: Proceedings of the 40th ACM SIGPLAN Conference on Programming Language Design and Implementation (PLDI 2019) (2019)
22. PhantomJS - Scriptable Headless Browser. Phantomjs.org

The Next Generation Web: Technologies and Services

Asoke K. Talukder[(⊠)]

SRIT India Pvt. Ltd., Bangalore, India
asoke.talukder@renaissance-it.com

Abstract. Tim Berners-Lee, invented the World Wide Web (WWW) or Web in short in 1989. The Web became so popular that for many, it is synonymous with the Internet or simply the Net. There were many flavors of the original Web like Web 2.0, Web 3.0, etc. All these versions of Web are different use-cases of the original Web, which is fundamentally Client-Server in nature. A client Web browser (or user-agent) makes a request for a document (or a transaction), and the server services that request – kind of a synchronized request-response service (Pull service). The Next Generation Web (NGW) will have a fundamental paradigm shift – it will be two-ways (full-duplex) asynchronous Peer-to-Peer (P2P) communication between two Web browsers using HTML5, Secured WebSocket (wss://), Server Sent Events (SSE), and JavaScript. The Next Generation Web (NGW) will be for human to human, and human to machineries (robots) interaction. Major technologies in Next Generation Web include WebRTC, Web Speech API, and WebUSB. WebRTC is already standardized by W3C and IETF. Web Speech API is at the draft a state. WebUSB is still evolving. NGW is a technology and not a solution – NGW does not need any additional downloads or plugins or any intermediate server. WebRTC supports real-time ultra-low latency audiovisual media and non-media arbitrary data with recording facility. Web Speech API includes speech recognition, speech synthesis, and audio processing on the Web browser. WebUSB will allow USB devices connected to the Web for Collaborative Robotics (Cobotics). Added with AI and IoT, Next Generation Web will revolutionize the Web application and digital transformation ecosystem from computers to mobile phones starting from simple Web page viewing to complex Health care applications and cobotics that will touch everybody's life.

Keywords: Digital transformation · Next Generation Web · NGW · HTML5 · WebSocket · Server Sent Events · JavaScript · WebRTC · Web Speech API · WebUSB · Nodes.js · IoT · Cyber Physical Systems · AI · Knowledge Graph · Secured communication · HIPAA · Cobotics

1 Introduction

On 4th of October 1957, the Union of Soviet Socialist Republics (USSR) launched Sputnik I – the first artificial Earth satellite into the space. The United States Department of Defense responded to Sputnik's launch by creating ARPA (Advanced Research Projects

© Springer Nature Switzerland AG 2020
L. Bellatreche et al. (Eds.): BDA 2020, LNCS 12581, pp. 209–229, 2020.
https://doi.org/10.1007/978-3-030-66665-1_14

Agency) in 1958 [1]. ARPA research played a central role in launching the "Information Revolution", through ARPANET. ARPANET's initial demonstration in 1969 led to the Internet that we use today [2].

World Wide Web (WWW) or Web in short was invented by Tim Berners-Lee in 1989. He invented it while he was working at CERN, European Laboratory for Particle Physics in Geneva, Switzerland. Originally World Wide Web was conceived and developed to meet the demand for information-sharing between scientists and researchers in universities and research institutes around the world working in CERN projects (Fig. 1). Tim Berners-Lee had the idea of enabling researchers from remote sites in the world to organize and pool together information and research publications [3].

Fig. 1. (a) The World Wide Web (WWW) inventor Tim Berners-Lee. Picture Taken: 11 Jul 1994 (c) CERN. Picture source: https://home.cern/science/computing/birth-web/short-history-web. (b) The first page of Tim Berners-Lee's proposal for the World Wide Web, written in March 1989 (c) CERN. Picture source: https://home.cern/science/computing/birth-web/short-history-web

In the Web's first generation, Tim Berners-Lee invented the HTML (Hyper Text Markup Language), the Hyper Text Transfer Protocol (HTTP), and the Uniform Resource Locator (URL) standards with Unix-based servers and character based browser version of the World Wide Web. In the second generation, Marc Andreessen and Eric Bina developed NCSA Mosaic at the National Center for Supercomputing Applications, University of Illinois. Several million people suddenly noticed the 'Web' [4]. Web became so popular that today for many 'Internet' or 'Net' became synonymous with Web.

The Clinton Administration's Telecommunications Act of 1996 made Internet free and launched the age of modern Internet policy – thrusting market forces and technological innovations. The authors of this act envisioned about what to happen in the future and that government could facilitate it best by letting private investments happen [5]. This accelerated the innovations in the Web technology space.

The first Web browser named 'WorldWideWeb' was constructed by Tim Berners-Lee in 1990. It was character mode software to access the Web. It did not have much impact outside of the scientific community. The Web became commercial on October

13, 1994 when the Netscape Navigator Web browser was released by the Mosaic (later Netscape) Corporation founded by Marc Andreessen and Jim Clark. Netscape Navigator used GUI (Graphical User Interface) and instantly became popular [6]. It can be stated that Netscape Navigator took Web out of the research lab and released it to the public. The dot-com boom started with Netscape going public on 9th August 1995. In 1995 Microsoft released their Internet Explorer Web browser and the browser war started. Within few years Internet Explorer became the market leader in the Web browser marketplace.

In 1995 few other important events happened in the Web Search Engine space. Jerry Yang and David Filo introduced the Yahoo directory based search engine called "Yahoo!". Crawler based search engine AltaVista was created by researchers at Digital Equipment Corporation and launched in 1995. Paul Flaherty conceived the idea of a crawler based search engine at Digital. Louis Monier and Michael Burrows wrote the Web crawler and indexer respectively.

Google was officially launched in 1998 by Larry Page and Sergey Brin. Google developed a search algorithm at first known as "BackRub" in 1996, with the help of Scott Hassan and Alan Steremberg [7]. The search engine soon became popular and Google went public in 2004. Google became one of the main players in the Internet and Web space with their Google Search Engine, Gmail, Google Map, Chrome Web browser and various AI (Artificial Intelligence) innovations. Google also developed Android operating system for smartphones and other devices.

Web came a long way and evolved from a simple document rendering mechanism on a user display terminal to the public-facing interface of all applications running anywhere in the world. Majority of intuitive interactive applications today have a Web interface. Users use these applications using a Web browser. During the "Severe Acute Respiratory Syndrome Coronavirus 2 (SARS-CoV-2)" commonly known as COVID-19 pandemic of 2020 the Web came to the rescue of the world. Starting from grocery to high school all became virtual through the Web during the COVID-19 pandemic and lockdown.

The era of Next Generation Web (NGW) started in 2008. During 2008 to 2012 many technologies were invented to transform the Web ecosystems – the main ones that will play a pivotal role in the NGW are HTML5, WebSocket, Node.js, and WebRTC along with JavaScript invented in 1995. These technologies helped a Web browser to transform from a passive client-server type application to a real-time peer-to-peer (P2P) active device. We will see the impact of this transformation through Next Generation Web (NGW) in the post-COVID era starting in 2021 when AI (Artificial Intelligence) and IoT (Internet of Things) join hands with peer-to-peer Web and mobile phones that will touch every industry and everybody's life.

2   Documents and Knowledge in the Web

Tim Berners-Lee thought of collaborative work where a researcher could actually link the text in the files themselves – kind of virtual embedding of one document within another. This is better than simply making a large number of research documents as files. This allowed scientists to arrange documents of related topics together simply through links. This would mean that while reading one research paper, a scientist could quickly display part of another paper that holds directly relevant text or diagrams or tables. Tim

Berners-Lee thought, multiple documents could be related or brought into the user's view by using some form of hypertext. Documentation of a scientific and mathematical nature would thus be represented as a 'web' of information held in electronic form on computers across the world. Tim Berners-Lee had worked on document production and text processing, and had developed his first hypertext system 'Enquire' in 1980 for his own personal use much before he invented World Wide Web. His prototype Web browser on the NeXT computer came out in 1990 [8].

Document presentation in computers was a challenge all along. There are different computers running various word-processing software; at the same time there are different printers from different vendors with proprietary formatting styles. How to design a document in word processing software that will print on a paper by all printers with exactly the same output? GML, SGML, runoff, nroff, and troff are the early versions of document formatting system. Word processing software will use any of these document formatting styles to send the document to the printer. UNIX group at Bell Labs were working on document presentation in UNIX. Roff is a descendant of the RUNOFF program developed by Jerry Saltzer. Douglas McIlroy and Robert Morris wrote runoff for Multics in BCPL based on Saltzer's program written in MAD assembler. Their program was "transliterated" by Ken Thompson into PDP-7 assembler language for his early UNIX operating system, circa in 1970 [9]. In 1973 Joseph Frank Ossanna, Jr authored the first version of document processing software troff for UNIX for Version 2 Unix at Bell Labs, in Assembly language and then ported to C. Another dominating document description language was PostScript developed for desktop publishing business at Adobe Systems by John Warnock, Charles Geschke, Doug Brotz, Ed Taft and Bill Paxton during 1982 to 1984. Another formatting software very popular in academic circle even today is LaTeX – it was written in the early 1980s by Leslie Lamport at SRI International.

While Unix group was building their document formatting software, IBM was working their own document formatting software generalized mark-up language focused on an alternative approach, whereby standard document structures such as headers, paragraphs, lists and so on were marked up by tags inserted into document text. In 1969, together with Ed Mosher and Ray Lorie, Charles F. Goldfarb, a lawyer by training invented Generalized Markup Language (GML) to solve the document representation problem for IBM. This work led to the Standard Generalized Mark-up Language (SGML), which became an international standard ISO 8879:1986 [10].

SGML enables a user to define a grammar for marked-up documents that defines the ways in which tags can be inserted into documents and printed on a printer in homogeneous fashion. Tim Berners-Lee chose SGML as a guide to define the HTML document format for the World Wide Web.

2.1 Semantic Web

The original WWW was "Web of documents", where a URL is used to fetch the document and present it to the user. In a business environment however WWW displays processed data or information from databases like your bank balance or ordering something from an e-commerce site. While presenting the data to the user, the browser uses some presentation styles such that the data is easy to read and ingested by the human user. The term "Semantic Web" conceived by Tim Berners-Lee refers to W3C's vision of the Web of linked data. The ultimate goal of the Web of data is to enable computers to do more useful artificial intelligent work and to develop systems that can support trusted interactions over the network. Tim Berners-Lee had an idea that WWW should not only fetch the data and display it blindly as a rendering agent, but WWW should read and understand the semantics (meaning) of the document or the data and present the meaning of the data for a layman who need not have the domain knowledge.

The goal of the Semantic Web is to make Internet data machine-understandable. Semantic Web technologies enable people to create data stores on the Web, build vocabularies, and write rules for transforming data into knowledge. Linked data are empowered by technologies such as RDF, SPARQL, OWL, and SKOS [11]. For example if you enter '*solar system*' in Google, along with the search results on the left side of the screen, you will also notice some sematic output on the right side of the screen extracted from multiple Web sites about the solar system including pictures of stars and a recommendation. John Mark of New York Times named this concept of semantic Web as Web 3.0 through an article 'Entrepreneurs See a Web Guided by Common Sense' in 2006. It was called as Web 3.0 because in 1999 Darcy DiNucci coined the term Web 2.0 that refers to websites that emphasize user-generated content, ease of use, participatory culture and interoperability for end users.

3   HTML and HTTP

It can be stated that HTML (Hyper Text Markup Language) is formally an application of SGML. The HTML Document Type Definition (DTD) formally defines the set of HTML tags and the ways that they can be inserted into documents. First version of HTML was created by Tim Berners-Lee in late 1991 but was never officially released. The first official version of HTML was HTML 2.0 which was published in November 1995 as IETF standard through RFC 1866.

HTML is a document structuring protocol. But we needed a communication protocol over Internet to fetch the HTML document from the server for us to display on a computer terminal. HTTP (Hyper Text Transfer Protocol) is the protocol for same. The initial HTTP protocol for the Web was simple. The client sent a request: 'GET this filename' and the other end sends back the file and closed the connection. There was no content type to tell the browser what kind of file was being sent. There was no status code for the browser to know if there was an error during fetch. MIME (Multipurpose Internet Mail Extensions) was the multimedia extension to email. MIME was adapted by HTTP so that when a browser receives a file from the server, it has a status code and a content type. This content type tells the browser or a user program whether or not the file is text/HTML, video/MPEG, image/GIF etc., which gives the browser a chance to call up the correct viewers to display the content of the file.

The HTTP we use today is the product of collaboration between CERN and a group at the National Center for Supercomputer Applications (NCSA) at the University of Illinois at Champaign-Urbana [12]. Hypertext Transfer Protocol standard HTTP/1.0 was released through RFC 1945 in May 1996. In May 2000 HTTPS (Hyper Text Transport Protocol Secured) was standardized through RFC 2818 that offered HTTP over TLS (Transfer Layer Security).

There are vast numbers of Web servers and Web services. How does HTTP locate just the file you need from somewhere on the Internet? The answer is through the use of the URL – the Uniform Resource Locator. This is like the telephone number of a file within a computer in the Internet. The URL does it by combining the domain name with the location (path) of the file to be sent to your machine. Uniform Resource Locators (URL) was published in December 1994 through RFC 1738.

HTML also evolved. HTML 3.0 released in 1995 was a major version of HTML with many additional features. Some of the important features are, support of CSS (Cascaded Style Sheets), Tables, and support for Non-Visual Media to cater for the visually challenged individuals.

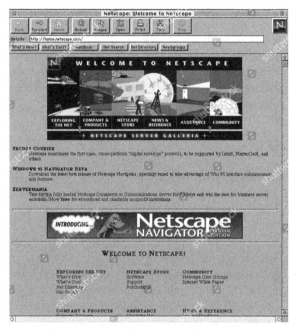

Fig. 2. World's first commercial graphic Web Browser the 'Netscape Navigator' homepage on August 9, 1995 the day Netscape went public and the Internet Boom started

Next major version of HTML was HTML 4.0. W3C recommendation of HTML 4.0 was published in December 1997 and revised in April 1998. HTML 4.0 offered many new features. Some of the notable features are Internationalization, to allow representation of the non-English languages through the support of Unicode. It added Frames, which allowed the ability to present multiple documents in one window. Another feature in HTML 4.0 was Advanced Tables. Another major enhancement of HTML 4.0 was scripting for the Web browser – to do some processing tasks to create dynamic Web pages and use HTML as a means to build networked applications.

4 Web Browsers

Tim Berners-Lee is credited with developing, both the first web server, and the first web browser. He developed the first browser called 'WorldWideWeb' in 1990. Later the WorldWideWeb program was renamed as Nexus [13]. The browser developed by Tim Berners-Lee was character mode and was not popular. Marc Andreessen and Eric Bina – two programmers working at NCSA (National Center for Supercomputing Applications) at the University of Illinois at Urbana–Champaign created a browser that used GUI (Graphical User Interface). The graphical browser was named as Mosaic. Mosaic is credited with sparking the Internet boom of the 1990s.

Mosaic was the first web browser aiming to bring multimedia content to non-technical general users, and therefore included images and text on the same page. Mosaic's creator Marc Andreessen, established the company Netscape in 1994. Netscape Navigator (Fig. 2) web browser from Netscape Communications Corporation (originally Mosaic Communications Corporation) dominated the Web market at the beginning; but lost to Internet Explorer from Microsoft and other competitors in the so-called first browser war. However, Netscape contributed towards the growth of Web technologies in many ways through many innovations. A major contribution of Netscape was the JavaScript – the most widely used programming language for client-side scripting of web pages. Netscape also developed SSL which was used for securing online communications before its successor TLS took over.

4.1 WAP Browser

At the end of the 20th century while Internet boom happened, wireless mobile hand phones were also becoming popular. Originally two dominant players in this space were Nokia and Ericsson with their mobile smartphones much before iPhone and Android were launched.

For smart mobile phones to access Web content over HTTP, WAP (Wireless Application Protocol) was introduced. WAP can be considered a scaled downed version of HTML designed for small wireless smartphone screen. WAP is a technical standard specifically designed for accessing information over a mobile wireless network.

In India WAP was introduced in 2000 by CellNext Solutions a company in wireless media space. CellNext collaborated with IBM to WAP enable IBM AS400 applications. IBM business users using AS400 could access their business applications through smartphones using WAP browsers in 2001. CellNext also introduced the first wireless email in India in 2001 that used WAP browser and WAP protocols. The WAP email system for CellNext was developed by Integra Micro Systems.

5 The Next Generation Web (NGW)

Web fundamentally works in Client-Server mode. It is a human to machine (H2M) interface, where a human being uses a Web browser to access a computer or an application or a service in a remote host. A client browser (or user-agent) makes a request using a URL for a document, data, or a transaction, and the server services that request. It is a kind of a synchronized request-response service. This is also referred to as a Pull service, where the user pulls a document from the remote computer using the Web Browser.

The Next Generation Web (NGW) will have a fundamental paradigm shift – it will be two-ways asynchronous Peer-to-Peer (P2P) communication. It is a conversation between two persons or two Web browsers – kind of human to human (H2H) communication using HTML5, JavaScript, WebSocket, and Node.js. In NGW there is no concept of predefined requester or responder or a sever. Anybody can become a requester or a responder with changing roles – either party can send messages or data to the other party in unsolicited fashion in real-time. It is a Push service where end-points do not Pull anything; they Push the data instantly. NGW offers all these peer-to-peer technologies without any intermediate server, plugin, or downloads.

The NGW will also be human to machineries (H2M) interaction, which is called cobotics. Cobotics is collaborative robotics where human beings collaborate with one or more machineries or robots and always is in charge. In robotics robots are autonomous and work independently without the direction of human being; whereas, in cobotics human beings are in charge and uses robotics for some function. A good example of cobotics will be robotic surgery where a human surgeon is in charge and guides a robotic surgeon to do the surgery. We will discuss then in following sections.

Next Generation Web fundamentally includes HTML5, along with WebSocket Secured (wss://), Server Sent Events (SSE), WebRTC, Web Speech API, WebUSB, JavaScript, and Node.js. A good repository to look for examples code and tutorials in HTML5 is https://www.html5rocks.com/en/. You may also like to refer Web Fundamentals at Google developers at https://developers.google.com/web.

Fig. 3. Official logos of HTML5, Node.js, and WebRTC. These are some of the important technologies in the Next Generation Web ecosystem.

5.1 HTML5

Following HTML 4.0, next major release of HTML was HTML5. Because it a new generation of HTML It is not called HTML 5.0 – instead it is called HTML5. HTML5 has a logo of its own (Fig. 3). HTML 2.0 to HTML 4.0 were enhancing the browser capabilities as a passive device for human-facing interfaces. In HTML5 there were many enhancements which make HTML5 browser an active device working like a computer. In HTML 4.0 scripting was added for local client-side processing. HTML5 is known to be a Living Standard and is maintained by a consortium of the major browser vendors (Apple, Google, Mozilla, and Microsoft), the Web Hypertext Application Technology Working Group (WHATWG). Some of the revolutionary features added in HTML5 to make it work as an active device are Video, Audio, Picture, Canvas, in combination with WebSocket, and Server Sent Events.

5.2 WebSocket Secured (WSS://)

In 1971 TCP Socket was formalized through RFC147 titled "The Definition of a Socket". The first paragraph of RFC147 standard states that "A socket is defined to be the unique identification to or from which information is transmitted in the network. The socket is specified as a 32 bit integer with even sockets identifying receiving sockets and odd sockets identifying sending sockets." TCP sockets are used for full-duplex transmission of data between two end-points in the Internet.

TCP sockets are the communication primitives of Internet and it requires lot of effort to develop any communication systems using socket. The avatar of TCP socket in the Web is WebSocket, which was first referenced as TCPConnection in the HTML5 specification. In 2008 first version of the WebSocket protocol was standardized. In December of 2009, Google Chrome 4 was the first browser to support WebSocket. Two new internet schemes "ws://", and "wss://" were introduced for WebSocket and WebSocket Secured respectively. WebSocket Secured uses SSL (Secured Socket Layer). The goal of WebSocket is to provide a mechanism for browser-based applications that need two-way full-duplex communication with servers that need not open multiple HTTP connections and can work without pooling (see below).

5.3 Server Sent Events

Server-Sent Event (SSE) is a mechanism for Web servers to initiate data transmission towards clients. It is generally known as Push service initiated by a server. They are commonly used to send unsolicited message, updates, alerts, notifications, or continuous data streams to a client Web browser. It is designed to enhance native, cross-browser streaming through a JavaScript API called EventSource, through which a client requests a particular URL in order to receive an event stream. The idea behind this may be equivalent to subscribe-publish in message queues [14]. A web application "subscribes" to a stream of updates generated by a server and, whenever a new event occurs, a notification is sent to the client.

To understand the importance of Server-Sent Events, we need to understand the limitations of its AJAX (Asynchronous JavaScript And XML) predecessors, which include Polling or Long Term Polling for alerts, notification, or unsolicited message to the user. Polling or a Pull service is a traditional technique used by the vast majority of AJAX applications for alerts. The basic idea is that the client application repeatedly polls a server for data or pulls some data from the server. Fetching data revolves around a periodical request-response format. The client periodically makes a request to the server and waits for the server to respond with data. If there is data ready to be sent to the client, the client fetches the data; if no data is available, an empty response is returned. This type of applications consumes lot of Internet resources.

Long polling (Hanging GET/COMET) is a slight variation of polling. In long polling, if the server does not have data available, the server holds the request open until new data is made available. Hence, this technique is often referred to as a "Hanging GET". When information becomes available, the server responds, closes the connection, and the process is repeated.

Server-Sent Events in HTML5 on the other hand have been designed from the ground up to be efficient. When communicating using SSE, a server can push data to your application whenever it wants, without the need to make an initial request. In other words, updates can be streamed from server to client as they occur. SSE opens a single unidirectional channel between server and client. The main difference between Server-Sent Events and long-polling is that SSE is handled directly by the browser and the user simply has to listen for the message.

One of the reasons SSEs is not very popular because WebSockets provide a richer protocol to perform bi-directional full-duplex data exchange. Having a two-way channel is more attractive for things like games, messaging applications, and for cases where you need near real-time updates in both directions. However, in some scenarios data does not need to be sent from the client very often. A few examples would be friends' status updates, stock tickers, news feeds, or other automated data push mechanisms.

SSEs are sent over traditional HTTP – they do not require a special protocol or server implementation. WebSockets on the other hand, require full-duplex connections and new Web Socket servers to handle the protocol. In addition, Server-Sent Events have a variety of features that WebSockets lack by design such as automatic reconnection, event IDs, and the ability to send arbitrary events.

5.4   JavaScript and Node.js

JavaScript is the scripting language that runs on a browser. JavaScript was invented by Brendan Eich while working with Netscape and introduced in 1995. In 1995 Microsoft released Internet Explorer and the browser war started. On the JavaScript front, Microsoft reverse-engineered the Navigator interpreter to create its own scripting language called JScript [15] and dominated the browser market. This started to change in 2004, when the successor of Netscape, Mozilla cofounded by Brendan Eich, released the Firefox browser. Firefox was well-received by many, taking significant market share from Internet Explorer.

In 2009 another major development happened. Node.js written initially by Ryan Dahl was introduced. Node.js is the same JavaScript but the execution engine sits outside of the browser – in a server. This in other words mean that a browser can now execute the JavaScript code preloaded in its memory and also continue executing JavaScript code at the server side through Node.js. With HTML5, client side JavaScript, and server side Node.js made a browser to be an active device. A combination of JavaScript and Node.js makes two browsers to communicate peer-to-peer and perform different functions which were not possible earlier.

5.5   Real-Time Communication Over Web (WebRTC and RTCWEB)

In May 2010, Google bought GIPS (Global IP Solutions) developing many real-time communication software components. Google open-sourced the GIPS technology and engaged with relevant standards bodies like the World Wide Web Consortium (W3C) and the Internet Engineering Task Force (IETF) to ensure industry consensus on real-time communication over Internet and the peer-to-peer Web.

WebRTC is a Web communication in real-time standardized by W3C [16]. The real-time communication is standardized by IETF (Internet Engineering Task Force) as RTCWEB. WebRTC being real-time browser to browser communication – it must be able to connect two browsers without a static IP. Even if two computers are within firewall with local IP address in two different domains in two different countries WebRTC should be able to connect these two browsers and communicate. How does WebRTC connect two computers with local IP address?

In standard Human to Computer pull service interface the web browser functions like a user-facing interface of the remote computer application. Let us take a simple example where you want to do some search using Google (www.google.com). If you do 'ping www.google.com' you will see the IP address of www.google.com shown on your terminal, which is a static IP address that is cataloged in the DNS (Domain Name Server) and does not change from time to time. If you use the command 'ifconfig' (ipconfig for Windows computer) you will see your own IP address. If you are inside a firewall, you will see a local IP address assigned to your computer by the DHCP (Dynamic Host Configuration Protocol) server something similar to 192.168.1.xxx. This dynamic IP address is unknown to the external world and changes from time to time. This in other words implies that a computer within a LAN has a local dynamic IP address which is not visible from outside of the LAN. NAT (Network Address Translation) does an address translation and connects you to Google.

The telephony network is peer-to-peer communication network – anybody can initiate a call as a caller and talk to a callee or the called party. In telephony network all telephones have a unique address that we call a 'telephone number'. This telephone number is unique in the world like the static IP of Google. For example the telephone number of SRIT in Bangalore, India is +91-80-4195-1999. Anybody anywhere in the world can use the telephone network to call this number to talk to someone at SRIT. For example, if you are in USA with your mobile number +1-408-504-2551 calling SRIT number using your mobile phone, the telephony signaling network will use the SS7 (Signaling System 7) signaling network to locate the callee (SRIT in this case) in Bangalore, India. Once the callee is located by the signaling network the call is established between you and the person at SRIT using the telephone network. The telephone network is separate from the signaling network. After the call is established, the signaling network disappears from the scene and does not play any other role.

Similar to the telephony network WebRTC uses a signaling function to locate the computers within a firewall. This is explained in Fig. 4. WebRTC consults with a signaling server to resolve this challenge of locating the callee computer (browser) within the firewall (with local IP address). Signaling is out of the scope of WebRTC. WebRTC system can use any signaling protocol. Signaling can use a STUN (Session Traversal Utilities for NAT) server, TURN (Traversal Using Relays around NAT) server, or even a SIP (Session Initiation Protocol) server to resolve the challenges related to NAT and local IP address.

It can be argued that if WebRTC uses a server why do we say that WebRTC communication is peer-to-peer without an intermediate server, plugin, or downloads? In reality WebRTC does not use any intermediate server, plugins, or downloads to communicate or exchange data. Once the connection between two browsers is established, similar to the telephony network the signaling server disappears from the scene – it does not have any role to play. After the addressing issue is resolved by the signaling server, WebRTC connects two Web browsers and they communicate directly peer-to-peer without any intermediate server, plugin, or downloads.

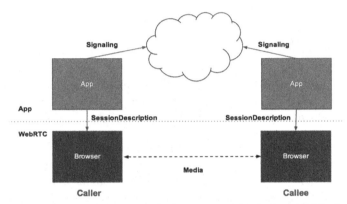

Fig. 4. WebRTC Architecture (Picture credit: Sam Dutton. (2013). WebRTC in the real world: STUN, TURN and signaling. Source of this architecture diagram is Sam Dutton. "WebRTC in the real world: STUN, TURN and signaling". Published: November 4th, 2013 (https://www.html5r ocks.com/en/tutorials/webrtc/infrastructure)

Moreover WebRTC uses HTTPS ensuring that the communication is end-to-end encrypted. WebRTC exchanges media directly using real-time audiovisual medium or arbitrary non-media data. In Fig. 4 upper part shows the caller connecting the callee through a signaling server; once the connection is established – WebRTC works as shown in the lower part of the figure where media (audio, video, and data) are transacted directly point-to-point without the signaling or any intermediate server.

WebRTC achieves all these complex functions through three simple APIs accessible through JavaScript in combination with WebSocket. These three APIs are:

1. `getUserMedia`: This API gives an access to turn real-time media streams like the microphone and the webcam in the device into basic JavaScript objects that can be easily manipulated.
2. `RTCPeerConnection`: This API is the core of peer-to-peer function in WebRTC to establish a peer-to-peer connection.
3. `RTCDataChannel`: This API is responsible for the exchange of real-time non-media data. Examples could be peer-to-peer file sharing, real-time chat, and other forms of IoT data.

WebRTC technology will transform IoT (Internet of Things) and PCS (Physical Cyber Systems) use-cases starting from healthcare to building automation. The real-time P2P feature of reliable secured data communication of WebRTC over the Web will ensures that any IoT device can be directly connected into the Web.

Below is a W3C code example taken from EXAMPLE 9 in Sect. 10.1 'Simple Peer-to-peer Example' from https://www.w3.org/TR/webrtc/#simple-peer-to-peer-exa mple of the WebRTC 1.0 standard [16]. The code assumes the existence of some signaling mechanism that is SignalingChannel. The STUN/TURN server can be used to get variables like their public IP address or to set up NAT traversal. They also have to send data for the signaling channel to each other using the same out-of-band mechanism. You can look at the live open source WebRTC application at https://appr.tc/.

```
const signaling = new SignalingChannel(); // handles JSON.stringify/parse
const constraints = {audio: true, video: true};
const configuration = {iceServers: [{urls:
'stun:stun.example.org'}]};
const pc = new RTCPeerConnection(configuration);

// send any ice candidates to the other peer
pc.onicecandidate = ({candidate}) => signal-
ing.send({candidate});

// let the "negotiationneeded" event trigger offer generation
pc.onnegotiationneeded = async () => {
  try {
    await pc.setLocalDescription();
    // send the offer to the other peer
    signaling.send({description: pc.localDescription});
  } catch (err) {
    console.error(err);
  }
};

pc.ontrack = ({track, streams}) => {
  // once media for a remote track arrives, show it in the re-
mote video element
  track.onunmute = () => {
    // don't set srcObject again if it is already set.
    if (remoteView.srcObject) return;
    remoteView.srcObject = streams[0];
  };
};

// call start() to initiate
function start() {
  addCameraMic();
}
```

```
// add camera and microphone to connection
async function addCameraMic() {
  try {
    // get a local stream, show it in a self-view and add it to
be sent
    const stream = await naviga-
tor.mediaDevices.getUserMedia(constraints);
    for (const track of stream.getTracks()) {
      pc.addTrack(track, stream);
    }
    selfView.srcObject = stream;
  } catch (err) {
    console.error(err);
  }
}

signaling.onmessage = async ({data: {description, candidate}})
=> {
  try {
    if (description) {
      await pc.setRemoteDescription(description);
      // if we got an offer, we need to reply with an answer
      if (description.type == 'offer') {
        if (!selfView.srcObject) {
          // blocks negotiation on permission (not recommended
in production code)
          await addCameraMic();
        }
        await pc.setLocalDescription();
        signaling.send({description: pc.localDescription});
      }
    } else if (candidate) {
      await pc.addIceCandidate(candidate);
    }
  } catch (err) {
    console.error(err);
  }
};
```

WebRTC is a technology and not a service or a solution; so, you can do almost anything using WebRTC – imagination is the limit. WebRTC offers many features and facilities. You can share screens (display device) by using `navigator.mediaDevices.getDisplayMedia`. You can do media processing like mixing of images, change color, add background etc. You can stream the canvas content or stream audio or video from external devices like a camera or microphone. You can even record and playback the media on a local disk and many more functions.

If you are interested in developing applications and services using WebRTC, I shall recommend that you please refer to sample code of WebRTC available at https://webrtc.github.io/samples.

5.6 Web Speech API (WSA)

Computer Telephony Integration (CTI) plays a significant role in many online telephony applications like Emergency services 911 in the US, Telephone banking, Customer support, Call center, Telephone triage and many more. These have been implemented using the IVR (Interactive Voice Response) systems [17]. IVR systems have been replaced by Visual IVR where the user interacts with a visual system in the Web and then connects to the IVR. This saved time and error – before the customer is connected to the service representative, all information about the customer is in front of the service center representative. In the Next Generation Web (NGW) Web Speech API removes these extra infrastructures. Web Speech API in the Web browser interfaces with a human and connects to the computer system directly. If you have SIP server, you can connect Web Speech API and the telephone network.

The Web Speech API (WSA) aims to enable web developers to provide, in a web browser, speech-input (speech-to-text) and speech output (text-to-speech) features that are typically not available when using standard speech-recognition software. The WSA itself is agnostic of the underlying speech recognition and synthesis implementation and can support both server-based and client-based embedded recognition and synthesis. The WSA specification is in draft state by W3C and available at https://wicg.github.io/speech-api/.

5.7 WebUSB

Every computer has a few USB (Universal Serial Bus) ports. This interface is used for many purposes starting from pen-drive to external hard disks or Webcam or mouse. The USB is the de-facto standard for wired peripherals. Most USB devices implement one of roughly a dozen standard device classes which specify a way for the device to advertise the features it supports and commands and data formats for using those features. Standard device classes include keyboard, mice, audio, video and storage devices. Operating systems support such devices using device-drivers provided by the device vendor. Many of these native codes have historically prevented these devices from being used by the Web. And that is one of the reasons the WebUSB API has been created: it provides a way to expose USB device services to the Web.

The WebUSB API provides a way to safely expose USB device services to the Web. It provides an API familiar to developers who have used existing native USB libraries and exposes the device interfaces defined by existing specifications. With this API you will have the ability to build cross-platform JavaScript SDKs for any device. The latest version of the W3C draft specification is available at https://wicg.github.io/webusb/.

6 Ancillary Technologies

There are other ancillary technologies that started emerging while Web was at its infancy. Java developed by James Gosling at Sun Microsystems was released in 1995 as a core component of Sun Microsystems' Java platform. Java like C++ is an object oriented procedural language. It was the language of choice for Web development. Java dominated the Web programing technology domain for a long time.

R language was created by Ross Ihaka and Robert Gentleman at the University of Auckland, New Zealand. The project was conceived in 1992, with an initial version released in 1995. R is an implementation of the S programming language created by John Chambers in 1976, while working at Bell Labs. R is a functional programing language and requires less coding and testing. Almost all statistical and mathematical algorithms and some AI libraries are ported in R. Soon R became the language of choice for data mining and statistical analysis.

Python was created by Guido van Rossum and released in 1991 at CWI (Centrum Wiskunde & Informatica) in the Netherlands. Python is a combination of object oriented with many features of functional programing. TensorFlow was developed by the Google Brain team for internal Google use. TensorFlow was made open source and released under the Apache License 2.0 on November 9, 2015. In no time TensorFlow was adopted by Python community to make it the de-facto standard and vehicle of AI. Python started gaining popularity when it started becoming the language for Artificial Intelligence (AI). All algorithms related to AI are available in Python.

While we talk about artificial intelligence we need to talk about human sensory organs like eye, ear, smell, taste, touch etc. Assuming there is a sensor to sense any of these signals, Python has a library for processing it. OpenCV for example – one of the leading computer vision libraries originally written in C++ has a port on Java and Python. For speech recognition or NLP (Natural Language Processing) you have many libraries in Python. For other AI functions like OCR (Optical Character Recognition), ANN (Artificial Neural Network) you have a Python package. Almost any algorithm you need for AI will be available either in Python or R.

Scratch, Python, HTML5, Java, C/C++ are the preferred languages for IoT (Internet of Thongs), CPS (Cyber Physical System), and IoE (Internet of Everything) driven by Raspberry Pi controller. Today's smartphones have many sensors. For example iPhone, Android operating systems on smartphones have Camera, Speaker, Motion Sensors, Position Sensors, and Environment Sensors. You can connect external sensors and wearables in these smartphones or your computers through the USB (Universal Serial Bus) port. Now you have sensors and the AI libraries with Next Generation Web. You just need to combine them and create innovative solutions.

As the price of smartphones are falling, and advanced wireless cellular networks like 4G and 5G with higher data bandwidths are becoming reality, the footprint of smartphones is increasing. Unlike a computer, smartphone is always powered on. Many people find it comfortable using the virtual keyboard or a soft-keyboard on a mobile smartphone compared to a keyboard in a computer. A computer will remain the platform of choice for development users and business users; but smartphones are likely to be the platform for general masses. Almost all web applications have a version for the smartphone or the mobile space. These applications in the mobile space are no longer called applications; they are called Apps. Google Android and iPhone iOS are the dominant players in the mobile space. They even have App stores where these mobile apps are generally hosted. There are many platforms and languages available in this space. However, Google Flutter is a very promising platform. Flutter is an open-source cross-platform graphical software development environment created by Google. Flutter is used to develop applications for

Android, iOS, Linux, Mac, Windows, Google Fuchsia, and the web from a single code-base written in Dart. Like Java – you write the code in Dart once and execute it in all environments. Dart even can compile to either native code or JavaScript.

In the Next Generation Web, CPS or IoE systems will be connected directly to a Web browser and send the sensor data real-time through WebRTC or Web Speech API, or WebUSB for server side processing. Eventually there will not be any Raspberry Pi controller at the sensor end. The sensor will collect the data and send it for real-time server side processing. NGW being real-time peer-to-peer with unlimited power, AI systems running at the server end can do many smart actions that were unthinkable earlier.

7 Next Generation Web Services

Next Generation Web is real-time peer-to-peer without any intermediate server or plugin, or downloads. This makes WebRTC end-to-end secure channel of communication without any third party software or systems in between. Any service needing high level of security can consider NGW as the carrier of technology and solutions. We already discussed ancillary technologies like CPS and AI. Here we will present two use-cases of NGW.

7.1 Use-Case I (HIPAA Compliant Remote Care)

In this use-case we discuss a HIPAA (Health Insurance Portability and Accountability Act) [18] compliant Contactless remote care. Healthcare organizations in the US require HIPAA compliance for patient privacy. A case in Oklahoma [19] about a doctor whose patient died has gained a lot of attention – the focus is on the fact that this doctor prescribed some restricted drugs and the entire doctor-patient interaction happened online, via Skype.

When somebody claims their system is "HIPAA Compliant", that usually means they have implemented their healthcare system to match the regulations of HIPAA as best as possible. The key technical points about HIPAA are that you must secure access to your system with strong authentication methods, and best practices to prevent hacking and data breaches. All patient data must be protected and conform to "Protected Health Information," or PHI. This means that information that can be used to identify individual patients should be encrypted when stored in your database, and encrypted when in-transit. The use of HTTPS URL's with SSL encryption is a must.

Figure 5 shows a Contactless ICU Care using WebRTC which is HIPAA compliant. In this picture there is a doctor and two patients (pictures are published with consent of individuals). In this figure a doctor is consulting with two patients from two different hospitals. The screen is split into two segments – each segment using separate peer-to-peer connection to these hospitals.

The dashboard in Fig. 5 shows Medical Chart with synthetic EHR (Electronic Health Records) data that shows the summary of all visits of the patient on a temporal scale with date, diagnosis, and prescribed medications. The Medical chart in Fig. 5 also includes at the top all comorbid diseases of the patient in machine understandable ICD10 codes.

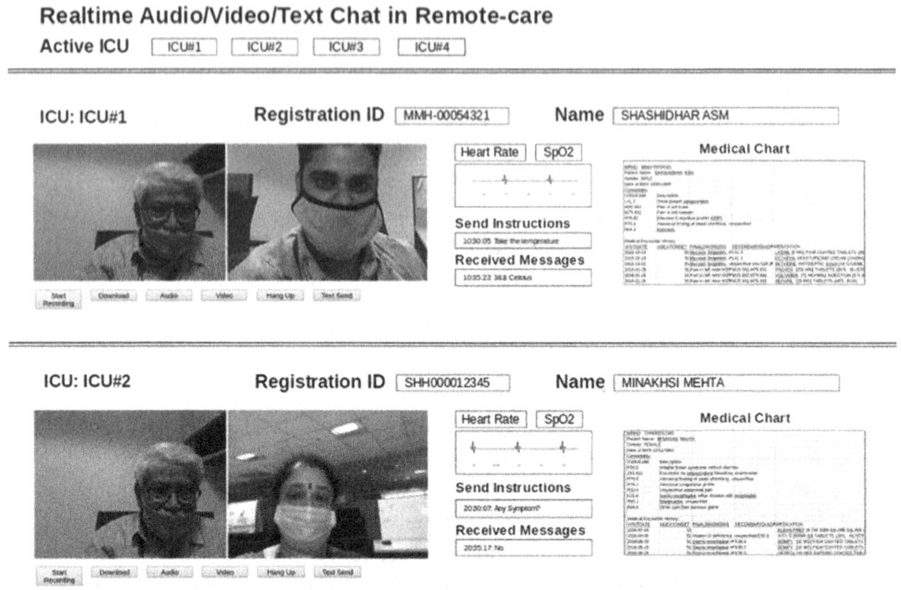

Realtime Audio/Video/Text Chat in Remote-care

Active ICU [ICU#1] [ICU#2] [ICU#3] [ICU#4]

Fig. 5. HIPAA compliant Contactless ICU Care. Here there are two patients and a doctor (picture are published with consent) with synthetic EHR data.

The audiovisual media channels between the ICU and the patient data is end-to-end encrypted through HTTPS and PKI (Public Key Infrastructure). Moreover, HTTPS on Chrome prevents CORS (Cross-Origin Resource Sharing). Because WebRTC is peer-to-peer end-to-end secured without any plugin, download, or intermediate server, there is no possibility of a man-in-the-middle or a hacker listening to the channel. Moreover, because HTTPS does not support CORS, the patient data in the Medical Chart is secured while on transit, though the doctor is consulting remote.

This is a follow-up session for the doctor to talk to the patients just to ensure that patients are fine. In this setup the doctor has one microphone which is used to talk to multiple patients. Unlike a conference call where one party can talk to all other parties or one can listen to everybody else, in a health care teleconsultation, each channel must be private. One patient should not be able to listen to the conversation the doctor is having with other patients. The voice channels between doctor and the ICU are point-to-point and independent of each other. Conversation with each patient is secured. Therefore, whenever the doctor selects one ICU, other voice channels to other ICUs are automatically muted. When the doctor selects a particular ICU the video channel of other ICUs are not muted – but the frame-rate is reduced.

Here along with the private audiovisual channels there is a full-duplex data channels for data chats between the doctor and nurse in-charge of the ICU. These data channels "Send Instructions" and "Received Messages" fields are also encrypted end-to-end. The heart rate is displayed on the display area that can show other real-time biomedical signals like ECG, SpO2 etc. These devices are connected through the USB (Universal Serial Bus). These streaming data through USB is directly transferred to the remote doctor encrypted end-to-end. The entire session of audiovisual interaction between the doctor and patient is recorded using the recording facility of WebRTC. This means, all interactions between the doctor and the patient are recorded at the doctor's end.

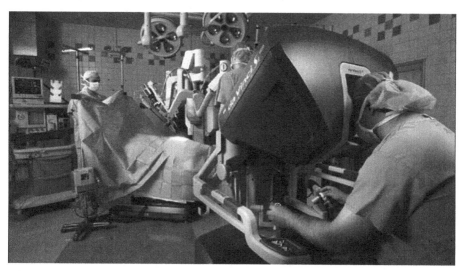

Fig. 6. Cobots in precision surgery where a human surgeon is controlling and supervising a robotic surgeon. Picture source: Robotic surgery has arrived [20].

7.2   Use-Case II (WebUSB in Cobotics)

In this use-case we discuss cobotics. Cobotics is collaborative robotics where humans are in charge and control remote machineries or robots. An autonomous robot performs tasks with a high degree of independence. WebRTC technology will transform IoT (Internet of Things) and PCS (Physical Cyber Systems) use-cases starting from building automation to cobotics. The real-time P2P feature of reliable data communication of WebRTC over the Web will ensures that any IoT device can be directly connected into the Web. These IoT devices will no longer need to be a standalone instrument driven by Raspberry Pi. WebRTC allows ultra-low latency connection for driving a robot. WebRTC offers end-to-end security; therefore, WebRTC can also be used as a VPN (Virtual Private Network).

Autonomous robots are expensive and have various constrains. An autonomous robot needs sufficient battery power, autonomous electromechanical flexibilities. In addition, robots must have sophisticated AI (Artificial Intelligence) for decision making along with synchronization of multiple sensors with AI functions. However, cobots are less expensive and easier to implement where cobots may function like robot coworkers or digital workers. An example could Flippy who flips and grills burgers in a California restaurant [21]. Figure 6 shows an example of cobotics where robotic surgeon is working as a coworker in an operating room along with nurses and human surgeons controlling the robots.

You can have multiple sensors and IoT devices with limited AI capabilities connected to the Web through WebRTC. An application in the Web managed by a person can control all these robots and implement a highly sophisticated cobot system. Some of the decision making can be shared between the human master controlling the robots and some decision making are delegated to some AI applications at the server end.

8   Conclusion

The United States Department of Defense responded to USSR's artificial satellite Sputnik's launch by creating ARPA in 1958. ARPA research played a central role in launching the ARPANET that laid the foundation in Internet that we use today.

In the Web's first generation, Tim Berners-Lee invented HTML, HTTP, and the URL standards. In the second generation, Netscape Communication created the world's first commercial graphic Web browser. On 9th August 1995 the Internet (Dot-Com) boom started with Netscape going public. The Clinton Administration's Telecommunications Act of 1996 made Internet free and helped private investors and innovation to drive the Internet.

While Web was an infant, Python was released in 1991. TensorFlow was developed by the Google Brain team for internal Google use. TensorFlow was made open source and released under the Apache License 2.0 on November 9, 2015. In no time TensorFlow was adopted by Python community to make it the de-facto standard and vehicle of AI.

Innovations between 2008 and 2018 transformed Web and Internet. Due to the active participation of many universities, standard bodies, governments, and business enterprises towards the cause of Open source, Next Generation Web (NGW) emerged. Next Generation Web is not simply a technology – it can be defined as an ecosystem of Web technologies. Along with Web technologies there were many other ancillary innovations in the Open domain software space that accelerated this journey. Some of the worth noting technologies are HTML5, JavaScript, HTTPS, WebSocket, Node.js, WebUSB, Web Speech API, WebRTC, Python3, Flutter, Dart, TensorFlow, ANN (Artificial Neural network), GAN (Generative Adversarial Network), Knowledge Graph, GNN (Graph Neural Network), and many other AI libraries.

During the COVID-19 pandemic and the lockdown of 2020, almost everything went online over the Web starting from online classes to online ordering of groceries. The technology roadmap of post-COVID-19 world will be the intersection of Artificial Intelligence, Cyber Physical Systems, Mobile Smartphone Apps, and the Next Generation Web. This will transform all industry verticals and touch everybody's life.

References

1. Van Sluyters, R.C.: Introduction to the Internet and World Wide Web. https://academic.oup.com/ilarjournal/article/38/4/162/656416
2. ARPANET. https://www.darpa.mil/about-us/timeline/arpanet
3. A short history of the Web. https://home.cern/science/computing/birth-web/short-history-web
4. Metcalfe, B.: Roads and Crossroads of Internet History Chapter 4: Birth of the Web. InfoWorld, vol. 17, no. 34, 21 Aug 1995

5. Thanks To Bill Clinton, We Don't Regulate The Internet Like A Public Utility. https://www. forbes.com/sites/realspin/2014/03/17/thanks-to-bill-clinton-we-dont-regulate-the-internet-like-a-public-utility/#564eee7c62e5

6. Netscape Navigator. https://en.wikipedia.org/wiki/Netscape_Navigator

7. History of Google. https://en.wikipedia.org/wiki/History_of_Google

8. A history of HTML. https://www.w3.org/People/Raggett/book4/ch02.html

9. roff. https://en.wikipedia.org/wiki/Roff_(software)

10. The Roots of SGML -- A Personal Recollection. http://www.sgmlsource.com/history/roots. htm

11. Semantic Web. https://www.w3.org/standards/semanticweb/

12. Introduction to the World Wide Web. https://www.w3.org/People/Raggett/book4/ch01.html

13. History of the web browser. https://en.wikipedia.org/wiki/History_of_the_web_browser

14. Message queue. https://en.wikipedia.org/wiki/Message_queue

15. JavaScript. https://en.wikipedia.org/wiki/JavaScript

16. WebRTC 1.0: Real-time Communication Between Browsers. W3C Editor's Draft 24 September 2020. https://w3c.github.io/webrtc-pc/

17. Talukder, A.K., Ahmed, H., Yavagal, R.: Mobile Computing Technology, Applications, and Service Creation. McGrawHill, New York (2010)

18. Summary of the HIPAA Security Rule. https://www.hhs.gov/hipaa/for-professionals/security/laws-regulations/index.html

19. Oklahoma Doctor Disciplined For Using Skype To Treat Patients?. https://telehealth.org/blog/oklahoma-doctor-disciplined-for-using-skype-to-treat-patients/

20. Robotic surgery has arrived. https://www.ucanaberdeen.com/robotic-surgery-has-arrived/

21. White Castle is testing a burger-grilling robot named Flippy. https://edition.cnn.com/2020/07/16/business/white-castle-flippy-robot/index.html

Adversarial Collusion on the Web: State-of-the-Art and Future Directions

Hridoy Sankar Dutta$^{(\boxtimes)}$ and Tanmoy Chakraborty

IIIT -Delhi, New Delhi, India
{hridoyd,tanmoy}@iiitd.ac.in

Abstract. The growth and popularity of online media has made it the most important platform for collaboration and communication among its users. Given its tremendous growth, social reputation of an entity in online media plays an important role. This has led to users choosing artificial ways to gain social reputation by means of blackmarket services as the natural way to boost social reputation is time-consuming. We refer to such artificial ways of boosting social reputation as *collusion*. In this tutorial, we will comprehensively review recent developments in analyzing and detecting collusive entities on online media. First, we give an overview of the problem and motivate the need to detect these entities. Second, we survey the state-of-the-art models that range from designing feature-based methods to more complex models, such as using deep learning architectures and advanced graph concepts. Third, we detail the annotation guidelines, provide a description of tools/applications and explain the publicly available datasets. The tutorial concludes with a discussion of future trends.

Keywords: Collusion · Blackmarkets · Online social networks · Social growth

1 Background

In recent years, we have seen an unprecedented popularity of online media, attracting a great number of people who want to share their thoughts and opinions. Gaining fame and reputation in online media platforms has become an important metric for several purposes – launching large-scale campaigns, promoting stocks, manipulating users' influence, and conducting political astroturfing, etc. The natural way of gaining appraisals and therefore attaining reputation can be a slow process. This has led to the creation of blackmarkets that help the online media entities to gain artificial appraisals such as followers, retweets, views, comments, etc. Gaining appraisals using blackmarket services violates the Terms and Service of online media platforms [2,3]. However, Google searches with keywords such as 'buy retweets', 'get YouTube views', etc. hit hundreds of websites offering 'easy', 'fast' and 'guaranteed' ways to increase appraisals [8,17]. This provides us with a clear picture on the popularity and impact of these services in providing appraisals across multiple online media platforms.

© Springer Nature Switzerland AG 2020
L. Bellatreche et al. (Eds.): BDA 2020, LNCS 12581, pp. 230–235, 2020.
https://doi.org/10.1007/978-3-030-66665-1_15

The problem is surprisingly prevalent; it is estimated that roughly 15% of Twitter accounts are operated by bots/collusive accounts [1], and more than 30% of the consumer reviews on online review/rating platforms are non-genuine [18]. Collusive entity detection is considered a challenging task as collusive users express a mix of organic and inorganic activities, and there is no synchronicity across their behaviors [4,8,10]. Thus, these users cannot be detected effectively by existing bot or fake user detection algorithms.

While the set of research for fake/fraud/spam entity detection in online media has expanded rapidly, there has been relatively little work that has studied the problem of collusive entity detection. The seminal work of [17] divided the black-market services into types based on their mode of operation: (i) **Premium** and (ii) **Freemium**. Premium blackmarkets provide appraisals when the customer pays money. On the other hand, in freemium services, customers themselves join the service and participate in providing collusive appraisals to other users of the service to gain (virtual) credits. In literature, it is seen that most of the existing studies on collusive entity detection are limited to online social networks and rating/review platforms. However, in reality, it is observed that collusive happens across multiple online media platforms such as video streaming platforms, recruitment platforms, discussion platforms, music sharing platforms and development platforms [9]. This further necessitates the development of advanced techniques to detect collusive entities across multiple online media platforms. Figure 1 shows the example of a blackmarket service which provides collusive appraisals to multiple online media platforms.

This tutorial presents a comprehensive overview of recent research and development on detecting the collusive entities using state-of-the-art social network analysis, data mining and machine/deep learning techniques. The topic is interdisciplinary, bridging scientific research and applied communities in data mining, human-computer interaction, and computational social science. This is a novel and fast-growing research area with significant applications and potential.

2   Objectives

The tutorial has three broad objectives:

1. To educate the community broadly about the problem of collusive activities on online media platforms.
2. To summarize advances made in the last few years in collusive entity detection with a description of annotation guidelines, publicly available collusive entity datasets, available tools and interfaces developed to detect collusive entities.
3. To encourage immediate adoption by researchers working in social network anomaly detection, as well as suggest future opportunities for short/long term investigations by researchers.

3   Organization and Outline

The tutorial will majorly cover the review progress in the area of collusive entity detection in different online media platforms. We will be presenting it in five

Fig. 1. Example of a blackmarket service which provides collusive appraisals to multiple online media platforms.

parts. Firstly, we start with an introduction to the problem with related definitions and concepts [9,10]. We briefly discuss the Terms and Service in online media platforms and the policy maintained by these platforms against fake and spam engagements. We also focus on how collusion is different from other relevant concepts such as fake, bot, sockpuppetry, malicious promotion, spam, content polluters, etc. We then show examples of how collusion happens across multiple online media platforms. We also highlight the challenges that we faced while designing models for collusive entity detection. We also shed some light on the limitations and restrictions of the APIs provided by the online media platforms and discuss the publicly available web scrapers and applications that helps to bypass the API rate limits. Secondly, we discuss the types of collusive activities: *individual collusion* [4–6,8,11,12,16] and *group collusion* [7,13,14,14,15,19]. We also highlight the differences between these types based on how they provide collusive appraisals. The third part is the core of our tutorial. Here, we address a number of models for collusive entity detection. Previous methods have employed different techniques ranging from feature-based methods, graph-based methods, topic modeling to complex structures using deep-learning based methods. We review the techniques that have been used for collusive entity detection and their different architectural variations. We also discuss the advantages and shortcomings of the models and their appropriateness to various types of problems. We then delve deeper into the characteristics of collusive entities by showing a variety of case studies. As case studies, we will show the effectiveness of the models on detecting collusive activities on Twitter and YouTube. In the fourth part, we focus on the annotation guidelines and description of publicly available datasets that are used for collusive entity detection. We first review the annotation guidelines for creating annotated datasets of collusive entities. Next, we provide pointers to existing studies with public links to the corresponding datasets, source codes and other related resources. A discussion of available tools and applications[1] for collusive entity detection will also be provided. In the final part, we spotlight some outstanding issues and discuss the future directions for collusive entity detection. These future research topics include: (i) collective collusion detection, (ii) understanding connectivity patterns in collusive network, (iii) event-specific studies, (iv) temporal modeling of collusive entities, (v) cross-lingual and multi-lingual studies, (vi) core collusive user detection, (vii) cross-platform spread of collusive content, (viii) multi-modal analysis of collusive entities, and (ix) how collusion promotes fake news. We believe our tutorial should interest the attending researchers and practitioners with interests in social network anomaly detection, user behavior modeling, graph mining, etc.

The outline of the tutorial is following:

1. **Introduction to collusion in online media.**
   1.1 Historical overview of online media platforms
   1.2 (Quick) Overview of collusion with definitions
   1.3 Example of collusive activities
   1.4 How collusion is different from other relevant concepts?
   1.5 Challenges in collusive entity detection
   1.6 How collusion happens across multiple online media platforms (social networks, rating/review platforms, video streaming platforms, recruitment platforms, discussion platforms, music sharing platforms, development platforms, other platforms)?
2. **Types of collusion**
   2.1 Individual collusion
   2.2 Group collusion
   2.3 How individual collusion differs from group collusion in providing collusive appraisals?
3. **Methods for collusive entity detection**
   3.1 Feature-based methods
   3.2 Graph-based methods
   3.3 Deep learning based methods
   3.4 Case studies
4. **Miscellaneous**
   4.1 How to annotate collusive entities?
   4.2 Description of publicly available datasets
   4.3 Tools, application and interfaces developed to detect collusion in online media
5. **Conclusion**
   5.1 Summary
   5.2 Open problems in collusive entity detection
   5.3 Future directions and discussion

---

[1] https://www.twitteraudit.com, https://followerwonk.com/analyze, https://botometer.iuni.iu.edu, https://www.modash.io/.

4 Target Audience

The intended audience for the tutorial are researchers of all levels seeking to understand the challenges, tasks and recent developments in collusive activities in online media. We will not assume any prerequisite knowledge and cover all the necessary concepts and definitions to ensure that the presentation is understandable to all tutorial attendees. Note that we will also cover some advanced topic materials as well.

We will also provide a detailed analysis of the existing works, a description of the publicly available datasets, tools/applications for collusive entity detection and an outline of future opportunities.

5 Presenters

Hridoy Sankar Dutta is currently pursuing his Ph.D. in Computer Science and Engineering from IIIT-Delhi, India. His current research interests include data-driven cybersecurity, social network analysis, natural language processing, and applied machine learning. He received his B.Tech degree in Computer Science and Engineering from Institute of Science and Technology, Gauhati University, India in 2013. From 2014 to 2015, he worked as an Assistant Project Engineer at the Indian Institute of Technology, Guwahati (IIT-G), India, for the project 'Development of Text to Speech System in Assamese and Manipuri Languages'. He completed his M.Tech in Computer Science and Engineering from NIT Durgapur, India in 2015. More details can be found at https://hridaydutta123.github.io/.

Tanmoy Chakraborty is an Assistant Professor and a Ramanujan Fellow at the Dept. of Computer Science and Engineering, IIIT-Delhi, India. He completed his Ph.D as a Google India PhD fellow at IIT Kharagpur, India in 2015. His primary research interests include Social Computing and Natural Language Processing. He has received several awards including Google Indian Faculty Award, Early Career Research Award, DAAD Faculty award. He leads a research group, LCS2 (http://lcs2.iiitd.edu.in), which primarily focuses on data-driven solutions for cyber-informatics. He is involved in mentoring several technology startups. For more details, please visit: http://faculty.iiitd.ac.in/~tanmoy/.

References

1. Politics and fake social media followers - lawsuit.org. https://lawsuit.org/politics-and-fake-social-media-followers/. Accessed 11 Aug 2020
2. Twitter: Platform manipulation and spam policy. https://help.twitter.com/en/rules-and-policies/platform-manipulation. Accessed 11 Aug 2020
3. Youtube: Fake engagement policy. https://tinyurl.com/yyvp68xh. Accessed 11 Aug 2020
4. Arora, U., Dutta, H.S., Joshi, B., Chetan, A., Chakraborty, T.: Analyzing and detecting collusive users involved in blackmarket retweeting activities. ACM Trans. Intell. Syst. Technol. **11**(3), 1–24 (2020)

5. Arora, U., Paka, W.S., Chakraborty, T.: Multitask learning for blackmarket Tweet detection. In: Proceedings of the 2019 IEEE/ACM International Conference on Advances in Social Networks Analysis and Mining, pp. 127–130 (2019)
6. Castellini, J., Poggioni, V., Sorbi, G.: Fake Twitter followers detection by denoising autoencoder. In: Proceedings of the International Conference on Web Intelligence, pp. 195–202 (2017)
7. Dhawan, S., Gangireddy, S.C.R., Kumar, S., Chakraborty, T.: Spotting collective behaviour of online fraud groups in customer reviews (2019)
8. Dutta, H.S., Chakraborty, T.: Blackmarket-driven collusion among retweeters–analysis, detection and characterization. IEEE Trans. Inf. Forensics Secur. **15**, 1935–1944 (2019)
9. Dutta, H.S., Chakraborty, T.: Blackmarket-driven collusion on online media: a survey. arXiv preprint arXiv:2008.13102 (2020)
10. Dutta, H.S., Chetan, A., Joshi, B., Chakraborty, T.: Retweet us, we will retweet you: spotting collusive retweeters involved in blackmarket services. In: ASONAM, pp. 242–249 (2018)
11. Dutta, H.S., Dutta, V.R., Adhikary, A., Chakraborty, T.: HawkesEye: detectingfake retweeters using Hawkes process and topic modeling. IEEE Trans. Inf. Forensics Secur. **15**, 2667–2678 (2020)
12. Dutta, H.S., Jobanputra, M., Negi, H., Chakraborty, T.: Detecting and analyzing collusive entities on YouTube. arXiv preprint arXiv:2005.06243 (2020)
13. Gupta, S., Kumaraguru, P., Chakraborty, T.: MalReG: detecting and analyzing malicious retweeter groups. In: CODS-COMAD, pp. 61–69. ACM (2019)
14. Kumar, S., Cheng, J., Leskovec, J., Subrahmanian, V.: An army of me: sockpuppets in online discussion communities. In: Proceedings of the 26th International Conference on World Wide Web, pp. 857–866 (2017)
15. Liu, S., Hooi, B., Faloutsos, C.: HoloScope: topology-and-spike aware fraud detection. In: Proceedings of the 2017 ACM on Conference on Information and Knowledge Management, pp. 1539–1548 (2017)
16. Shah, N.: FLOCK: combating astroturfing on livestreaming platforms. In: WWW, pp. 1083–1091 (2017)
17. Shah, N., Lamba, H., Beutel, A., Faloutsos, C.: OEC: open-ended classification for future-proof link-fraud detection. CoRR abs/1704.01420 (2017). http://arxiv.org/abs/1704.01420
18. Streitfeld, D.: The Best Book Reviews Money Can Buy, Tulsa, Okla (2012). http://www.todroberts.com/USF/BookReviews_for_Sale.pdf
19. Wang, Z., Gu, S., Zhao, X., Xu, X.: Graph-based review spammer group detection. In: KAIS, pp. 1–27 (2018)

Comparing Performance of Classifiers Applied to Disaster Detection in Twitter Tweets – Preliminary Considerations

Maryan Plakhtiy[1], Maria Ganzha[1] (ID), and Marcin Paprzycki[2](✉) (ID)

[1] Warsaw University of Technology, Warsaw, Poland
M.Ganzha@mini.pw.edu.pl
[2] Systems Research Institute Polish Academy of Sciences, Warsaw, Poland
marcin.paprzycki@ibspan.waw.pl

Abstract. Nowadays, disaster "detection", based on Twitter tweets, has become an interesting research challenge. As such it has even found its way to a Kaggle competition. In this work, we explore (and compare) multiple classifiers, applied to the data set from that challenge. Moreover, we explore usefulness of different preprocessing approaches. We experimentally establish the most successful pairs, consisting of a preprocessor and a classifier. We also report on initial steps undertaken towards combining results from multiple classifiers into a meta-level one.

Keywords: Tweet analysis · Text preprocessing · Classifiers · Performance comparison · Meta-classifiers

1 Introduction

Currently, social media like Facebook, Twitter, Instagram, and others, are one of the most followed sources of news. Almost all news agencies, organizations, political parties, etc., post their information/news/advertisements there, as they are the most visited web sites. However, each of these news outlets has its own purpose and/or target audience. For example, Facebook is known to be a political discussion arena, Instagram is used by bloggers and businesses, to advertise their products. Finally, Twitter, because of its short messages, has become one of the fastest, and widely used, news/events "spreaders". Thus it became one of the most important sources of communication in the case of an emergency. Particularly, this is the case, when almost every person has a smartphone, which lets them "immediately" announce an observed emergency. As a result, a growing number of news agencies, emergency organizations, local government branches, and others, became interested in monitoring Twitter for emergency announcements. This is to help them to act faster and, possibly, save lives.

In this context, a very interesting question arises: how to distinguish an actual emergency-announcing tweet, from a "fake news tweet", clickbait, or a non-related tweet. While relatively easy for humans, making this distinction

© Springer Nature Switzerland AG 2020
L. Bellatreche et al. (Eds.): BDA 2020, LNCS 12581, pp. 236–254, 2020.
https://doi.org/10.1007/978-3-030-66665-1_16

is quite challenging for computers. Therefore, event detection, on the basis of tweets' content, is a popular research area. While the most popular seems to be the sentiment analysis, ongoing research involves also fake news detection, text classification, or disaster detection. To stimulate work, related to the latter one, a competition was created on the Kaggle site [10]. Unfortunately, before the submission deadline, a leakage of correct predictions occurred, and the competition had to be canceled. Nevertheless, this work is based on the data set originating from this competition, and its aim is to apply (and compare performance of) well-known classifiers, to analyse content of tweets, to detect disasters. Moreover, interactions between data preprocessing that can be applied to the tweets and the classifiers is investigated. Interestingly, there is not much work that would delve into this topic. Finally, preliminary results, related to combining "suggestions" from multiple classifiers, and creating a meta-classifier, are presented.

In this context, we proceed as follows. In Sect. 2 we summarize most pertinent related work. Next, in Sect. 3, we describe the Kaggle data set and present summary of basic statistical analysis that was applied to it. We follow, in Sect. 4, with the description of data preprocessing that have been experimented with. Moreover, in Sect. 5, the experimental setup is outlined. Material presented up to this moment provides the foundation and allows to report, in Sect. 6, the experimental results obtained when different classifiers have been paired with different preprocessors and applied to the prepared data. Finally, in Sect. 7, initial results concerning combining multiple classifiers are introduced.

2   Related Work

Let us now summarize the state-of-the-art of classifiers that can be (and have been) used in context of disaster detection (or other forms of tweet content analysis). We focus our attention on most recent, successful approaches that may be worthy further examination.

In [13], authors summarized effective techniques for data analysis in the context of Twitter data. They also suggested that Logistic Regression [25] is best suited for the disaster management applications. Obtained results supported this claim. In a somewhat similar research, reported in [2], authors analysed damages caused by the disaster (e.g. how many people were injured, homes destroyed, etc.). They have created a Tweedr (Twitter for Disaster Response) application, which consisted of: classification, clustering, and extraction. Among used classifiers, the Logistic Regression was proven to be the most effective, with the $\sim 88\%$ accuracy for the disaster detection. However, the data set used for tests, was really small and skewed (by over abundance of "positive examples").

In [24], authors leveraged a Hybrid CNN-LSTM classifier architecture, for the Fire Burst Detection. Unfortunately, they did not provide details of achieved accuracy, only claim that this approach was successful. Hybrid CNN-LSTM was chosen, based on the [1], where LSTM, LSTM-Dropout, and Hybrid CNN-LSTM, were used for the Fake News Identification on Twitter. Here, 82.29% accuracy was achieved with the LSTM, 73.78% with the LSTM-Dropout, and 80.38%

with the Hybryd CNN-LSTM (on the PHEME data set [28]), which consisted of ∼ 5, 800 tweets. These classifiers are promising, since they do not need the context, and the considered data set involves different types of disasters.

In [26], authors proposed a Hierarchical Attention Networks (HAN), for the document classification, which outperformed other considered classifiers. They compared several popular methods, testing them on six classification tasks, and stated that HAN was the best in each case. Note that authors worked with complete documents, whereas in Twitter, an entry is limited to 140 characters. Another paper supporting use of the attention mechanism is [27], where authors address the hyperpartisan (highly polarized and extremely biased) news detection problem. Authors compared different classifiers on a data set consisting of million articles, and claimed that the Bidirectioanal LSTM model, with self attention (Bi-LSTM-ATTN), performed best. Similarly to [26], they worked with large documents. However, they reduced them to 392 tokens (during preprocessing), removing stopwords, and selecting a 40000 token vocabulary. In this setup (using Bi-LSTM-ATTN), they obtained 93,68% accuracy for the hyperpartisan news detection, while other models reached accuracy of approximately 89.77–91.74%.

One of interesting approaches, to the problem at hand, is the Bidirectional Encoder Representations from Transformers (BERT; [5,8]), which obtained state-of-the-art results in a wide array of Natural Language Processing (NLP) tasks. In [14], BERT was used for classification of tweets related to disaster management. There, performance of BERT-default, and its modifications, was compared to the baseline classifier, when applied to the combination of CrisisLex [3] and CrisisNLP [4] data sets, consisting jointly of 75800 labeled tweets. Overall, all BERT classifier outperformed the baseline one, and BERT-default had the best recall score (∼ 64.0% accuracy).

Separately, work reported in [7,23], where authors combine suggestions from multiple sources to deliver improved results, may be worthy considerations. In this context note that different classifiers independently determine, if a given tweet is (or is not) reporting an actual disaster. Hence the question: can suggestions from multiple classifiers be combined (e.g. using a simple neural network), resulting in a better accuracy of classification?

In the mentioned researches, data preprocessing is applied in each case, but is not comprehensively considered. Therefore, besides testing known classifiers, we have decided to study effects of preprocessing on the performance of the preprocessor+classifier pairs. This is one of important contributions of this paper. The list of preprocessing methods that have been tried, can be found in Sect. 4.

Overall, based on considered literature, it was decided to test six "standard classifiers": Logistic Regression, Ridge, SVC, SGD, Decision Tree and Random Forest, and eleven NN-based classifiers: LSTM, Bi-LSTM, LSTM-Dropout, CNN-LSTM, Fast Text, RNN, RCNN, CNN, GRU, HAN, and BERT. Classifiers were tested using Count [19] and Tfidf [22] vectorizers. Ten NN-based classifiers (excluding BERT) were tested using Keras Embeddings [11] and, independently, Global Vectors for Word Representation (GloVe [16]) Embeddings. To combine results from multiple classifiers, a standard backpropagation neural network was used.

3   Data Set and Its Preliminary Analysis

Let us now describe the data set used in our work. As noted, it originates from the Kaggle competition: *Real or Not? NLP with Disaster Tweets. Predict which Tweets are about real disasters and which ones are not* [10]. The complete data set contains 10,876 tweets that were *hand classified* into: (1) *True*, i.e. notifying about actual disasters, and (2) *False*. This data set was (within Kaggle) divided into two files: *train.csv*, and *test.csv*. The *train.csv* contains 7613 rows, and 5 columns: *id, keyword, location, text, target*. The first 10 rows from the *train.csv* file are presented in Table 1.

Table 1. First rows from the *train.csv* file.

id	Keyword	Location	Text	Target
1	NaN	NaN	Our Deeds are the Reason of this #earthquake M ...	1
4	NaN	NaN	Forest fire near La Ronge Sask. Canada	1
5	NaN	NaN	All residents asked to 'shelter in place' are ...	1
6	NaN	NaN	13,000 people receive #wildfires evacuation or ...	1
7	NaN	NaN	Just got sent this photo from Ruby #Alaska as ...	1
8	NaN	NaN	#RockyFire Update => California Hwy. 20 closed ...	1
10	NaN	NaN	#flood #disaster Heavy rain causes flash flood ...	1
13	NaN	NaN	I'm on top of the hill and I can see a fire in ...	1
14	NaN	NaN	There's an emergency evacuation happening now ...	1
15	NaN	NaN	I'm afraid that the tornado is coming to our a ...	1

Here, the individual columns have the following meaning:

- *id* - does not contain any relevant information, so it will not be used;
- *keyword* - contains keywords, or *NaN* if the tweet does not include a keyword;
- *location* - contains location, or *NaN* if the tweet does not include location;
- *text* - contains text of the tweet,
- *target* - contains correct prediction for the tweet; *1* means that the tweet notifies about a disaster, and *0* if the tweet does not announce an emergency.

File *test.csv* contains *test data* that was to be used within the competition. This file contains 3263 rows and 4 columns: *id, keyword, location* and *text*, with the same meaning as above. Obviously, it does not include the *target* column.

Following the link to the Kaggle Leaderboard [9], it is easy to notice a number of submissions that have a perfect score. This is because of the data leakage, i.e. correct predictions for the "test tweets" appeared on the Internet. Hence, it can

be reasonably claimed that some participants used this data to submit their
"predictions". Moreover, it is **not possible** to establish, which results that are
not 100% accurate, have been obtained by training actual classifiers. This is
because it is easy to envision that some participants could have "produced"
results "close to perfect", but not "stupidly perfect". Obviously, due to the
leak, Kaggle cancelled this competition. Nevertheless, analyzing the Leaderboard
it can be stipulated that the first "realistic result" is 0.86607 submitted by
the "right_right_team" team. However, for obvious reasons, this claim is *only* a
"reasonable speculation".

Overall, in what follows, all classifiers were trained using data from the
train.csv file. Moreover, performance is reported in terms of the actual Kaggle score, calculated by applying the trained classifiers to the *test.csv* file. In
this way, keeping in mind all, above mentioned, limitations of this approach, a
baseline performance can be formulated.

3.1 Statistical Analysis of the Data Set

A Python program, inspired by [6], was developed to analyze key characteristics
of the data set. We report these aspects of data that are known (from the literature; see, Sect. 2) to, potentially, influence the performance of the classifiers. To
start on the most general level, in Table 2, class distributions for the *train data
set* and the *test data set* are presented. It is easily seen that there are more "non
disaster tweets" than "disaster tweets", in both data sets. However, their ratio
(number of non disaster tweets divided by number of disaster tweets) is almost
the same.

Table 2. Basic statistics of the *train* and *test* data sets

Distribution	*Train data set* (7613–100%)			*Test data set* (3263–100%)		
Type	Disaster	Non Disaster	Ratio	Disaster	Non Disaster	Ratio
Class	3243 (42,6%)	4370 (57,4%)	0,7421	1402 (43,0%)	1861 (57,0%)	0,7534
Keyword	3201 (42,0%)	4351 (57,1%)	0.7357	1386 (42,5%)	1851 (56,7%)	0,7488
Location	2177 (28,6%)	2903 (38,1%)	0,7499	936 (28,7%)	1222 (37,5%)	0,7660

Each data set, has a *keyword* column. In Table 2, it is shown how many
tweets of each class have a keyword data (not all tweets were provided with
keywords). It can be noticed that trends, for both data sets, match trends of
class distribution. It thus can be concluded that there is no visible correlation
between tweet keyword data and its class. Overall, data from the *keyword* column
will be used in the training process.

Also, in Table 2, it is visible that a large number of tweets do not have the
location data. However, the general trend is very similar to the class distribution
(ratio of non disaster tweets to disaster tweets). Hence, it can be concluded that

there is no immediate correlation between location data and tweets' class. This hypothesis was experimentally confirmed.

Data from Table 2 supports the conjecture that *train* and *test* data sets originated from the same data set (see, also [17]). This conjecture has been further experimentally verified and thus, in the remaining parts of this Section, we depict properties only for the *train data set*.

Let us start form the distribution of the number of characters in the disaster and non-disaster tweets, which is depicted in Fig. 1.

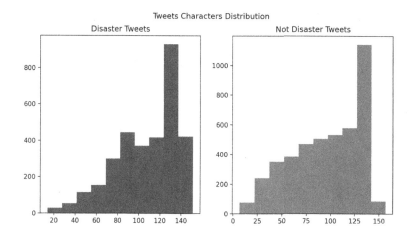

Fig. 1. Characters distribution for disaster and non-disaster tweets.

It can be seen that disaster tweets and the non-disaster tweets have quite similar character distributions. But, it is also fair to notice that, in general, non disaster tweets have more characters than the disaster tweets. This fact, however, has not been used explicitly in model training.

Next, let us compare distributions of numbers of words in disaster and non-disaster tweets (presented in Fig. 2).

Here, it can be easily seen that the number of words distributions have similar shapes. Moreover, note that the scale for both graphs is the same. It should be kept in mind that the fact that non disaster tweets have slightly "taller bars" is caused by the fact that there are more tweets in this class.

Thus far we have not found substantial statistical differences between disaster and non disaster tweets. Situation is quite similar when the word length distribution is considered (see, Fig. 3). While the scales of both figures are slightly different, the overall shape of the distribution is very much alike.

Let us now consider the top 10 stopwords found in disaster and non-disaster tweets. They have been captured in Fig. 4.

In both cases, *the* stopword is the most popular one. The fact that it seems to be more popular in non disaster tweets is "normal", because there are more tweets of this class. In the non disaster class, the next most popular stopwords

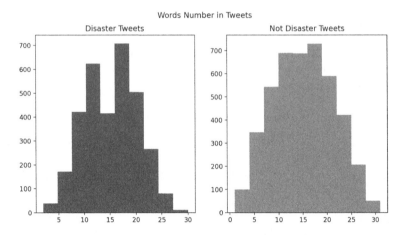

Fig. 2. Word number distribution for disaster and non-disaster tweets.

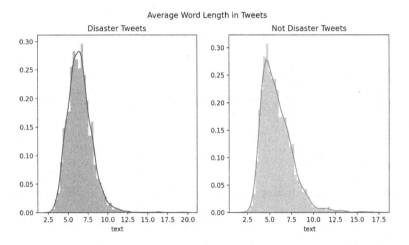

Fig. 3. Word length distribution for disaster and non-disaster tweets.

are: *i, you, my*. Interestingly, these stopwords are not present in the disaster tweets class. There, the most popular stopwords (other than '*the*') are: *in, of, on, for, at*. These are descriptive stopwords that can be connected with the description of the details of the disaster. Therefore, stopwords may be useful in recognizing the nature of a tweet (and should not be removed).

As the last statistical comparison, let us consider punctuation symbols in both tweet classes (as illustrated in Fig. 5).

It can be seen that both classes of tweets have the same order of top 3 punctuation signs ('-', '—', ':'). Hence, punctuation symbols do not differentiate between classes (and can be removed during the preprocessing phase of data preparation).

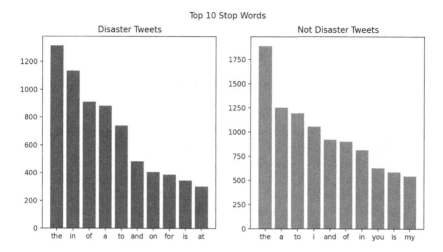

Fig. 4. Top 10 stopwords in disaster and non-disaster tweets.

Finally, after analysing occurrences of the top 10 hashtags in the data set, it was noticed that there is a fairly small number of hashtags embedded in tweets, as compared to the total number of tweets. For example, in the *train data set* the top hashtag *#news* occurred less than 60 times, while the top not disaster tweet hashtag *#nowplaying* occurred around 20 times (in total). Thus, it can be conjectures that hastags will not bring much value to the classifier training. Therefore, it was decided to explicitly tag tweets that include a hashtag, but the hashtag itself will not be considered as a "different word" (e.g. *#news* and *news* will be treated as the same word).

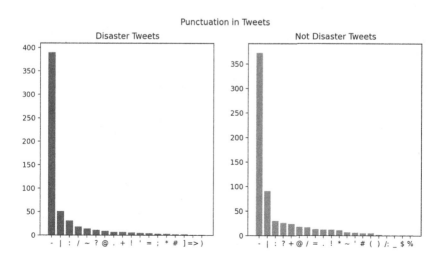

Fig. 5. Punctuation characters in distribution for disaster and non-disaster tweets.

4   Data Preprocessing

Data preprocessing is a standard initial step of all data analytics/machine learning procedures. While the NLTK [15] is a popular natural language processing library, it had to be slightly augmented to deal with the tweets. Specifically, there are 17 configurable parameters in the *TweetPreprocessing.preprocess* method, created for the purpose of preprocessing. Therefore, 2^{17} different preprocessors can be created. Hence, it would be extremely time consuming to check the performance of all possible preprocessors, especially since the majority of them will perform poorly. This latter statement has been (partially) experimentally verified. Thus, it was decided to select, and "freeze" a set of operations performed by all of them, and thus substantially reduce the total number of possible preprocessors. Operations that have been selected to be used in all preprocessors were: (1) make all words lower letter, (2) remove slang words, (3) remove links, (4) remove Twitter users, (5) remove hashtags, (6) remove numbers. These operations were selected experimentally (it was found that they "always" improved the performance of classifiers). The last frozen parameter was (7) to not to perform stemming. This was an unexpected discovery, as stemming is one of key preprocessing techniques in NLP. After extensive experimentation with the remaining 10 parameters, 10 preprocessors have been selected. They are summarized in Table 3 (names of parameters are self-explanatory).

Note that six operations add flags to the string (e.g. *Hash Flag* and *Link Flag* set to "T" mean that *"My #string has a link* https://mylink.com*"* will be transformed into *"My #string has a link* https://mylink.com *< hashtag > < link >"*). Here, it was found that the content of these items might not be important. For example, the link itself, usually has no useful information. However, the fact that the tweet contains a link, user, hashtag, numbers, or location, might be important. Note that the Global Vectors for word Representation (GloVe, [16]) has similar flags included. It has vector representations for *Link*, *User*, *Hash* and *Number* flags. While analysis reported in Sect. 3.1, suggested that *keyword* and *location* columns are not likely to be useful, these flags have been included for the testing purposes.

Preprocessors PA01, PA02, PA03, PA04 and PA05 were selected to test how the presence of *Link*, *User*, *Hash*, *Number*, *Keyword* and *Location* flags, influences performance of classifiers. Here, the remaining parameters are set to *T*.

Preprocessors PA06, PA07, PA10, and PA01, have similar configuration for the "flag presence parameters", but different configurations of the remaining parameters. The four parameters *Link Flag*, *User Flag*, *Hash Flag* and *Number Flag*) are set to *T* to match the vector representations in GloVe. These preprocessors test how *Split Word*, *Remove Punctuation*, *Remove Stopwords*, and *Remove Not Alpha* parameters influence the performance of classifiers.

Finally, preprocessors PA08 and PA09, and PA05, have the "flag presence parameters" turned to *F*. They test how *Split Word*, *Remove Punctuation*, *Remove Stopwords*, and *Remove Not Alpha* parameters influence the performance of classifiers.

Table 3. Selected preprocessing algorithms

ID	Link Flag	User Flag	Hash Flag	Number Flag	Keyword Flag	Location Flag	Remove Punct	Remove Stpwrds	Remove NotAlpha	Split Words
PA01	T	T	T	T	F	F	T	T	T	T
PA02	F	T	T	F	F	F	T	T	T	T
PA03	T	T	T	T	T	T	T	T	T	T
PA04	T	T	T	F	F	F	T	T	T	T
PA05	F	F	F	F	F	F	T	T	T	T
PA06	T	T	T	T	F	F	F	F	F	T
PA07	T	T	T	T	F	F	F	F	F	F
PA08	F	F	F	F	F	F	F	F	F	F
PA09	F	F	F	F	F	F	T	F	F	F
PA10	T	T	T	T	F	F	T	F	F	F

5   Experimental Setup

Based on literature review, summarized in Sect. 2, 17 classifiers were selected: Logistic Regression, Ridge, SVC, SGD, Decision Tree and Random Forest, from the scikit-learn [20]; LSTM, Bi-LSTM, LSTM-Dropout, CNN-LSTM, Fast Text, RNN, RCNN, CNN, GRU, HAN and BERT utilizing Keras [12].

Apart from pairing classifiers with preprocessors, classifiers were tested for different parameter configurations, within each pair. Six Sklearn based classifiers were tested with the *Count* ([19]) and *Tfidf* ([22]) vectorizers, with different *ngrammar* ranges ([1,1], [1,2], [1,3]). The 10 NN-based classifiers (excluding BERT) were tested with Keras and GloVe Embeddings, with different hyper parameters. Experiments with BERT were conducted separately because it is a pre-trained classifier, requiring an embedding mechanism different from other NN-based classifiers. It should be noted that additional experiments have been performed to further explore possibility of building a "perfect classifier", but they were found to be much worse that these reported here.

We proceeded with exactly the same data split as in the Kaggle competition. The original data set (10,876 tweets) was split into standard 70–30% and 10-fold cross-validation was used. Specifically, for the cross-validation we split the *train data* into 90/10% (6447/716 tweets). Here, experiments were executed using StratifiedKFold [21] with 10 splits, and the seed equal to 7.

6   Experimental Results

Let us now report and discuss obtained results. Note that the complete set of results, for all classifiers and preprocessors that have been experimented with, can be found at [18]. This is provided since the actual volume of generated data is such, that it cannot be properly depicted in the article itself.

6.1   Base Classifiers and Preprocessors

Let us start from considering combining preprocessing algorithms, with Ridge, Linear Regression, Logistic Regression, Random Forest, Decision Tree, SVC and SGD classifiers. They were tested with ten different preprocessing algorithms using *Tfidf* and *Count* vectorizers, with different grammar ranges ([1, 1], [1, 2], [1, 3]). Overall, 360 (6 classsffiers * 10 preprocessing algorithms * 2 Vectorizers * 3 ngrammars) tests cases have been completed, exploring all possible combinations of classifier and preprocessor pairs. In Table 4 best results obtained for each of the six classifiers are presented.

Table 4. Best results for classifiers not based on neural networks.

Classifier	Vectorizer [grammar range]	Algorithm ID	Score
Ridge	Tf-idf [1, 2]	PA08, PA09	0.80444
Logistic Regression	Count [1, 2]	PA04	0.80245
SVC	Count [1, 1]	PA03	0.80251
SGD	Tf-idf [1, 3]	PA08	0.80156
Random Forest	Tf-idf [1, 1]	PA02	0.78881
Decision Tree	Count [1, 2]	PA05	0.76332

As can be seen, the Ridge classifier performed the best. Moreover the same result was obtained in combination with two different preprocessors (PA8 and PA9). Reaching beyond results presented in Table 4, when considering all experimental results, no specific preprocessing algorithm has been found to be a "definite champion". Hence, a "relative strength" of preprocessors was calculated. For each preprocessor, for each classifier, the best three results were considered. If a given preprocessor was the best for a given classifier, it received 3 points, the second place scored 2 points, while the third place 1 point (see, Table 5).

Table 5. Preprocessing algorithms relative score

Algorithm	Positions Points	Score
PA09	3 + 1 + 1 + 3	8
PA04	3 + 2 + 2 + 1	8
PA02	1 + 2 + 3 + 2	8
PA08	3 + 3	6
PA03	1 + 3	4
PA01	2 + 1	3

Using the proposed scoring method (while keeping in mind its limitations), the best scores were achieved for **PA09**, **PA04** and **PA02** preprocessors. However, it is the **PA09** that can be considered a winner, as it earned two "first places". The second best was **PA04** (one first place, when combined with the Logistic Regression classifier). This preprocessor adds link, user, hash flags and removes punctuation and stopwords. The third best was **PA02**, which is very similar to **PA04**, except the **PA04** adds link flag but **PA02** does not.

To complete the picture, let us now present the 15 best preprocessor+classifier pairs (in Table 6).

From results presented thus far it can be concluded that the three "leaders among non-NN-based classifiers" were: Ridge, SVC, and Logistic Regression. They performed best, but their best results were obtained when they were paired with different preprocessing algorithms. Henceforth, there is no preprocessor that could be said that it performs best with every classifier. This may indicate that, in general, for different problems, different classifiers require different preprocessors to reach full potential.

6.2   Preprocessors and Neural Network Based Classifiers

Let us now report on the results of the tests that combined preprocessors and the ten different NN-based classifiers: LSTM, LSTM-Dropout, Bi-LSTM, LSTM-CNN, Fast Text, RCNN, CNN, RNN, GRU and HAN. All these classifiers were tested with Keras and GloVe embeddings, with different hyper-parameters. Note that results of BERT classifier are reported separately, because BERT requires an embedding mechanism different from other NN-based classifiers.

Table 6. Top 15 preprocessor+classifier pairs

Classifier	Algorithm ID	Score
Ridge	PA08	0.80444
Ridge	PA09	0.80444
SVC	PA03	0.80251
Logistic regression	PA04	0.80245
SVC	PA04	0.80233
Ridge	PA02	0.80211
Logistic regression	PA04	0.80211
Ridge	PA09	0.80196
Ridge	PA08	0.80196
SVC	PA10	0.80165
SVC	PA07	0.80162
SGD	PA08	0.80156
Ridge	PA02	0.80153
Logistic regression	PA02	0.80153
SVC	PA06	0.80147

Let us start with the NN-based classifiers utilizing Keras embeddings. In Table 7 presented are top pairs scores of classifier + preprocessor.

Table 7. Top scores for the neural network based classifiers and Keras embeddings

Model	Preprocessor	Score
LSTM	PA09	0.79684
LSTM_DROPOUT	PA01	0.79782
BI_LSTM	PA04	0.79746
LSTM_CNN	PA04	0.79384
FASTTEXT	**PA07**	**0.80374**
RCNN	PA10	0.79418
CNN	PA07	0.79994
RNN	PA01	0.79859
GRU	PA06	0.80055
HAN	PA06	0.80172

It can be easily noticed that the average scores in Table 7 are quite similar to each other. Moreover, as in the case of non-NN approaches, different classifiers performed best when paired with different preprocessors. As a matter of fact, not a single preprocessor occurs in this table more than two times. Again, there is no pair of preprocessor and NN-based classifier, which stands out significantly among others. However, the best average score was achieved with the PA07 preprocessor and the Fast Text NN-based classifier.

Let us now consider NN-based classifiers utilizing GloVe [16] embeddings. In Table 8 top pairs of classifiers and preprocessors are presented.

Table 8. Top average scores for the neural network based classifiers and GloVe embeddings

Model	Preprocessor	Score
LSTM	PA04	0.80996
LSTM_DROPOUT	PA02	0.81257
BI_LSTM	PA04	0.81039
LSTM_CNN	PA04	0.80610
FASTTEXT	PA04	0.81229
RCNN	PA09	0.80478
CNN	PA10	0.80622
RNN	PA02	0.80788
GRU	**PA04**	**0.81444**
HAN	PA06	0.80975

In general, it can be seen that average scores, in Table 8, are higher than these in Table 7. This suggests that using GloVe embeddings increased the performance of considered classifiers. Also, it can be seen that the PA04 preprocessor was at the top for 5 out of 10 models. Moreover, the GRU classifier, paired with the PA04 preprocessor, had the top average score. It is safe to state that the PA04 preprocessor stands out as the best pair for classifiers, which use the GloVe embeddings. One of possible reason, why this preprocessor was "the best", might be that it adds three flags to the tweets (user, link, and hashtag, which also have representation in GloVe). Moreover it removes "noise", such as stopwords, punctuation, and non-alphabetic words. In this way it it may be the best match with the GloVe embeddings used in this series of experiments. However, this is just a stipulation, which has not been further verified and as such, needs to be taken with a grain of salt.

6.3 Preprocessors and BERT Classifier

In this section, experiments with the BERT-Large model, from the TensorFlow Models repository, on GitHub [8], are discussed. Here, BERT uses $L = 24$ hidden layers of size of $H = 1024$, and $A = 16$ attention heads. BERT was tested with the ten preprocessors, and without the preprocessing phase. The average scores are presented in Table 9.

It can be easily noticed that the average scores obtained by BERT classifier are much higher than the average scores obtained with all other tested classifiers. The best average score is achieved when BERT was combined with the PA08 preprocessor. It should be noted that the best individual score, achieved for any experiment, was 84,064% accuracy. This result would have landed at the 79th place, out of 3402 submissions, in the Kaggle Leaderboard (if scores, which are better than 98%, were excluded, as potentially fraudulent). While, the average score is much better in capturing the overall strength of the approach, it is also obvious that the scores reported for the competition were the best ones actually

Table 9. BERT averages scores for different preprocessors

Model	Preprocessor	Score
BERT	None	0.82991
BERT	PA01	0.82109
BERT	PA02	0.82216
BERT	PA03	0.82182
BERT	PA04	0.82155
BERT	PA05	0.81649
BERT	PA06	0.83154
BERT	PA07	0.83252
BERT	**PA08**	**0.83469**
BERT	PA09	0.82832
BERT	PA10	0.82581

achieved in any single run. Therefore, it can be stated that the 79th place, claimed above, is somewhat realistic. This is also the best score we managed to obtain my using any single classifier (with its best sidekick preprocessor).

7   Combining Predictions Form Multiple Classifiers

Let us now consider the possibility of "combining suggestions" from multiple classifiers (paired with their best preprocessors). The goal, here, is to deliver results that would be more accurate than these from any single classifier+preprocessor pair. Specifically, the top six classifiers were taken from those listed in Table 4. Moreover, top ten NN-based classifiers, based on the Keras embeddings, were taken from those listed in Table 7, as well as top ten NN-based classifiers, based on GloVe embeddings, from Table 8. Finally, the best BERT pair from Table 9 was selected. As a result, 27 pairs of classifier+preprocessor were selected. For clarity, they are listed in Table 10.

Table 10. Top 27 pairs of classifier and preprocessor

Model/Classifier	Preprocessor
BERT	PA08
LSTM-GloVe	PA04
LSTM_DROPOUT-GloVe	PA02
BI_LSTM-GloVe	PA04
LSTM_CNN-GloVe	PA04
FASTTEXT-GloVe	PA04
RCNN-GloVe	PA09
CNN-GloVe	PA10
RNN-GloVe	PA02
GRU-GloVe	PA04
HAN-GloVe	PA06
LSTM	PA09
LSTM_DROPOUT	PA01
BI_LSTM	PA04
LSTM_CNN	PA04
FASTTEXT	PA07
RCNN	PA10
CNN	PA07
RNN	PA01
GRU	PA06
HAN	PA06
RIDGE	PA08
SVC	PA03
LOGISTIC_REGRESSION	PA04
SGD	PA08
DECISION_TREE	PA05
RANDOM_FOREST	PA02

Each classifier+preprocessor pair, from Table 10, was trained using 70% of the (original Kaggle) *train data set*, and validated against the remaining 30%. Note that, for each of the 27 individual classifiers, the same data split (exactly the same data used in both the training and the validation sets) was used, during the training phase.

Each of the trained classifiers produced its own predictions, for the tweets in the validations set, resulting in 27 independent lists of "suggestions". These 27 suggestion vectors were transformed into input for the neural network: $[v_1, v_2..., v_n]$, where v_i – is the list of 27 predictions per tweet; $i = 1 \ldots n$; n – is the number of tweets in the validation set (30% of the train data set). Here, a standard backpropagation neural network, with 27 input neurons, 14 neurons in the hidden layer and a single output neuron, was used. After training the neural network on this input (with the same split 70%/30%), the average results did only barely beat the score of the best standalone BERT+ preprocessor pair. Specifically, BERT classifier (combined with PA08 preprocessor), had the best recorded accuracy of 0.83604. This has to be compared with the best single result of the meta-classifier, which was better by 0.004. This improvement would not have boost the score in the Kaggle competition. The average score of the best BERT+preprocessor pair was 0,83469, while the improvement (of the average score) of the meta-classifier was 0.00071. Overall, it can be concluded that combining predictions from multiple classifiers has potential, but requires further investigation.

8   Concluding Remarks

Let us now summarize the main conclusions for the research discussed within the scope of this paper.

– Based on literature analysis, and comprehensive experimentation, ten preprocessors have been selected. All of them had (the same) 7 out of 17 possible tweet preprocessing parameters/operations "frozen". Frozen parameters have "always" improved the performance of classifiers.
   Preprocessors PA01, PA02, PA03, PA04 and PA05, tested how the presence of *Link, User, Hash, Number, Keyword* and *Location* flags, influences performance of classifiers. The remaining parameters were set to *T*.
   Preprocessors PA06, PA07, PA10, and PA01, have similar configuration for the "flag presence parameters", but different configurations of the remaining parameters. Parameters *Link Flag, User Flag, Hash Flag* and *Number Flag*) were set to *T*. among others, to match the vector representations in GloVe. These preprocessors test how *Split Word, Remove Punctuation, Remove Stopwords*, and *Remove Not Alpha* parameters influence the performance of classifiers.
   Finally, preprocessors PA08 and PA09, and PA05, have the "flag presence parameters" turned to *F*. They test how *Split Word, Remove Punctuation, Remove Stopwords*, and *Remove Not Alpha* parameters influence the performance of classifiers in this context.

- Six "base" classifiers, using different vectorizers and ngrammars, combined with selected preprocessors, were tested. It was discovered that the best average of 10-fold cross validation accuracy of 0.80444 was achieved by the Ridge classifier, utilizing Tfidf vectorizer with [1, 2] ngrammar range, independently with two preprocessors: PA08 and PA09 (Table 6). These two preprocessors are almost the same, except that PA08 does not remove punctuation from tweets, while PA09 does. Overall, the top three classifiers were Ridge, SVC and Logistic Regression. Moreover, the worst results were achieved by the Decission Tree and Random Forest classifiers (regardless of the applied preprocessing). Top three preprocessors were PA09, PA04 and PA02, as they obtained the same relative score (Table 5). However, it cannot be said that one of them is the best one overall, and that every classifier performs the best with it.
- Ten NN-based classifiers, using Keras embedddings, were combined with selected preprocessors and tested. It was found out that Fast Text NN-based classifier, combined with PA07 preprocessor, had the best average score of 0.80374 (Table 7). Average scores of all classifiers were "close to each other". However, they were achieved when combined with different preprocessors. Hence, again, no preprocessor was found to perform best with all NN-based classifiers.
- The same ten NN-based classifiers, combined with sepected preprocessors and using GloVe embeddings were tested. All NN-based classifiers with GloVe embeddings produced better scores than same classifiers with Keras embeddings. The top average score of 0.81444 was achieved by GRU NN-based classifier combined with PA04 preprocessor (Table 8). Again, all classifiers obtained similar results. However, in this case, the PA04 preprocessor produced 5 top averages results out of 10 NN-based classifiers, and was a clear "winner".
- BERT classifier had the best average score among all classifiers. The best average score 0.83469 was achieved with the PA08 preprocessor (Table 9). The best single score with BERT classifier was achieved with the same preprocessor, and was 0.84064. This score would result in reaching 79th place, out of 3402 submissions, in the Kaggle Leaderboard (excluding scores better than 98% that can be assumed to be compromised).
- Finally, the 27 pairs of best classifiers + preprocessors (from Table 10) were combined, using a simple neural network, into a meta-classifier. This approach did overcome best single BERT model accuracy by maximum of 0.004.

The single most interesting research direction, following from the above-reported results, could be to further investigate possibility of combining predictions of multiple classifiers, to achieve better overall accuracy.

References

1. Ajao, O., Bhowmik, D., Zargari, S.: Fake news identification on twitter with hybrid CNN and RNN models. In: Proceedings of the International Conference on Social Media and Society. SMSociety (2018). https://doi.org/10.1145/3217804.3217917

2. Ashktorab, Z., Brown, C., Nandi, M., Mellon, C.: Tweedr: mining twitter to inform disaster response. In: Proceedings of the 11th International ISCRAM Conference - University Park, May 2014. http://amulyayadav.com/spring19/pdf/asht.pdf

3. CrisisLex: Crisislex data set. https://github.com/sajao/CrisisLex/tree/master/data/CrisisLexT26

4. CrisisNLP: Crisisnlp data set. https://crisisnlp.qcri.org/

5. Devlin, J., Chang, M.W., Lee, K., Toutanova, K.: BERT: pre-training of deep bidirectional transformers for language understanding, May 2019. https://arxiv.org/pdf/1810.04805.pdf, arXiv: 1810.04805

6. Es, S.: Basic EDA, cleaning and glove. https://www.kaggle.com/shahules/basic-eda-cleaning-and-glove

7. Ganzha, M., Paprzycki, M., Stadnik, J.: Combining information from multiple search engines - preliminary comparison. Inf. Sci. **180**(10), 1908–1923 (2010). https://doi.org/10.1016/j.ins.2010.01.010

8. Google: BERT GitHub. https://github.com/google-research/bert

9. Kaggle: Kaggle competition leaderboard. https://www.kaggle.com/c/nlp-getting-started/leaderboard

10. Kaggle: Kaggle competition: real or not? NLP with disaster tweets. Predict which tweets are about real disasters and which ones are not. https://www.kaggle.com/c/nlp-getting-started/overview

11. Keras: Keras embedding layer. https://keras.io/api/layers/core_layers/embedding/

12. Keras: Keras library. https://keras.io/

13. Kursuncu, U., Gaur, M., Lokala, K.T.U., Sheth, A., Arpinar, I.B.: Predictive analysis on Twitter: techniques and applications. Kno.e.sis Center, Wright State University, Jun 2018. https://arxiv.org/pdf/1806.02377.pdf, arXiv: 1806.02377

14. Ma, G.: Tweets classification with BERT in the field of disaster management. Department of Civil Engineering, Stanford University (2019). https://web.stanford.edu/class/archive/cs/cs224n/cs224n.1194/reports/custom/15785631.pdf

15. NLTK: NLTK library. https://www.nltk.org/

16. Pennington, J., Socher, R., Manning, C.D.: GloVe: global vectors for word representation. https://nlp.stanford.edu/projects/glove/

17. Plakhtiy, M.: Applying machine learning to disaster detection in twitter tweets. https://drive.google.com/file/d/1k2BGDn3t76rQjQIMA2GaRIzSFLXwnEIf/view?usp=sharing

18. Plakhtiy, M.: Results on google drive. https://docs.google.com/spreadsheets/d/1eP0DdEMxzNLT6ecfdN5Kf5ctK1BSYNgoq1YHylSVDLc/edit?usp=sharing

19. SkLearn: count vectorizer. https://scikit-learn.org/stable/modules/generated/sklearn.feature_extraction.text.CountVectorizer.html

20. SkLearn: Scikit-learn library user guide. https://scikit-learn.org/stable/user_guide.html

21. SkLearn: StratifiedKFold. https://scikit-learn.org/stable/modules/generated/sklearn.model_selection.StratifiedKFold.html

22. SkLearn: TFIDF vectorizer. https://scikit-learn.org/stable/modules/generated/sklearn.feature_extraction.text.TfidfVectorizer.html

23. Stadnik, J., Ganzha, M., Paprzycki, M.: Are many heads better than one - on combining information from multiple internet sources. Intelligent Distributed Computing, Systems and Applications, pp. 177–186 (2008). http://www.ibspan.waw.pl/~paprzyck/mp/cvr/research/agent_papers/IDC_consensus_2008.pdf, https://doi.org/10.1007/978-3-540-85257-5_18

24. Thanos, K.G., Polydouri, A., Danelakis, A., Kyriazanos, D., Thomopoulos, S.C.: Combined deep learning and traditional NLP approaches for fire burst detection based on twitter posts. In: IntechOpen, April 2019. https://doi.org/10.5772/intechopen.85075
25. Wikipedia: Logistic regression. https://en.wikipedia.org/wiki/Logistic_regression
26. Yang, Z., Yang, D., Dyer, C., He, X., Smola, A., Hovy, E.: Hierarchical attention networks for document classification. Carnegie Mellon University, Microsoft Research, Redmond, June 2016. https://www.cs.cmu.edu/~./hovy/papers/16HLT-hierarchical-attention-networks.pdf
27. Zhang, C., Rajendran, A., Abdul-Mageed, M.: Hyperpartisan news detection with attention-based BI-LSTMS. Natural Language Processing Lab, The University of British Columbia (2019). https://www.aclweb.org/anthology/S19-2188.pdf
28. Zubiaga, A., Hoi, G.W.S., Liakata, M., Procter, R.: PHEME data set. https://figshare.com/articles/PHEME_dataset_of_rumours_and_non-rumours/4010619

Business Analytics

Applying Machine Learning to Anomaly Detection in Car Insurance Sales

Michał Piesio[1], Maria Ganzha[1] , and Marcin Paprzycki[2]([✉])

[1] Warsaw University of Technology, Warsaw, Poland
M.Ganzha@mini.pw.edu.pl
[2] Systems Research Institute Polish Academy of Sciences, Warsaw, Poland
marcin.paprzycki@ibspan.waw.pl

Abstract. Financial revenue, in the insurance sector, is systematically rising. This growth is, primarily, related to an increasing number of sold policies. While there exists a substantial body of work focused on discovering insurance fraud, e.g. related to car accidents, an open question remains, is it possible to capture incorrect data in the sales systems. Such erroneous data can result in financial losses. It may be caused by mistakes made by the sales person(s), but may be also a result of a fraud. In this work, research is focused on detecting anomalies in car insurance contracts. It is based on a dataset obtained from an actual insurance company, based in Poland. This dataset is thoroughly analysed, including preprocessing and feature selection. Next, a number of anomaly detection algorithms are applied to it, and their performance is compared. Specifically, clustering algorithms, dynamic classifier selection, and gradient boosted decision trees, are experimented with. Furthermore, the scenario where the size of the dataset is increasing is considered. It is shown that use of, broadly understood, machine learning has a realistic potential to facilitate anomaly detection, during insurance policy sales.

Keywords: Anomaly detection · Fraud detection · Insurance sector · Clustering algorithms · Ensemble methods · Gradient boosted decision trees

1 Introduction

According to the 2018 data, available form the Polish Central Statistical Office, in the sector of compulsory motor vehicle insurance, the gross premium amounted to PLN 14.779 billion [1]. Therefore, for the enormous volume of transactions, even small errors in premium calculation can lead to substantial losses. It should be relatively obvious that, broadly understood, machine learning methods may be used to minimise or even eliminate problems, and at least to some extent protect against sales of "dangerous" policies.

In this work we consider possibility of applying *anomaly detection*, during the process of concluding a car insurance policy. In this context, we apply machine learning techniques to an actual dataset from an insurance company located in

L. Bellatreche et al. (Eds.): BDA 2020, LNCS 12581, pp. 257–277, 2020.
https://doi.org/10.1007/978-3-030-66665-1_17

Poland. This dataset has been hand tagged to indicate, which policies contain anomalies and which are correct.

In what follows we focus our attention on two classes of anomalies, and both involve values of parameters actually used to calculate the premium of the insurance. Any possible errors in their values could severely affect the financial gain from concluding such a policy and even lead to potential losses.

It should be relatively obvious that the considered anomaly detection problem is actually a *classification problem*, which can be stated as follows. Assuming that a new policy contract is introduced to the computer system (of an insurance company), should it be classified as "safe", or as "dangerous". In a perfect situation, if the considered methods were accurate, the "system" would autonomously conclude all "safe" contracts, while contracts "flagged as potentially anomalous" would be sent to "someone" to be double checked.

One more interesting issue is as follows. Let us assume that an insurance company has a certain amount of data that is used to train "classifier(s)". Over time, subsequent contracts will be concluded (or, rejected). Thus, how to deal with a systematically increasing volume of data. It would be possible to "every Saturday, re-train the classifier(s) with complete data set". Should all available data be actually used, or a "sliding window-type" approach applied? We will present initial results of our attempt at investigating this issue.

In this context, we proceed as follows. We start (in Sect. 2) with a more detailed conceptualization of the problem at hand. Next, in Sect. 3, we summarize state-of-the-art found in related literature. We follow with the description of the dataset and the preprocessing that was applied to it (in Sect. 5). These sections provide the background that is used in the following Sect. 6, where results of application of clustering, ensemble and decision tree based methods, to untagged data, are presented and analysed. This section should be seen as the core contribution of this work. Lastly, the problem of anomaly detection, considered from the perspective of a system where data is being systematically accumulated is considered, in Sect. 7.

2   Conceptualization of the Anomaly Detection Problem

The presence of abnormal data, can be caused by many factors, and can lead to a significant financial loss, damage to the reputation, and/or customer trust. In the real-world practices of insurance companies, the main reasons for appearance of incorrect data are as follows.

- Human error – insurance agents deal with a steady stream of customers; with large number of tasks, and the time pressure, mistakes can happen;
- deliberate falsification of data – customers may try to deceive the company, in order to reduce the price of the insurance;
- incorrect data deliberately provided by the agent – agents can, dishonestly, increase their revenue, by falsifying customer data, to reduce the premium, to encourage customers to purchase the insurance; this, in turn, can provide the agent with a financial benefit, in the form of the margin on sales;

- errors in the internal computer system – unforeseen errors, in the sales system, may lead to an incorrect calculation of the premium;
- external system errors – connectivity with external systems that are (often) asked about the data directly related to issuing the policy, may be lost and this may result in errors and inaccuracies in the sales system.

Therefore, it would be extremely beneficial to have an effective method of verification of the entered data. The optimal solution would a "module", in the software supporting sales, that would analyse the data and "flag" transactions deemed as "suspicious". Next, it could inform both the agent and her/his supervisor about potential anomaly. If such module could not work in "real-time", it could examine recently sold policies. This would allow the discovery of problematic contracts that could be examined by human assessors. This could result in annulment of the most "problematic" ones, and could still be beneficial to the company. Moreover, it is important to capture changes that may occur over time to the sales process. Such changes may lead to an "evolution of anomalies". For example, agents can discover new methods of cheating, to increase sales.

Let us also stress that, with the amount of available data, and the complexity of the sales process, a thorough human oversight is impossible. Therefore, the only reasonable solution is to try to apply, broadly understood, data analytics approaches to facilitate detection of potential anomalies. The aim of this work is to investigate if such approach can actually work.

3 State-of-the-Art

Let us now summarize anomaly detection methods that are potentially pertinent to the insurance sales case. Discussed are also methods that, although used in the insurance industry, were not a good choice for the issues covered here.

The first group of methods are the clustering algorithms. They can be used alone, or combined with more complex algorithms, to improve their results. It has been shown that, in datasets with a fairly low number of anomalies, in relation to the "normal data elements", algorithms based on the choice of the leader [2] may be used. Such methods are quite fast, even for large datasets, but do not work well when anomalies occur with high frequency. Therefore, this family of clustering algorithms was omitted.

Work reported in [3] uses clustering to audit claims payments. Proposed algorithm is based on the *k-means method* [4]. Analysis of obtained clusters can support the work of an expert, in the analysis of problematic cases. However, there is still the issue of automating operation of such algorithm, in order to check the policy before its conclusion. Cited work does not provide a solution.

A promising approach, to anomaly detection, is the use of *team methods*, which combine multiple estimators. This allows focusing on data locality, rather than on the case-by-case examination, taking into account the complete dataset [5]. It uses simple base estimators, *kNN* (the k nearest neighbors), and *LOF* (local anomaly coefficient). Both approaches are derived from clustering methods and, therefore, can be naturally compared with those described above.

The insurance industry, in [6], has explored use of *logistic model* and *naive Bayesian classifier* as base estimators. Moreover, *neural networks, decision trees* and *support vector machines* have been tried (see, [7]). However, this work dealt with (car insurance) claims reporting, not policy purchase. Another slightly problematic aspect was the use of base estimators, with relatively long training time.

Another group of algorithms, used in the insurance industry, are methods based on the support vector machines (SVM) [8]. SVM-based algorithms try to find decision boundaries, by mapping input data into a multidimensional kernel space. Meanwhile, the [9] demonstrates the use of *spectral analysis*, as an approximation of an SVM, to point to fraudulent insurance claims. SVMs, however, require accurate selection of hyperparameters, which can be problematic, given the dynamics of the sales system. It can be difficult to adapt the algorithms to work effectively during the sales process, which discourages their use.

Labeled training dataset allows use of "supervised learning". Here, the leading methods are *random forests* [10] and *gradient boosted decision trees* [11] (belonging to the category of decision tree based methods). Particularly promising results, in the context of detecting damage-related financial fraud, were reported in [11]. Additionally, the authors note that decision trees may successfully detect anomalies in newly arriving data, which is very useful in insurance industry.

In the context of detecting fraudulent claims, studies have been carried, reaching beyond insurances and claims. The focus of [12] is exploring claimants data, searching for suspicious relationship cycles that indicate the possibility of deals between agents and customers. However, this method was not investigated, due to lack of needed data, and difficulty of collecting it, to facilitate research.

In summary, various methods have been applied to detect anomalies in insurance industry. However, to the best of our knowledge, while detection of anomalies in reported claims is often investigated, detecting fraud in policy sales has not been discussed so far.

4   Selected Algorithms

Ultimately, it was decided to investigate use of three approaches to anomaly detection: clustering, ensemble methods, and gradient boosted decision trees. Let us now discuss them in more detail, and state reasons for their selection.

4.1   Clustering Algorithms

Clustering algorithms have been selected because of their simplicity and ease of visualization of results. Here, work reported in [3] applies the centroid algorithm, an extension of the k-means algorithm, based on working with mini-batches [13], which reduces the computation time. However, while being the most popular clustering method, it has problems when working with large datasets. Therefore, it was decided to try also different clustering algorithms, which may improve the results and/or reduce the computation time.

The first of the selected algorithms is clustering applying *Gaussian mixture model clustering* [14]. It is based on the assumption that the dataset has a Gaussian distribution of observation parameters. Moreover, while the centroid algorithm makes a "hard classification" (assigning each point to exactly one cluster), the mixed Gaussian method expresses cluster assignment as a probability, allowing to capture the uncertainty of classification. For the problem at hand, the mixed Gaussian model expresses how likely a given policy will end up in a cluster of "dangerous contracts". This can be advantageous, as it is often difficult to decide with certainty if a policy is problematic or not.

Second, clustering algorithms belonging to the family of hierarchical algorithms, were selected. The first is *BIRCH algorithm* (Balanced Iterative Reducing And Clustering Using Hierarchies; [15]), which was chosen for its efficiency for very large datasets. The last clustering algorithm is the (also hierarchical) Ward method [16]. It usually works well for data with a balanced number of data elements across clusters, as it is based on the sum of squares criterion.

Research reported in [3] operated on the full available dataset, which allowed for greater precision, but involved large computational costs. The selected algorithms require calculating the distance between each pair of points, which was too costly due to the size of the set considered here. In the case of the k-means algorithm, even the use of the optimized (mini-batch) version is associated with the time complexity of $\mathcal{O}(knmt)$, where k is the number of clusters, n is the number of records in the dataset, m number of features of a single record, and t number of iterations of the algorithm. Therefore, a decision was made to reduce the data to two or to three dimensions. This, in addition to the cost reduction, offers easier visualization and interpretation of results. For this purpose, we chose the t-distributed Stochastic Neighbor Embedding *t-SNE* algorithm [17]. This algorithm builds a probability distribution over pairs of points, so that similar points are close to each other, while points with large differences are spaced far apart in the graph. Considering the relatively high percentage of anomalies in the dataset, this approach was expected to allow for easier division of the set into fragments with high, or low, anomaly count.

It is also crucial to note that the information whether fraud/error occurred is not available during the clustering process. It is only considered after the set is clustered, to check how the anomalies are distributed within individual clusters. Overall, the algorithms try to divide the entire dataset in such a way that some clusters contain mostly policies that are fraudulent, while the remaining clusters consist mostly of correct ones. Note that the point is not to split the dataset into two clusters (bad and good), but into groups of clusters with mostly anomalous data and mostly correct data. Achieving such clustering could allow assessment whether a policy is "safe" (or not). Specifically, upon the arrival of a new policy, it is mapped to a single (two- or three-dimensional) point, and assign to the corresponding cluster. If anomalies dominate in this cluster, the new policy becomes "suspicious". Here, note that the mapping of new records, which can use the t-SNE algorithm (see, [18]), has not been tested as focus of this work was on clustering data into groups dominated by correct and bad policies.

4.2 Dynamic Combination of Classifiers

The ensemble-based approach often addresses the problem of incorrectly detecting large numbers of false positives, or false negatives. It has been tried in the insurance industry (see, [6,7]). Therefore, it was decided to use Dynamic Combination of Detector Scores for Outlier Ensembles (DSO, [5]), an innovative framework that uses base estimators that, so far, have not been explored in the insurance sector.

4.3 Gradient Boosted Decision Trees

A common problem in the detection of anomalies is the manual tagging of suspicious elements, which would allow for supervised learning. In the insurance industry, however, there are approaches that take advantage of labeled datasets for high precision anomaly detection [11]. Here, with very good results, the gradient boosted decision trees were used, to detect fraud in insurance claims. It should be also noted that decision trees are characterized by high speed and optimal use of resources.

5 Dataset

The dataset examined in this study was obtained from an insurance companies operating in Poland. It contains actual data on automobile insurances for individual clients. Here, data was collected during three years (June 2016–19). Obviously, data was anonymized and business sensitive information was removed. While a much wider set of parameters is collected during insurance policy sales, based on the literature and experience of the lead author, who worked in the insurance company, the following parameters have been selected:

- age of the car owner,
- age of the car co-owner,
- driving permit years,
- number of claims in the last year,
- years since the last insurance claim,
- total number of years of policy ownership,
- bonus-malus,
- car age,
- car mileage,
- difference (in days) between the policy conclusion date and the insurance protection beginning date,
- insurance premium price.

During preprocessing, records with null values, in any of the columns, have been deleted. The only exception was *age of the co-owner*, which was set to 0, when it was not defined. Only a small proportion (7.3%) of policies contain co-owner data, but the experience has shown that this information is valuable in the fraud detection (so the data was kept). The dataset included policies that were not purchased, and they were removed. However, in the future they could be considered. Originally, the collection consisted of 39805 policies. After preprocessing, 35350 were left. Finally, the entire dataset has been normalized.

Experience of the lead author indicates that two popular types of anomalies, which directly affect the insurance price, should be considered. (1) Incorrect values of the parameter: *total number of years of policy ownership*. Here, when a value higher than the real one is entered in the system, it can lead to a reduction in the premium, and a financial loss to the company. Normally, the value provided by the customer should be verified by an external system, the Insurance Guarantee Fund [20], but this may not always be successful. Here, the service may experience interruptions in availability, or the personal data may be incorrect. If the verification fails, the value entered by the agent is used, which may lead to the abuses described above. (2) Incorrect values of the parameter *years since the last insurance claim*. Here, again, entering a value higher than the actual one may lead to an unwarranted reduction of the premium. Here, the verification by the Insurance Guarantee Fund may fail, again, for the same reasons.

The status of each element in the dataset was tagged by querying the Insurance Guarantee Fund, and setting two binary flags, if data is correct, it is a first, or second, type anomaly. The number of type one anomalies was 12568 (35.6%), and the number of type two anomalies was 9479 (26.8%). Moreover, 2985 (8.4%) of data elements belong simultaneously to both subsets of anomalies. However, this fact was not further explored; though it may be an interesting research direction to be undertaken. To represent the occurrence of anomalies, the t-SNE algorithm was used, resulting in a "flattened" representation depicted in Fig. 1.

(a) Anomalies of the total number of years   (b) Anomalies of the years since the last
of policy ownership   insurance claim

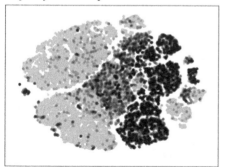 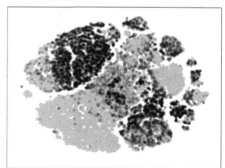

Fig. 1. Distribution of anomalies (black) in the dataset, reduced to two dimensions with the t-SNE algorithm.

As can be seen, policies that contain incorrect data constitute a large part of all contracts. In many places they are "very close to each other" (appear in large clusters), i.e. they have similar values of other parameters. However, in other areas, problematic policies are scattered across clusters of valid contracts.

6   Analysis of Results

Let us now analyze the results delivered by the selected, and above described, anomaly detection algorithms.

6.1   Clustering Algorithms

One of important questions, when applying clustering algorithms is the optimal number of clusters. Here, it was experimentally established that the optimal number is between 8 and 20 clusters. Too few clusters did not deliver an appropriate separation of (good/bad) policies. Too many clusters, on the other hand, caused an undesired combination of normal and abnormal data elements within clusters and resulted in unnecessary division of relatively homogeneous policies into separate clusters. Based on multiple experiments, it was decided to report, in this Section, results obtained for 12 clusters. This number brought about satisfactory clustering of policies and understanding of cluster content.

Moreover, operation of algorithms on data, reduced to three dimensions, was verified in order to check whether further flattening of data to two dimensions does not cause a large loss of information, and non-optimal clustering.

Finally, it should be remembered that the clustering algorithms work with untagged data. Hence, information about anomalies was imposed on the results of clustering, to understand and analyse the them. Specifically, each policy, where fraud or an error was committed, was plotted as a point in a different color. As an example, in Fig. 2, results of applying BIRCH algorithm to the data with type one anomaly are presented.

It can be seen that BIRCH tends to cluster groups of data elements that are "far away", compared to neighboring groups (separated, green cluster circled with a blue ellipse), while this phenomenon is rare in other algorithms. Other, uniform areas were similarly clustered in the remaining three clustering methods.

All methods adequately separated the entire clusters into the area of the main occurrence of anomalies (marked with a green ellipse). This means that the anomalies are relatively similar, in terms of parameter values. There are as many as 8626 policies in the "black cluster", while there are only 1022 "correct contracts". Such clustering of anomalies, should facilitate their effective detection. If a new policy would be marked as belonging to this cluster it is likely to be "problematic".

In Fig. 3, results of applying k-means algorithm to the data with type two anomaly are presented.

As can be seen, data with most anomalies (marked with the green ellipse) materializes in several clusters. Recall, that the goal is to split the data in such

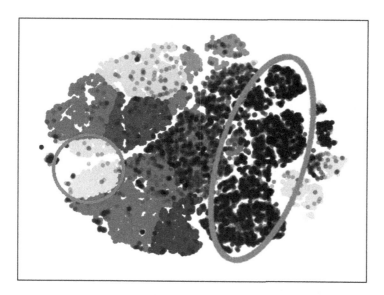

Fig. 2. Clustering results of BIRCH algorithm; 12 clusters; black points are anomalies; colored fragments represent separate clusters; type one anomaly; data reduced to/mapped into two dimensions. (Color figure online)

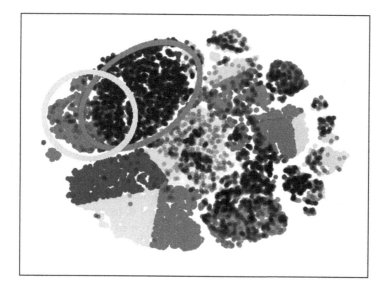

Fig. 3. Clustering with k-means algorithm; 12 clusters; black points are anomalies; colored fragments represent separate clusters; type two anomaly; data reduced to/mapped into two dimensions. (Color figure online)

a way that normal and abnormal data elements dominate separate clusters. Therefore, it is not a problem that the largest number of anomalies dominates multiple (smaller) clusters. In total, there are 4,577 anomalies in the circled area. However, one can notice that part of this "abnormal area" involves clusters (fragments) with very low number of anomalies (marked with a yellow circle). Therefore, if a new policy that would be assigned to the area covered by the yellow ellipse, it is not obvious if it is problematic or not.

Let us now present results of clustering when data is mapped into three dimensions. This approach preserves more information, but makes it more difficult to interpret the visualization. We start, in Fig. 4, with type one anomaly, and BIRCH algorithm.

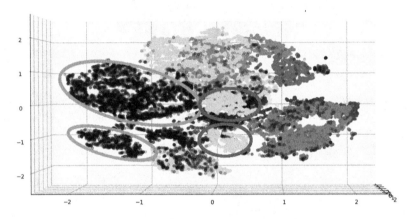

Fig. 4. Clustering results of BIRCH algorithm; 12 clusters; black points are anomalies; colored fragments represent separate clusters; type one anomaly; data reduced to/mapped into three dimensions.

As can be seen, mapping data into three dimensions allowed for better detection of clusters with a very low (maximum of 4.3%) anomaly count (marked with blue ellipses). In addition, the algorithms still managed to find clusters with a large number of abnormal policies (marked with orange ellipses). The separation of anomalies is at a satisfactory level, as 98.1% of dangerous policies have been located. Hence it is a good candidate for actual use in the insurance industry.

Next, three dimensional clustering of data with the type two anomaly, using k-means algorithm, has been depicted in Fig. 5.

As can be seen, type two anomalies are more sparsely distributed across the dataset than type one anomalies. This suggests that policies with type two anomalies differ more from each other (as they are not closely grouped). This is also evidenced by the statistics on the "density of anomalies" in the clusters. The maximum density of anomalies in a single cluster was lowest for the k-mean's method, and it was 71.3% of data elements of the particular cluster. However, this method recorded the most significant improvement when data was reduced only into three, and not into two, dimensions.

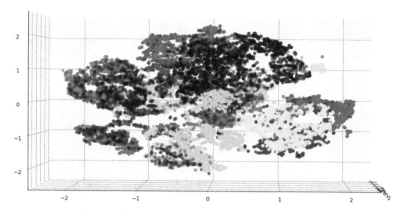

Fig. 5. Clustering results of k-means algorithm, 12 clusters, black points are anomalies, colored fragments represent separate clusters; first type anomaly; data reduced to/mapped into three dimensions.

6.2 Dynamic Combination of Classifiers

In the case of a dynamic combination of classifiers, two yet unexplored in the industry base estimators were selected – k-Nearest Neighbours and Local Outlier Factor.

In order to achieve a clearer visualization of the results, the dataset was reduced to two dimensions, using the t-SNE method. The algorithm, on the other hand, operates on the full available dataset – in case of this method, the dataset reduction method was not used to optimize the computation time, as it was the case for the clustering algorithms. Such procedure was not necessary, because the algorithm used in this approach works with the time complexity $\mathcal{O}(nm + n\log n + s)$, where n number of records in the dataset, m number of features of a single record, and s is the number of combined detectors. As unsupervised detectors are used, the algorithm does not use information as to whether fraud or an error has occurred in the policies tested.

We start with the presentation of results obtained using kNN base detectors for the first anomaly type (in Fig. 6).

Here, recall that the kNN base estimators mark data elements as anomalous if they are further away from their closest neighbors than other points in the "nearby area". Hence, majority of data elements marked as abnormal, have been found on the edges of the clusters, and in the lower density groups (marked with a pink ellipse, in Fig. 6). In the larger and more clustered areas "on the left hand side of the figure" (marked with green ellipses), the method was less likely to find anomalies, but still managed to point to a few "suspicious ones". Nevertheless, numbers of policies incorrectly marked as anomalies (3,492) and undetected anomalies (4,833) are high. This makes it doubtful if this approach can be successfully used in practice.

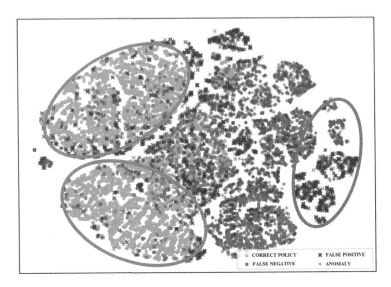

Fig. 6. Prediction of type one anomaly using kNN base detectors. Correctly identified anomalies are denoted as green, false negatives – red, false positives – blue, correct policies – yellow. (Color figure online)

Next, results of application of the kNN base estimators for type two anomaly are presented (in Fig. 7).

The majority of type two anomalies appear in a large, relatively dense area, in the upper left corner of Fig. 7 (marked with a green ellipse). However, the algorithm only occasionally marks the policies from this part of the dataset as anomalous. Very low precision can also be observed in other areas. There, the main problem is incorrect marking of normal data as erroneous (cyan ellipses). Despite attempts to diversify the parameterization of the algorithm, it was not possible to obtain more satisfactory results. Moreover, only 19.4% of actual anomalies were correctly detected, while 23.3% of correct policies were marked as frauds. Likely, this indicates that the kNN base detectors should not be applied to this problem in practice.

Interestingly, the results based on the LOF method differ from these based on the kNN. This can be seen when comparing Figs. 6 and 8. Here, a much higher number of points inside larger, more dense clusters is labeled as abnormal. However, the effectiveness of this approach is also low, which can be seen in large number of false positives. The fragment marked by the cyan ellipse, which has a lower density of points, very rarely has any policies marked as anomalies. Only 18.6% of data elements in this area have been flagged as anomalies when, in fact, this is precisely where the anomalies are located.

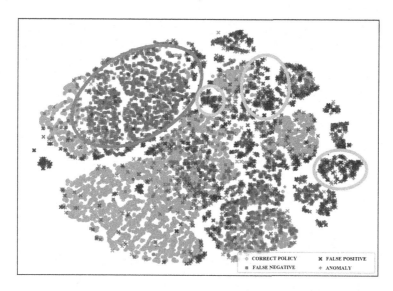

Fig. 7. Prediction of type two anomaly using kNN base detectors. Correctly identified anomalies are denoted as green, false negatives – red, false positives – blue, correct policies – yellow. (Color figure online)

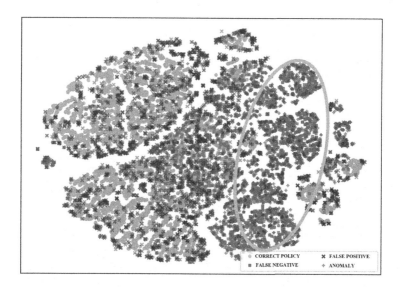

Fig. 8. Anomaly prediction for type one anomaly using LOF base detectors.

The LOF approach worked better for type two anomaly, since it detected at least some problematic policies (in area marked with a cyan ellipse, in Fig. 9). Nevertheless, the efficiency of the algorithm remained low. Only 17.5% of anomalies were detected and false positives were also common (9.3% of correct policies were marked as anomalies).

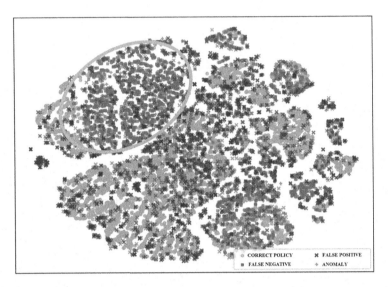

Fig. 9. Anomaly prediction of the parameter of years since the last insurance claim using LOF base detectors.

6.3   Gradient Boosted Decision Trees

To complete the overview of experiments clustering data in two dimensions, experiments with gradient boosted decision trees are reported. We start with the type one anomaly (in Fig. 10). Here, the XGBOOST library was used.

Analysing the Fig. 10, one can observe that decision trees rarely misclassify anomalies when they are clustered in one area (for example, in the area marked by the red ellipse), Here, false positives appear only occasionally (99.8% of data was marked correctly). The most problematic area is the "center of data" (marked with a cyan ellipse). Here, normal and abnormal policies are mixed. However, still, 79.3% of policies have been correctly classified. Note that, in this case, the number of false positives and false negatives is similar (difference of only 4%).

The algorithm also coped relatively well (97.1% precision) with predicting the type two anomaly. However, in the main cluster of anomalies (indicated by the pink ellipse, in Fig. 11), 6.4% of anomalies were not detected. Nevertheless, one can also observe areas practically free from mislabeled data elements (marked

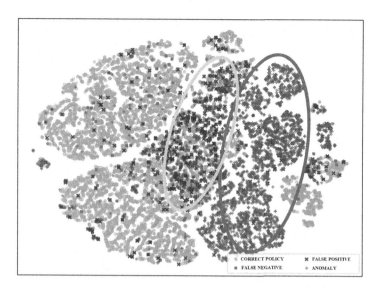

Fig. 10. Prediction of the type one anomaly using gradient boosted decision trees.

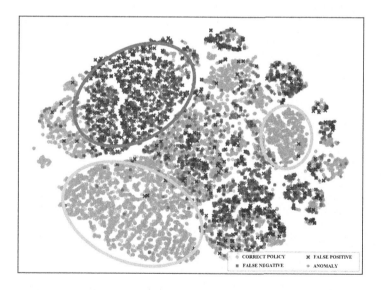

Fig. 11. Predicting anomaly of type two using gradient boosted decision trees.

with cyan ellipses). This is difficult to achieve even for supervised algorithms. In these areas, erroneous predictions constituted only 0.002% of policies.

6.4   Comparing Effectiveness of Different Classifiers

Let us now present the combined results achieved by all approaches. Here, the accuracy was measured with the Area Under Curve (AUC) of the Receiver Operating Characteristic [19]. Specifically, the AUC was calculated as follows. Each of the clusters was assigned a numerical value equal to $\frac{ra}{ra+n}$, where r is the ratio of the number of normal to abnormal policies in the entire set, a is the number of anomalies in a given cluster, and n the number of normal data elements in a given cluster. When preparing the ROC plot, all elements belonging to the clusters were marked as anomalies if the value of the cut-off point was lower than the value assigned to the cluster.

Let us start form type one anomaly. As represented by the results in Table 1 clustering data reduced to two dimensions had achieved worse precision than clustering data reduced to three dimensions, albeit the difference was minimal for some clustering algorithms. Dividing the dataset into more clusters tended to produce better results. Overall, best result achieved by the gradient boosted decision trees (AUC = 0.9907). However, all three dimensional approaches, also achieve accuracy, which would be satisfying for majority of insurance companies.

Table 1. Type one anomaly; accuracy measured with AUC; best results obtained with the indicated parameters.

Algorithm	Parameters	AUC
Clustering BIRCH 2D	14 clusters	0.9052
Clustering Ward 2D	16 clusters	0.9273
Clustering Gauss 2D	18 clusters	0.9334
Clustering k-means 2D	18 clusters	0.9224
Clustering BIRCH 3D	20 clusters	0.9381
Clustering Ward 3D	18 clusters	0.9284
Clustering Gauss 3D	20 clusters	0.9312
Clustering k-means 3D	16 clusters	0.9313
Ensemble kNN	Maximum of averages pseudo-truth	0.6884
Ensemble LOF	Average of maximums pseudo-truth	0.5766
Decision trees	Tree depth 8	0.9907

Precision of the algorithms, for the type two anomaly, was significantly lower across all tested methods (see, Table 2). The smallest decrease of precision was observed for the gradient boosted decision trees (decline of only 0.03). The decline for the remaining clustering approaches was of order of 0.1. Finally, accuracy of the ensemble methods, was barely better, or even worse, than random selection (i.e. one could as well flip a coin to decide is a policy was an anomaly or not).

Table 2. Type two anomaly; accuracy measured with AUC; best results obtained with the indicated parameters.

Algorithm	Parameters	AUC
Clustering BIRCH 2D	14 clusters	0.7954
Clustering Ward 2D	20 clusters	0.8304
Clustering Gauss 2D	20 clusters	0.8230
Clustering k-means 2D	18 clusters	0.8060
Clustering BIRCH 3D	18 clusters	0.8113
Clustering Ward 3D	20 clusters	0.8163
Clustering Gauss 3D	18 clusters	0.8332
Clustering k-means 3D	20 clusters	0.8234
Ensemble kNN	Maximum of averages pseudo-truth	0.5164
Ensemble LOF	Average of maximums pseudo-truth	0.4743
Decision trees	Tree depth 8	0.9607

7 Accumulating Data in Insurance Company System

Let us now consider a typical situation when data, concerning sold policies, is being accumulated in the computer systems(s) of the insurance company. As noted above, it is possible to use some (possibly very simple) method to compare each new policy to the existing data model and establish "where it belongs". As a result the policy can be judged to be normal or potentially problematic. Next, every N days (e.g. every weekend) the complete existing dataset can be used to create a new data model. However, in this approach, an important question needs to be answered, in order to achieve the highest possible anomaly detection precision. How to deal with "older" data. Should all data be kept and used to train the model or, maybe, older data should be removed, as it may have adverse effect on detecting anomalies in the dynamically changing world.

There is one more aspect of this problem. While the dataset examined in this study is relatively large, this is not going to be the case for companies that are just entering the insurance industry. Here, they will have only a small initial dataset available as a source for model training. Note that it is not important how large and how small are the respective datasets. The key point is that one of them is considerably smaller than the other. Therefore, it was decided that by limiting the size of the (currently available) dataset it will be possible to simulate a situation where a relatively new insurance company tries to create a business tool to be used with a small (initial) dataset.

Therefore, as an extra contribution, it was decided to perform some preliminary investigations, to lay foundation of understanding of what is happening in the considered scenario. Thus, while thus far we have ignored the chronology of data elements, this information was available in the dataset. Therefore, the data was sorted chronologically, and only the "initial part" was allocated to the training set. First, the dependence of the precision of anomaly detection, on the size of the initial training set was tested. Models built on the initial training sets consisting of 6, 12, 18, 24, and 30 thousand data elements were prepared and their effectiveness was compared on the corresponding fragments of the test set. Each fragment of the set, for which the effectiveness was checked, consisted of 2000 policies. Results for type one anomaly have been depicted in Fig. 12.

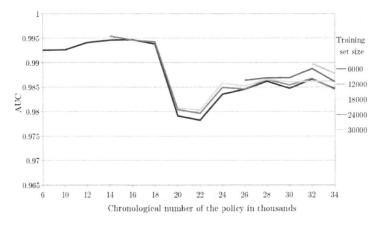

Fig. 12. Precision of prediction depending on the training set size; type one anomaly; AUC; gradient boosted decision trees.

The first observation is obvious, the lowest accuracy occurs for the smallest training set. However, even then, accuracy remains acceptable for a real-life insurance company. Sudden changes to the precision around the 18 thousandth policy can be historically tracked to the changes to the tariff used to determine the final insurance premium sum, which changed the structure of the dataset. They could have also resulted in agents finding different methods of cheating.

The second considered scenario was, what would happen if one used a "sliding window" and kept only certain volume of most current data, and discarded the older ones, in order to capture the latest trends. This scenario was investigated as follows – results from the previous experiment were compared to the precision obtained with the training set where the older data was omitted. Sliding window of 12,000 data elements was used. Results have been depicted in Fig. 13.

As can be seen, across the board, omitting old data did not improve the results delivered by the trained models. Nevertheless, in sporadic cases, there were areas where the models trained with the reduced training sets obtained better results. However, the differences were not significant.

In summary, while it is obvious that a number of open questions remains, and is worthy investigation, it was established that: (1) for the size of the dataset that was available for this study, it seems that the best approach it to keep all the available data; and (2) even for relatively small datasets, gradient boosted decision trees provide reasonable accuracy of prediction of anomalies of both types.

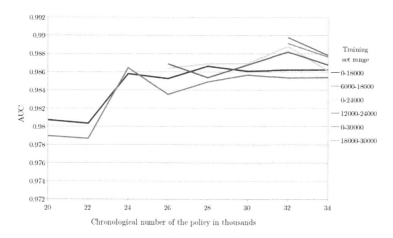

Fig. 13. Precision of prediction depending on the training set range; type one anomaly; AUC; gradient boosted decision trees.

8   Concluding Remarks

In this work we have considered a novel problem of detecting anomalies in policy sales. Experimental work was based on a car insurance dataset, obtained from an actual insurance company. To establish anomalies in an unsupervised learning scenario, three classes of algorithms have been applied: clustering algorithms, dynamic classifier selection, and gradient boosted decision trees.

The gradient boosted decision trees algorithm dominated all the tested methods in terms of the anomaly prediction accuracy. The favorable characteristics of this approach is also that it allows for easy and quick analysis of new policies, already during their conclusion, allowing the company to be adequately secured against financial losses. Further analysis of the algorithm also showed that it performs well when the size of the dataset, available for training, is small. This may be a significant advantage for start-ups, wanting to use machine learning to guarantee optimal protection against potential frauds and mistakes. The study of the deviation of the effectiveness of the algorithm considering the influx of new data showed that this method is quite resistant to potential changes in the nature of anomalies. However, especially in this area, a number of open questions remain. Nevertheless, to address them properly, a much larger (especially from the time of collection perspective) dataset should be used.

References

1. Polish Central Statistical Office: Polish Insurance Market in 2018 (2019). https://stat.gov.pl/en/topics/economic-activities-finances/financial-results/polish-insurance-market-in-2018,2,8.html
2. Talagala, P.D., Hyndman, R.J., Smith-Miles, K.: Anomaly detection in high dimensional data. J. Comput. Graph. Stat. (2020)
3. Thiprungsri, S., Vasarhelyi, M.A.: Cluster analysis for anomaly detection in accounting data: an accounting approach. Int. J. Digit. Account. Res. **11**, 69–84 (2011)
4. MacQueen, J.: Some methods for classification and analysis of multivariate observations. In: Symposium on Mathematical Statistics and Probability (1967)
5. Zhao, Y., Hryniewicki, M.K.: DCSO: dynamic combination of detector scores for outlier ensembles. In: ACM KDD Workshop on Outlier Detection De-Constructed (ODD v5.0) (2018)
6. Viaene, S., Derrig, R.A., Baesens, B., Dedene, G.: A comparison of state-of-the-art classification techniques for expert automobile insurance claim fraud detection. J. Risk Insur. **69**, 373–421 (2002)
7. Hassan, A.K.I., Abraham, A.: Modeling insurance fraud detection using ensemble combining classification. Int. J. Comput. Inf. Syst. Ind. Manag. Appl. **8**, 257–265 (2016)
8. DeBarr, D., Wechsler, H.: Fraud detection using reputation features, SVMs, and random forests. In: Proceedings of the International Conference on Data Science (2013)
9. Niana, K., Zhanga, H., Tayal, A., Coleman, T., Li, Y.: Auto insurance fraud detection using unsupervised spectral ranking for anomaly. J. Financ. Data Sci. **2**, 58–75 (2016)
10. Anton, S.D.D., Sinha, S., Schotten, H.D.: Anomaly-based intrusion detection in industrial data with SVM and random forests. In: International Conference on Software, Telecommunications and Computer Networks (2019)
11. Dhieb, N., Ghazzai, H., Besbes, H., Massoud, Y.: Extreme gradient boosting machine learning algorithm for safe auto insurance operations. In: IEEE International Conference of Vehicular Electronics and Safety (2019)
12. Bodaghi, A., Teimourpour, B.: Automobile insurance fraud detection using social network analysis. In: Moshirpour, M., Far, B.H., Alhajj, R. (eds.) Applications of Data Management and Analysis. LNSN, pp. 11–16. Springer, Cham (2018). https://doi.org/10.1007/978-3-319-95810-1_2
13. Béjar, J.: K-means vs Mini Batch K-means: A comparison, KEMLG - Grup d'Enginyeria del Coneixement i Aprenentatge Automàtic - Reports de recerca (2013)
14. McLachlan, G.J., Basford, K.E.: Mixture models. Inference and applications to clustering (1988)
15. Zhang, T., Ramakrishnan, R., Livny, M.: BIRCH: an efficient data clustering method for very large databases (1996)

16. Ward Jr., J.H.: Hierarchical grouping to optimize an objective function. J. Am. Stat. Assoc. **58**, 236–244 (1963)
17. van der Maaten, L., Hinton, G.: Visualizing data using t-SNE. J. Mach. Learn. Res. **9**, 2579–2605 (2008)
18. van der Maaten, L.: Learning a parametric embedding by preserving local structure. In: Proceedings of the Twelth International Conference on Artificial Intelligence and Statistics (2009)
19. Fawcett, T.: An introduction to ROC analysis. Pattern Recognit. Lett. **27**, 861–874 (2006)
20. Insurance Guarantee Fund. https://www.ufg.pl/infoportal/faces/pages_homepage

Authorship Identification Using Stylometry and Document Fingerprinting

Shubham Yadav[1], Santosh Singh Rathore[1(✉)], and Satyendra Singh Chouhan[2]

[1] Department of IT, ABV-IIITM Gwalior, Gwalior, India
shubham.iiitmgwl@gmail.com, santoshs@iiitm.ac.in
[2] Department of CSE, MNIT Jaipur, Jaipur, India
sschouhan.cse@mnit.ac.in

Abstract. Authorship identification focuses on finding the particular author of an anonymous or unknown document by extracting various predictive features related to that document. It helps to predict the most probable author of articles, messages, codes, or news. This task is generally viewed as a multi-class, single labeled categorization of text. This topic is important and among the most interesting topics in the field of Natural Language Processing. There are many applications in which this can be applied such as identifying anonymous author, supporting crime investigation and its security, also detecting plagiarism, or finding ghost-writers. Till now, most of the existing works are based on the character n-grams of fixed length and/or variable length to classify authorship. In this work, we tackle this problem at different levels with increasing feature engineering using various text-based models and machine learning algorithms. The propose work analyses various types of stylometric features and define individual features that are high in performance for better model understanding. We evaluate the proposed methodology on a part of Reuters news corpus. It consists of texts of 50 different authors on the same topic. The experimental results suggest that, using document finger printing features enhance the accuracy of classifier. Moreover, PCA (Principal Component Analysis) further improves the results. In addition, we compare the results with other works related to the authorship identification domain.

Keywords: Natural Language Processing · Stylometry · Document Finger Printing · Authorship identification · Machine learning classifiers

1 Introduction

Authorship identification has been a very important and practical problem in Natural Language Processing. The problem is to identify the author of a specific document from a given list of possible authors [1,2]. A large amount of work exists on this problem in the literature. It has become even more important with the widespread use of various forums where anonymity is allowed. A good system for authorship identification can handle cases of misuse of anonymity [3].

© Springer Nature Switzerland AG 2020
L. Bellatreche et al. (Eds.): BDA 2020, LNCS 12581, pp. 278–288, 2020.
https://doi.org/10.1007/978-3-030-66665-1_18

Moreover, such systems can also be used for plagiarism detection [4]. Analysis of authorship of historical texts has already been done many times using techniques for authorship attribution. Various spoofing e-mails can also be detected by using these techniques [5]. In the traditional problem of authorship identification, one candidate author is assigned a text of unknown authorship, provided a collection of candidate authors for whom samples of unquestioned authorship are available. From a machine learning perspective, this can be seen as a method of categorizing single-label text in multiclass.

The key concept behind recognition of authorship is to quantify certain textual features, these features contribute to the author's style of writing called stylometry. For solving this problem, conventional text classifiers like naive bayes, logistic regression and SVM have been used, and various deep learning models like LSTM, LDA have also been tried [2,6]. Different models perform differently depending upon the type of training and testing data available. The recognition of authors is assisted by methods of statistics and computation. This research field leverages developments in practice in the areas such as machine learning, information extraction and the analysis of natural languages.

To the best of our knowledge, stylometric features along with Document Finger Printing solution has yet not explored for the problem of Authorship Identification. The work presented in this paper uses a combination of the above two mentioned techniques, which acts as a novel approach to tackle this problem of multi-class single label text categorization task. We also implemented other feature extraction techniques for comparison purpose. For the feature modelling part, *TfIdf Vectorizer* along with the extracted textual features and for the predictive modelling part models like SGD Classifier, Improved-RFC and Naive bayes have been implemented. The experimental evaulation is done on various performance matrices.

The remainder of the paper is structured as follows. Section 2 discusses the related work in authorship identification. Section 3 presents the methodology used to extract all of the necessary features and classification algorithms. Section 4 discuss the dataset used, performance evaluation measures and experimental results. Finally, conclusions and future directions are given in the last section.

2   Related Works

The problem of authorship identification has been studied extensively. Many hand-designed features have been developed in order to tackle this problem. In the literature, some researchers have presented a few solutions to this problem. Most previous works seems to follow a common paradigm: a feature extraction step followed by a classification step. This common classification architecture has proven to work effectively for the problem at hand. However, the reported performance has been measured on small and relatively easy datasets.

In [7], researchers focused on the Reuters news articles dataset comprising of texts related to the news topics from 50 different authors. They suggested a variable length n-gram approach in their research work, which was inspired by prior work to choose variable length word sequences. Author(s) proposed a method for the selection of variable length n-gram function of mainly 3, 4 and 5 g was based on extracting multiword terms. The training dataset consists of 2,500 texts (50 per author), and 2,500 testing texts were also included in the test data collection. Used baseline information gain, and SVM is trained using the important set of features, with linear kernels given C equal to 1. Finally, accuracy is calculated which came to be 72.56% using information gain (IG) parameter. Recent work related to this domain was cross-lingual authorship identification [8]. The aim of cross-lingual identification of authorship is to predict the author of a random document written in one language by taking the identified documents into other languages. The key challenge of this task is the design features not available to other languages from one language in the dataset. In their research, they analyzed various kind of stylometric features and find the better performing language independent features for analysis. Authors partitioned the documents in various fragments where each and every fragment can be further broken into chunks of fixed size. The dataset they worked on multi language corpus of 400 authors with about 820 documents in six different languages. Used three different distance measures that include commonly known partial, standard and modified Hausdorff distance. Used probability k-NN classifier, to make predictions which are further merged to make a final prediction for the document. There were 196 monolingual, 204 bilingual, 25 multilingual authors (more than two languages). Dataset was extracted from an online book archive in languages from English to Portuguese. Calculation of accuracy was done in two metrics, Fragment accuracy and Document accuracy with results like maximum of 94.65% and 96.66% respectively.

In [9], Deep learning based technique proposed for Authorship Identification. In this research work, Several specialized deep learning models are used at different levels to test authorship recognition solution output through a deep learning approach. For the data-processing tasks two techniques are used, one is word representations which used GloVe vword vectors as pre-trained word embeddings, they used the GloVe vectors as the already trained word embeddings in which it contains 4 lakh tokens and the second technique was Input Batch Alignment to use GPU's parallel processing advantage to speed up the training cycle. In their work, they used 2 labelled datasets which were adopted to train, test and evaluate models. Authorship detection was conducted on both an article and event-level news dataset and a narrative data set. In addition, the GRU and LSTM network were analyzed, fine-tuned, and used for output asses. Also, the siamese network is attempted to analyze the similarities between pf two posts, which looked solid. The narrative dataset consists of 50 writers with different articles. There were some 45,000 paragraphs contained in the corpus. Article-level GRU performed better on authorship recognition, offering 69.1% accuracy on C50 and 89.2% accuracy on gutenberg data collection. Furthermore, on all datasets, the Siamese network reached authentication precision of 99.8%. In [10], a dissertation work based on the recognition of authorship from unstructured texts is proposed. The impetus for pursuing this was the rise in vast volumes of anonymous texts on the Internet, which indicated a strong and urgent need to recognize authorship. In this study, they suggested a semantic association model on language, non-subject stylistic terms and word-dependence relationships to reflect the writing style of unstructured texts by different writers, develop an unsupervised approach to extract stylistic features and enforce PCA and LDA alogrithms [11] for accurate recognition. For the feature construction process, ten feature sets are generalised broadly categorised as structural, lexical, syntactic, semantic, syntactic and semantic combined among which only three were proposed in the work mentioned earlier. Classification models like kNN, SVM and LDA were implemented on different feature sets with proper model parameters tuning achieving the highest accuracy of about 78.44% on the RCV1 corpus.

3   Methodology

The aim of the proposed methodology is to check the effectiveness of stylometry feature with document fingerprinting for authorship identification. The overall methodology is shown in Fig. 1. We have used four different scenarios by using four different feature extraction approach for performance evaluation. In first case, we extracted TF-IDF features of the dataset only and apply the machine learning algorithms. In second case, we used only stylometry features to identify authorship. In third case, we use combination of stylometry feature with document fingerprinting for identification purpose. Finally, we apply PCA (Principle Component Analysis) over stylometry feature with document fingerprinting for identification. The detailed discussion is as follows.

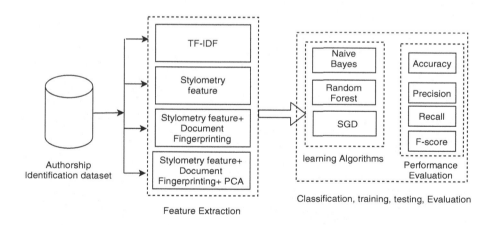

Fig. 1. Overview of the experimental methodology

3.1   TF-IDF (Term Frequency Inverse Document Frequency)

TF-IDF Feature model is used to generate numeric features from the text, which will be further applied to the model for prediction. For 'D' total documents and 'd' be a particular document in the dataset. Mathematically, it is defined as:

$$tfidf_{i,d} = tf_{i,d} \cdot idf_i \tag{1}$$

$$tf_{i,d} = \frac{n_{i,d}}{\sum_k n_{k,d}} \tag{2}$$

$$idf_i = \log\frac{|D|}{|d : t_i \in d|} \tag{3}$$

The TF-IDF Vectorizer tokenizes documents, learn the vocabulary and reverse weighting of document frequencies, and allowed to encode new documents [12].

3.2   Stylometric Features

Stylometric features or style markers try to capture the writing style of the author. We have used a large number of stylometric features in our model. The motivation behind these features comes from the survey paper on authorship attribution [7]. List of features extracted is as follows:

Token number, Number of unique tokens, Sentence number, Punctuation fraction, Average token length, Standard token-length deviation, Average sentence length (wrt words), Average sentence length (wrt characters), Number of digits

The features pertaining to types and token are: number of types (unique words), number of tokens, type-token ratio. Some authors have a tendency to write long sentences whereas some authors prefer shorter sentences. In order to capture this, we have included various features in our model such as average length of a sentence (both in terms of words and characters), standard deviation of sentence lengths. Some authors pose more questions as compared to others.

For assimilating such ideas in our model we have introduced features like relative number of declarative, interrogative, imperative and exclamatory sentences. Other style markers include relative number of punctuations, relative number of digits in the document etc. Natural Language Tool Kit (NLTK) has been used to extract these style markers from the text.

3.3   Document Finger-Printing

The detection of fingerprints is a well-known technique in forensic sciences. Also applicable to other contexts can be the basic concept of defining a subject based on a collection of features left by the subject acts or behavior. Another such method is the detection of authorship of text based on an author fingerprint. Fingerprinting has mainly been used in the past for plagiarism detection.

Basically, the idea is to generate a compact signature for each document and then those signatures can be matched for similarity. We used 2 types of document finger-printing, character level finger-print generation and word n-gram level fingerprinting. In character level finger-printing, we first create a set of all $k(k \geq 5)$ length character n-grams. Then we select a small subset of these character n-grams based on their hash values using the Winnow's algorithm [13] implemented in this python fingerprint library [14] The hash value of the sequence of chosen character n-grams is returned as the signature of the document. To make the process efficient we used fingerprint library. Similarly word level features can be used and similarity between 2 documents can be computed as the number of matches found. This word-level analysis is very slow if implemented normally. So, we used DARMC Intertextuality Analysis Library for effective computation of the similarity function [7].

3.4 Dimensional Reduction Using Principle Component Analysis (PCA)

Dimension Reduction process in order to eradicate non-required features and working on those features that contribute much to the identification. One of the most common methods of linear dimensional reduction is principal component analysis (PCA). It can be used separately or can also be used as a starting element for other methods of reduction. It is a widely used and accepted dimension reduction technique. Next author identification using classification algorithms are implemented resulting in expected outcome and later comparison of the results obtained with the results of past similar works is done.

3.5 Learning Algorithms

We have used three different learning techniques, random forest, naive Bayes, and SGD classifier to build and evaluate the performance of presented approach. A detailed description of these techniques is given by Ali et al. [15]. All learning techniques are implemented using Python *sklearn* library.

4 Results and Discussion

To evaluate the performance of presented methodology, we performed experiments on a benchmark dataset (Sect. 4.1). The implementation was done in Python using the commonly used ML and NLP libraries like scikit learn, NLTK and so on.

Parameters of *TfIdfVectorizer* are tuned as the ngram range is set to (1,2) which indicates that both unigrams and bigrams are considered, lowercase attribute is set to False and stopwords are ignored by setting the value of parameter to 'english'. Further, parameters of SGD Classifier used are set such as the log loss value depicts the working as logistic regression. Constant multiplying the regularization term α is set to 0.000001 with penalty defaults to l2 which is the normal regularizer for linear models, thus the same is used. N iter value, the actual number of iterations needed to achieve the stop criterion. For multiclass fits, the limit is specified to be 5 over each binary fit. Shuffle: Whether to shuffle the training data after each epoch or not. Defaults to True, and here too, labeled True.

4.1   Dataset Used

The RCV1 corpus was already used in research into authorship identification. However in the sense of the AuthorID project, the top 114 RCV1 authors with at least 200 available text samples were selected. Like other methods, in this review the criteria for selecting the authors were the subject of the available text samples. Thus, after removing all duplicate texts found using the R-measure, the top 50 text writers classified with at least one subtopic of the CCAT class (corporate/industrial) were chosen. In this way the subject dimension is sought to be minimized when distinguishing between the texts. Subsequently, as steps have been taken to minimize the dominance of the genre, differences in authorship can be considered to be a more significant factor in the differentiation of the texts. Consequently, differentiating between writers is more difficult because all the examples of text deal with similar topics rather than when some writers deal mainly with economics, others deal with foreign affairs etc. Each document was written by one person, and only one. The training corpus consists of 2,500 texts (50 per author), and the test corpus contains 2,500 additional texts (50 per author) which do not overlap with the training texts.

4.2   Performance Matrices

The performance of the presented models has been evaluated in terms of their accuracy scores. Also, for the best fit model among all the average macro F-score is calculated as harmonic mean of average Precision and average Recall. For this multi-class single label classification task, the equations for Average Precision, Recall, and F-score are defined in the Eqs. 4, 5, and 6, respectively[1].

$$AvgPrecision = \frac{\sum ClassesPrecisions}{\text{Number Of Authors}} \tag{4}$$

$$AvgRecall = \frac{\sum ClassesRecalls}{\text{Number Of Authors}} \tag{5}$$

$$AvgFscore = 2 * \frac{\text{Avg Precison * Avg Recall}}{\text{Avg Precision + Avg Recall}} \tag{6}$$

In general, Precision gives the measure of correctness (True positives) among the predicted positives (True Positive + False Positive).

---

[1] TP = True Positive, FP = False Positive, FN = False Negative.

Recall gives the measure of retrieval. Correctly predicted Positive class (TP) with total Positive class (TP+FN).

F-score gives measure of Precision and Recall combined together. If precision of a model is good but not recall or vice-versa, the overall f-score for the model will not be good. F-score for performance evaluation gives good result if both Precision and recall for the model are good.

4.3   Experimental Results and Comparative Analysis

We started out with the simple naive bayes classifier which gave a decent accuracy. Then we tried various other classifier models of which SGD Classifier performed the best for us. Therefore, we selected SGD classifier for further evaluation. Then we added stylometric features mentioned in the list of features extracted, to our model. This led to a considerable increase in accuracy. After that we went further to add the document finger-printing features which again led to a significant increase in accuracy.

After applying successfully Principal Component Analysis, some features which are highly or lightly correlated with each other are reduced retaining the variation provided with all the stylometric features together upto the maximum extent. After doing this, the features namely Number of sentences, Average token length, Number of digits are found to be some of those features that are providing the information the least of a particular author and can be termed as not so good features. Further, performance of implemented models is evaluated on various metrics. These are tabulated in Table 1.

Table 1. Accuracy results of implemented models

Experiments	Accuracy	Avg Precision	Avg Recall	Avg f-score
Naive Bayes	69.0	0.663912	0.834237	0.739392
SGD Classifier	76.4	0.724207	0.812304	0.765729
SGD + Stylometry	79.5	0.783021	0.782225	0.782622
RFC + Stylometry + Document Fingerprinting	79.8	0.800073	0.793102	0.796572
SGD + Stylometry + Document Fingerprinting	81.1	0.820042	0.792351	0.805958
After PCA	81.4	0.830517	0.787058	0.808204

Figure 2 shows the comparison results of the presented models with other similar works in terms of accuracy measure. We have included works performed on the same RCV1 corpus C50 dataset. From the figure, it can be observed that except for the naïve Bayes classifier, all other presented models yielded better prediction accuracy than the similar previous works. SVM with sentences performed better than the SGD classifier alone. However, all other models significantly outperformed SVM with sentences work. Further, it can be observed that

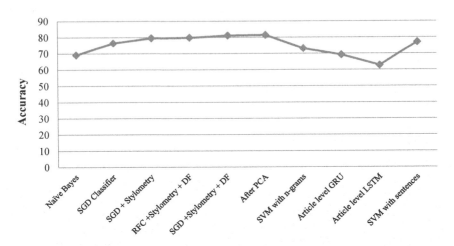

Fig. 2. Comparison of the presented models with other similar works in terms of accuracy measure (*DF = Document Fingerprinting)

combining document fingerprinting and stylometry resulted in a better performance than sequential advance models like LSTM. Thus, the presented models helped in getting better classification results for the authorship identification.

4.4 Other Methods Explored

Knowledge of Most Confused Authors: Due to data insufficiency, we restricted ourselves to simpler models that use more number of features. For performing error analysis on our model, we constructed confusion matrix for the authors. From there we tried to use the knowledge of most confused authors for improving the performance of our model. One key observation was that the confusion matrix was a sparse matrix. It means an author is confused with only a small number of other writers, not nearly any other author. In particular, there are certain pairs of authors who are confused with each other. For disambiguating between such authors we trained smaller models made exclusively for those authors. First a confusion matrix is created for the predictions on the training set. Then from the confusion matrix we pick top-k most confused pairs of authors (depending on the total number of mispredictions among them). We train smaller models on each of these selected pairs. Now for prediction on a test example, if the predicted author is found to be one of the confused authors, we use the majority of the outputs from the smaller models. This method increased accuracy by a small amount. However, this method is computationally expensive and therefore we could not use it effectively.

Lemmatization and POS Tagging: We considered replacing the words with their root forms using Porter Stemmer lemmatization. But it did not result in an increase in accuracy. We also tried the syntactic feature of appending the words with their Part-Of-Speech (POS) tags. Again this idea did not work out in our favour. These features try to hide the actual word information. Perhaps due to this they did not improve the accuracy.

5   Conclusion and Future Work

In the presented work, we have shown that stylometric features and finger-printing features are crucial to the task of authorship identification. Moreover, traditional classifiers like naive Bayes, SVM and logistic regression are performing better than neural network models such as LSTM on the same dataset. With the help of comparison between the implemented solution and the approach used in earlier works, it can be concluded that there is a high scope of inventing new solution approaches to tackle this problem of Authorship Identification. In the future, There is also a lot of practical research worth studying. Some possible experiments, such as attempting various sizes of word embedding and ways of representing a text or sentence rather than simply averaging, could be undertaken. One promising approach is to use LSTMs with more amount of data and to build on the previous work added with all the stylometric features and document finger-printing ideas. In addition, an ensemble of models aims to give rise to the accuracy of attribution of authorship.

References

1. Peng, F., Schuurmans, D., Keselj, V., Wang, S.: Language independent authorship attribution with character level n-grams. In: 10th Conference of the European Chapter of the Association for Computational Linguistics, Budapest, Hungary, April 2003. Association for Computational Linguistics (2003)
2. Gamon, M.: Linguistic correlates of style: authorship classification with deep linguistic analysis features. In: COLING (2004)
3. Stamatatos, E.: A survey of modern authorship attribution methods. J. Am. Soc. Inf. Sci. Technol. **60**(3), 538–556 (2009)
4. Zhang, S., Lin, T.Y.: Anautomata based authorship identification system. In: Workshops with the 12th Pacific-Asia Conference on Knowledge Discovery and Data Mining, vol. 1, pp. 134–142 (2008)
5. Brocardo, M., Traore, I., Woungang, I.: Authorship verification of e-mail and tweet messages applied for continuous authentication. J. Comput. Syst. Sci. **81**, 12 (2014)
6. Bajwa, I.S., Anwar, W., Ramzan, S.: Design and implementation of a machine learning-based authorship identification model. Sci. Program. 14 (2019)
7. Houvardas, J., Stamatatos, E.: N-gram feature selection for authorship identification. AIMSA **4183**, 77–86 (2006)
8. Sarwar, R., Li, Q., Rakthanmanon, T., Nutanong, S.: A scalable framework for cross-lingual authorship identification. Inf. Sci. **465**, 323–339 (2018)
9. Chen, Q., He, T., Zhang, R.: Deep learning based authorship identification. Department of Electrical Engineering Stanford University, Stanford (2017)
10. Zhang, C., Xindong, W., Niu, Z., Ding, W.: Authorship identification from unstructured texts. Knowl.-Based Syst. **66**, 08 (2014)

11. Jafariakinabad, F., Tarnpradab, S., Hua, K.: Syntactic recurrent neural network for authorship attribution, February 2019
12. Stein, R., Jaques, P., Valiati, J.: An analysis of hierarchical text classification using word embeddings. Inf. Sci. **471**, 216–232 (2019)
13. Schleimer, S., Wilkerson, D., Aiken, A.: Winnowing: local algorithms for document fingerprinting. In: Proceedings of the ACM SIGMOD International Conference on Management of Data, vol. 10, April 2003
14. Marton, Y., Wu, N., Hellerstein, L.: On compression-based text classification. In: Losada, D.E., Fernández-Luna, J.M. (eds.) ECIR 2005. LNCS, vol. 3408, pp. 300–314. Springer, Heidelberg (2005). https://doi.org/10.1007/978-3-540-31865-1_22
15. Almazroi, A.A., Mohamed, O.A., Shamim, A., Ahsan, M.: Evaluation of state-of-the-art classifiers: A comparative study

A Revenue-Based Product Placement Framework to Improve Diversity in Retail Businesses

Pooja Gaur[1]([✉]), P. Krishna Reddy[1], M. Kumara Swamy[2],
and Anirban Mondal[3]

[1] IIIT, Hyderabad, India
pooja.gaur@research.iiit.ac.in, pkreddy@iiit.ac.in
[2] CMR Engineering College, Hyderabad, India
m.kumarswamy@cmrec.ac.in
[3] Ashoka University, Haryana, India
anirban.mondal@ashoka.edu.in

Abstract. Product placement in retail stores has a significant impact on the revenue of the retailer. Hence, research efforts are being made to propose approaches for improving item placement in retail stores based on the knowledge of utility patterns extracted from the log of customer purchase transactions. Another strategy to make any retail store interesting from a customer perspective is to cater to the varied requirements and preferences of customers. This can be achieved by placing a wider variety of items in the shelves of the retail store, thereby increasing the *diversity* of the items that are available for purchase. In this regard, the key contributions of our work are three-fold. First, we introduce the problem of concept hierarchy based diverse itemset placement in retail stores. Second, we present a framework and schemes for facilitating efficient retrieval of the diverse top-revenue itemsets based on a concept hierarchy. Third, we conducted a performance evaluation with a real dataset to demonstrate the overall effectiveness of our proposed schemes.

Keywords: Utility mining · Patterns · Diversity · Concept hierarchy · Retail management · Itemset placement

1 Introduction

Over the past decade, we have been witnessing the prevalence of several popular medium-to-mega-sized retail stores with relatively huge retail floor space. Examples of medium-sized retail stores include Walmart Supercenters, while Macy's Department Store at Herald Square (New York City, US) and Dubai Mall (Dubai) are examples of mega-sized retail stores. Typically, such mega-sized stores have more than a million square feet of retail floor space. Such large retail stores typically have a huge number of aisles, where each aisle contains items that are stocked in the slots of the shelves.

© Springer Nature Switzerland AG 2020
L. Bellatreche et al. (Eds.): BDA 2020, LNCS 12581, pp. 289–307, 2020.
https://doi.org/10.1007/978-3-030-66665-1_19

In such large retail stores, tens of thousands of items need to be placed in the typically massive number of slots (of the shelves) of a given store in a way that improves the sales revenue of the retailer, thereby posing several research challenges. In this regard, existing works have focused on scalable supply chain management [2], inventory management [35], stock-out management [35], identification of frequent and/or high-utility itemsets [6] and strategic placement of items (products) as well as itemsets for improving the revenue of retail stores [9]. In essence, given the importance and relevance of the retail sector in today's world, this is an active area of research with several ongoing efforts.

Incidentally, another strategy of improving the revenue of the retailer is to provide customers with a wider *diversity* of items so that customers have more choice in terms of variety of the items that are available to them for purchase [19, 36]. For example, a customer, who wishes to buy soft drinks, would generally prefer a wider and more diverse range of soft drinks (e.g., Coke, Coke-Light, Pepsi, Sprite, Limca, etc.) to choose from. As another example, there could also be different brands and varieties of soaps, shampoos and tomato sauce in a retail store. It is a well-known fact in the retail industry that customers belong to different market segments e.g., Alice prefers a specific brand of tomato sauce or a specific brand of a soft drink, while Bob may prefer another brand of tomato sauce as well as another brand of a soft drink.

Given that each market segment is relatively large in practice, *increasing the diversity* in a retail store generally attracts customers from *multiple market segments*, thereby in effect increasing the overall foot-traffic of the retail store. For example, if a retail store were to only stock Coke in its shelves, it would likely miss out on the customers, whose preference could be Coke-Light or Pepsi. In fact, research efforts are being made to improve diversity for real-world retail companies by collecting data about sales, customer opinions and the views of senior managers [13, 18, 27, 33]. In essence, there is an opportunity to improve the revenue of the retailer by incorporating the notion of diversity (of items) during the placement of itemsets.

In the literature, itemset placement approaches have been proposed in [6–9] to improve the revenue of the retailer by extracting the knowledge of high-revenue itemsets from specialized index structures. These index structures are populated based on the information in the log of customer purchase transactions. However, these works do not address itemset placement based on concept hierarchy based diversity of items. Moreover, the notion of diverse frequent patterns have been proposed in [20, 21] to extract the knowledge of diverse items based on the customer purchase transactional dataset as well as a given concept hierarchy of the items. Furthermore, it has also been demonstrated that the knowledge of diverse frequent patterns can potentially be used for improving the performance of recommender systems. Observe that while the works in [20, 21] address the diversity issue, they do not consider the context and issues associated with itemset placement in retail stores.

In this paper, we have made an effort to improve the diversity of itemset placement in retail stores by proposing the notion of concept hierarchy based diverse revenue itemsets. In particular, based on the notion of diverse revenue patterns, we have proposed a framework, which improves the diversity of itemset placement, while not compromising the revenue. To this end, we have proposed two itemset placement schemes.

The main contributions of this paper are three-fold:

- We introduce the problem of concept hierarchy based diverse itemset placement in retail stores.
- We present a framework and schemes for facilitating efficient retrieval of the diverse top-revenue itemsets based on a concept hierarchy.
- We conducted a performance evaluation with a real dataset to demonstrate the overall effectiveness of our proposed schemes.

To the best of our knowledge, this is the first work that incorporates the notion of concept hierarchy (of items) for increasing the diversity in itemset placement for retail stores in order to improve the revenue of the retailer.

The remainder of this paper is organized as follows. In Sect. 2, we discuss existing works and background information. In Sect. 3, we present the context of the problem. In Sect. 4, we present our proposed itemset placement framework. In Sect. 5, we report the results of our performance evaluation. Finally, we conclude the paper in Sect. 6 with directions for future work.

2   Related Work and Background Information

In this section, we discuss an overview of existing works as well as some background information.

2.1   Related Work

The placement of items on retail store shelves considerably impacts sales revenue [4,5,11,17]. The placement of products in the shelves of a retail stores is an important research issue, which has motivated several efforts [10,11,17,34,35]. In [10,34,35], it has been studied that how the visibility of the items impacts the total revenue generated by the store.

Several research efforts have also been made in the area of utility mining [23,29,30]. The goal of these efforts is to identify high-utility itemsets from transactional databases of customer transactions. The work in [30] proposed the Utility Pattern Growth (UP-Growth) algorithm for mining high-utility itemsets. It keeps track of the information concerning high-utility itemsets in a data structure, which is designated as the Utility Pattern Tree (UP-Tree). Moreover, it uses pruning strategies for candidate itemset generation. The work in [23] proposed the HUI-Miner algorithm for mining high-utility itemsets. It used a data structure, designated as the *utility-list*, for storing utility and other heuristic information about the itemsets, thereby enabling it to avoid expensive candidate itemset generation as well as utility computations for many candidate itemsets. The work in [29] proposed the CHUI-Miner algorithm for mining closed high-utility itemsets. In particular, the algorithm in [29] is able to compute the utility of itemsets without generating candidates. While these existing works have focused primarily on identifying high-utility itemsets, none of these works has addressed the itemset placement problem for retail stores.

Placement of itemsets in retail stores has been investigated in [4–9,14,17] The work in [5] has shown that the profits of retail stores can be improved by making efficient use of the shelf space. The work in [17] proposed a model and an algorithm for maximizing the total profit of a given retail store. In particular, the algorithm proposed in [17] allocates shelf space by considering the available space and the cost of the product. The work in [4] exploited association rules to address the item placement problem in retail stores. In particular, it used frequent patterns to select the items for placement and demonstrated the effectiveness of frequent patterns towards item placement. Furthermore, the work in [14] investigated the mixed-integer programming model towards retail assortment planning and retail shelf space allocation for maximizing the revenue of the retailer.

A framework has been presented in [6,9] to determine the top-revenue itemsets for filling a required number of (retail store) slots, given that items can physically vary in terms of size. A specialized index structure, designated as the STUI index has been proposed in [6,9] for facilitating quick retrieval of the top-revenue itemsets. Furthermore, an approach has been proposed in [7] for placing the itemsets in the premium slots of large retail stores for achieving diversification in addition to revenue maximization. Notably, the notion of diversification in the work in [7] is based on reducing the placement of duplicate items. This facilitates towards improving long-term (retail) business sustainability by attempting to minimize the adverse impact of macro-environmental risk factors, which may dramatically reduce the sales of some of the retail store items. As such, the emphasis of the work in [7] is orthogonal to the focus of this work. Moreover, the work in [7] proposed the kUI (k-Utility Itemset) index for quickly retrieving top-utility itemsets of different sizes. Additionally, the work in [8] addresses the problem of itemset placement by considering slots of varied premiumness. Observe that none of these existing works considers concept hierarchy based diversity in the context of itemset placement for retail stores.

The issue of improving the diversity in item placement for retail stores has been investigated in [13,18]. The work in [18] aimed at increasing diversity in retail items by mixing the food items and other retailer items. The work in [13] differentiated between two types of retailers, namely (a) limited-diversity retailers specializing in either food or non-food product categories and (b) extended-diversity retailers offering both food and non-food product categories. However, these works do not consider a concept hierarchy based approach for improving the diversity of retail items in a more fine-grained manner.

Interestingly, the work in [3] has proposed an approach to improve the diversity of recommendations. In [12,28], efforts have been made to improve the diversity of search engine recommendations. In the area of recommender systems [20,21,31,37], a model of concept hierarchy based diversity has been introduced and it has been demonstrated that it improves the variety of recommendations in e-commerce recommendation systems.

2.2   About Concept Hierarchy Based Diversity

In [20–22], an approach has been proposed to compute the diversity of a given pattern (itemset) by using concept hierarchies. In a given domain, a set of items can be grouped into a category and a pattern may contain items belonging to different categories. The items and the categories form a concept hierarchy. The concept hierarchy [15,16,24–26,32] is a structure representing the relationship between items and their categories in a tree, where the items occupy the leaf nodes. Here, the intermediate nodes represent the categories of the items. The *root* of the concept hierarchy is a virtual node. A sample concept hierarchy is depicted in Fig. 1.

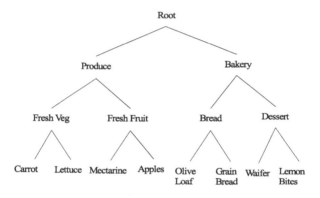

Fig. 1. A sample concept hierarchy from Instacart Market Basket Analysis dataset

The notion of diversity captures the extent of items in the pattern belonging to multiple categories. The notion of diversity of a pattern captures the merging behavior, i.e., how the items in lower level merge to the items at higher level. If the items of the pattern merge quickly to the higher level node, the pattern has low diversity. Similarly, if the items of the pattern merge slowly, that pattern has relatively high diversity. As an example, consider the items and concept hierarchy depicted in Fig. 1. Here, the pattern {Carrot, Lettuce} has low diversity as it merges quickly, whereas the pattern {Carrot, Grain Bread} has the maximum diversity as the items merge at the *root*.

Given a pattern and a concept hierarchy [20–22], the diversity is computed using the Diversity Rank (*DRank*) metric. Given a set of items and the concept hierarchy C, the formula to compute the *DRank* of a pattern Y is $DRank(Y) = \left(\frac{P - (|Y| + h - 1)}{(h - 1)(|Y| - 1)} \right)$. Here, P is the number of edges in the sub-tree of the concept hierarchy C formed by the items of the pattern Y, and h is the height of C.

2.3 About k-Utility Itemset index

The work in [6] proposed an itemset placement scheme based on efficiently determining and indexing the top high-utility itemsets. The itemsets are organized in the form of the kUI (k-Utility Itemset) index to retrieve top-utility itemsets quickly. We shall now provide an overview of the kUI index.

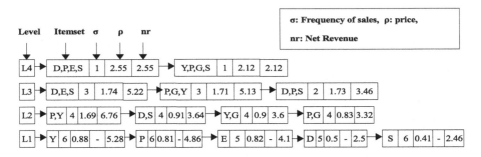

Fig. 2. Example of the kUI Index

The kUI is a multi-level index, where each level corresponds to a particular itemset size. It optimizes the task of getting potential itemsets of a given size from a large number of itemsets. The kUI index is organized into N levels, where each level contains a fixed number λ of itemsets. At the kth level, kUI index stores top-λ ordered itemsets of size k.

Figure 2 shows an illustrative example of the kUI index. Each node in the linked list at the kth level contains a tuple of $\langle itemset, \sigma, \rho, nr \rangle$, where ρ is the price of the a given itemset $itemset$, σ is the frequency of sales of the itemset and nr is the net revenue, which is computed as $(\rho \times \sigma)$. The itemsets at each level of the kUI index are sorted in descending order based on net revenue to facilitate quick retrieval of the top-revenue itemsets.

3   Context of the Problem

Consider a finite set γ of m items $\{i_1, i_2, \ldots, i_m\}$. We assume that each item of the set γ is physically of the same size. Moreover, we assume that all slots are of equal size and each item consumes only one slot. Each item i_j is associated with a price ρ_j and a frequency σ_j. The revenue of a given itemset is the product of its price and its frequency of sales. Observe that σ_j is the number of transactions in which i_j occurs in the transactional database. The price of a given itemset is the sum of the prices of all of the items in that itemset. The net revenue (nr) of a given itemset is the price of the itemset multiplied by the number of times that the itemset was purchased. Furthermore, recall our discussion in Sect. 2.2 concerning the notion of diversity.

Problem Statement: Given a set γ of items, a transactional database D over set γ, price ρ_i of the ith item in set γ, a concept hierarchy C over the items in set γ and a given number of slots to be filled, the problem is to determine the placement of itemsets in the slots such that the total revenue of the retailer as well as the diversity should be improved.

4   Proposed Framework

In this section, we first discuss the basic idea of the proposed framework. Next, we explain the proposed schemes.

4.1   Basic Idea

Transactional data of retail customers provide rich information about customer purchase patterns (itemsets). Given a set of transactions, an approach to identify high-revenue itemsets has been explained in [6], as previously discussed in Sect. 2.3. We designate this approach as the Revenue-based Itemset Placement (RIP) approach. Under the RIP approach, it is possible to maximize the revenue of the retailer by extracting and placing high-revenue itemsets. However, the RIP approach does not consider diversity of items, while performing itemset placement. Recall that in Sect. 2.2, we have discussed an approach [21,22] to compute the $DRank$ (Diversity Rank) of a given itemset, given a set of items and a concept hierarchy.

   The issue is to develop a methodology to identify potential high-revenue itemsets with relatively high $DRank$ values. The basic idea is to propose a new ranking mechanism to rank itemsets based on both revenue and diversity of the itemsets. Hence, given an itemset X, we propose a new measure, designated as *Diverse Net Revenue (dnr)*, of X, which is equal to the product of the net revenue (nr) and the $DRank$ value of X. Given a pattern X, $dnr(X) = DRank(X) \times nr(X)$. In essence, the implication of the dnr measure is that the selected itemsets for placement should have either a high value of nr or a high value of $DRank$ or both.

   Based on the notion of diverse revenue itemsets, we propose a scheme, which we shall henceforth refer to as the Diverse Rank Itemset Placement (DRIP) scheme. In particular, the DRIP scheme comprises the following: (i) a methodology to identify top itemsets with high dnr value and building an index called Concept hierarchy based Diversity and Revenue Itemsets ($CDRI$) index and (ii) a methodology to place the itemsets by exploiting the $CDRI$ index, given the number of slots as input.

   Observe that the proposed DRIP scheme prioritizes itemsets with high net revenue and high $DRank$ values. Consequently, it may miss some of the top-revenue itemsets with low $DRank$ values. As a result, even though DRIP improves both revenue and diversity, it may not meet the performance of the RIP approach [7] on the net revenue aspect. Recall that the RIP approach exploits the kUI index, as previously discussed in Sect. 2.3. Hence, we propose another approach called the Hybrid Itemset Placement (HIP) approach, which considers the top-revenue itemsets as in the RIP approach·and the top itemsets with high revenue and high $DRank$ values as in the DRIP approach. We shall now discuss the details of our two proposed approaches.

4.2   The DRIP Approach

We shall first discuss our proposed DRIP approach and use an illustrative example to explain the approach. The DRIP approach comprises two main components: 1. *Building of the CDRI index* and 2. *Itemset placement scheme*. The input to the proposed approaches is a transactional database consisting historical transactions performed by the users at a retail store, item prices and a Concept hierarchy of the items in the transactions.

1. Building of the CDRI index

$CDRI$ stands for *Concept hierarchy based Diversity and Revenue Itemsets index*. $CDRI$ is a multi-level index containing N levels. At each level of the $CDRI$ index, λ itemsets are stored. The kth level of the $CDRI$ index corresponds to itemsets of size k.

Each level of the $CDRI$ corresponds to a hash bucket. The data is stored as a linked list of nodes at each level. The node is a data structure having required information for an itemset. Each node in the linked list at the kth level contains a record consisting of the following fields: $\langle itemset, \sigma, \rho, DRank, dnr \rangle$. Here, ρ is the price of the given *itemset*, σ is the frequency of sales of the itemset. Here, $dnr = DRank \times \sigma \times \rho$.

The $CDRI$ index is built level by level. The level 1 of the $CDRI$ index is formed by inserting a record for each item. The following process is repeated to compute the itemsets at level k. We first extract candidate itemsets by concatenating the itemsets of level $k-1$ with the items of level 1. Next, we remove the duplicate items in each itemset and exclude all itemsets, whose size does not equal k. We compute the support for the itemsets by scanning the transactional database. For each itemset X, price of X is calculated. The price of the itemset is equal to the sum of price of the items in it and $DRank(X)$ is computed using the formula to compute $DRank$. Next, we compute dnr for each itemset. The itemsets are then sorted w.r.t. dnr and the top-λ itemsets are placed at the kth level of the $CDRI$ index.

Figure 3 depicts an example of the $CDRI$ index as used in the proposed approach. Each node in the linked list at k^{th} level contains a tuple of $(itemset, \sigma, \rho, DRank, dnr)$, where ρ is the price of the itemset *itemset*, σ is the frequency of sales of the itemset, $DRank$ is the diversity rank of the itemset and dnr (diverse Net Revenue) is $DRank \times \sigma \times \rho$.

Algorithm 1 contains the pseudo-code to build the $CDRI$ index. The algorithm essentially has two parts i.e., first building the level 1 of index and then building level 2 to max_level of the index. All items (i.e., one-itemsets) are picked from the transactional dataset T and placed in a temporary list A (Line

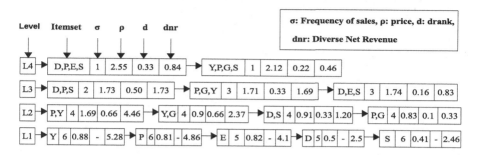

Fig. 3. Illustrative example of the CDRI index

Algorithm 1: Building of $CDRI$ Index

Input: T: Transactional database, price ρ_i for each item i, C: Concept
Hierarchy, max_level: Number of levels in $CDRI$, λ: Number of entries
in each level of $CDRI$, st: support threshold

Output: $CDRI$ /* CDRI index */

1 Initialize $CDRI[max\_level][\lambda]$ to NULL

2 Scan T and compute support $s(X)$ for each item X in T and store it in array A
/* Level 1 of CDRI */

3 Sort A based on the support and remove all items which have the support less
than the st

4 Select top-λ items from A and for each item X, insert $\langle X, s(X), \rho(X)\rangle$ in
$CDRI[1][\lambda]$ in the descending order of the support*price

5 **for** $lev = 2$ to $lev = max\_level$ **do**

6   Create combination of itemsets in lev - 1 with the items in level 1 and
generate all itemsets of length lev and store in A

7   Scan T and compute support $s(X)$ for each itemset X in A

8   Sort A based on the support and remove all items which have the support
less than the st

9   Compute price ρ_i for each itemset X in A by combining the price values of
the items in X

10   Compute $DRank(X)$ for all X in A

11   Sort all itemsets in A in descending order based on diverse net revenue
(dnr) values of itemsets. i.e., $dnr(X) = DRank(X) \times Support(X) \times \rho(X)$

12   Select λ itemsets from A in the descending order of dnr value of the itemsets
and for each itemset X in A, insert $\langle X, s(X), \rho(X), DRank(X), dnr(X)\rangle$ in
$CDRI[lev][\lambda]$ in the descending order of dnr value of the itemsets in A

13 **end**

14 **return** $CDRI$

Algorithm 2: DRIP Placement Algorithm

Input: $CDRI$ index, Total Number of Slots (Ts)
Output: slots (Placement of itemsets in Slots)

1 Initialise slots[Ts] to NULL
2 $CAS \leftarrow Ts$ /* CAS indicates the number of currently available slots */
3 **while** $CAS > 0$ **do**
4   Starting from level=2, select one itemset with top dnr from each level of $CDRI$ index and place it in A
5   Choose the itemset X from A which has the highest revenue per slot
6   Remove X from $CDRI$ index
7   If $(CAS - |X|) < 0$ then break
8   Place X in slots[Ts]
9   CAS = CAS - $|X|$
10 **end**
11 **return** $slots$

---

2). For the items in A, the number of times an itemset is found in T is computed and the items with a frequency less than *support_threshold* are removed from A (Line 3). The top-λ items are picked from A now and placed in level 1 of $CDRI$ index (Line 4). For each level lev in 2 to max_level, all possible candidate itemsets of size lev are generated using itemsets placed in level 1 and level $lev - 1$ and placed in list A (Line 6). For each itemset in A, support (also called frequency of sales of the itemset) is found from T and all the itemsets with support less than *support_threshold* are removed from A (Line 8). Now, price for each itemset in A is calculated as sum of prices of items it has (Line 9) and $DRank$ is calculated for each itemset in A using *Concept Hierarchy* (Line 10) with the method discussed earlier. All the itemsets in A are sorted based on *diverse revenue* $(d_i \times s_i \times \rho_i)$ (Line 11). From this sorted list, top-λ itemsets are picked and the nodes are placed in $CDRI$ index (Line 12). Once this process is done for each level, the formed $CDRI$ index is returned.

2. Itemset Placement Scheme in DRIP
Given the number of slots, we present the itemset placement scheme for DRIP approach. It can be noted that $CDRI$ index contains itemsets of different sizes. We only consider itemsets with the size greater than 1. Given itemsets of different sizes and net revenue values, we first pick the itemset which maximizes revenue contribution per slot. The notion of *Revenue per slot* is defined as follows.

Revenue Per Slot: Consider an itemset X of size k which consumes k slots. Revenue per slot refers to the per slot revenue contribution of X itemset.

$$net\ revenue\ per\ slot\ for\ X = \frac{price\ of\ X \times support\ of\ X}{size\ of\ X} \tag{1}$$

The placement approach keeps picking the itemset with maximum *revenue per slot* and placing it in the available slots.

Algorithm 2 depicts the steps for proposed DRIP scheme. The algorithm stops when CAS becomes 0. The CAS is initialized as number of slots to fill (Line 2). While there are slots left to fill, get the top node of each level of the $CDRI$ index and place it in a temporary list A (Line 4). Now, the best node (node with highest *per slot revenue*) is picked from A and also removed from $CDRI$ as it is chosen to be placed (Line 5, 6). Then, append the itemset of this node in the slots and reduce the number of slots currently available as we have placed an itemset (Line 8,9). When the loop ends, the itemsets have been placed in the *slots* and the algorithm returns *slots* (Line 11).

3. Illustrative Example of DRIP Scheme

Now, we will explain the proposed DRIP scheme in the following with an example. The Fig. 4 shows the whole approach.

We consider a *transactional dataset* which has 10 transactions of 8 items. The hierarchical relation between the items and categories as *Concept hierarchy*. Total number of slots is 25. Default value of parameters are provided in Table 1. We explain the placement of itemsets in the given slots using DRIP Scheme.

The first step is to build a $CDRI$ index for the given data using Algorithm 1. The generated $CDRI$ can be observed in the figure. After the $CDRI$ index is generated we proceed with Algorithm 2 for placement. First we compare revenue per slot of top-nodes from each level of $CDRI$ i.e [2.23, 0.57, 0.21]. Since the top revenue for level 2 (index 1 here) is highest, we place the itemset from level 2 in slots and remove is from $CDRI$ index. The updated top per slot revenues for comparison becomes, $[1.18, 0.57, 0.21]$ and we pick node from level 2 again as it has highest per slot revenue. This process is continues till all slots are filled. The step wise placement is shown in Step 2 in the figure.

4.3   Hybrid Itemset Placement Approach (HIP)

The HIP approach comprises of two main components: 1. *Identifying itemsets* and 2. *Itemset placement scheme*. We briefly explain these components.

1. Identifying itemsets

Given transactional database over a set of items, concept hierarchy over a set of items, price details of items, we build both revenue index (RI) index and $CDRI$ index. The process to build $CDRI$ index is explained in the preceding section. The process to build RI index is similar to the process of building $CDRI$ index, except, the itemsets of indexed based on net revenue (as in kUI index, refer Sect. 2.3).

2. Itemset placement scheme

The HIP scheme uses a combination of RI and $CDRI$ to get itemsets for placement. During placement, it maintains vectors to top pointers and their *revenue_per_slot* for both RI and $CDRI$. We introduce a variable called RD_-

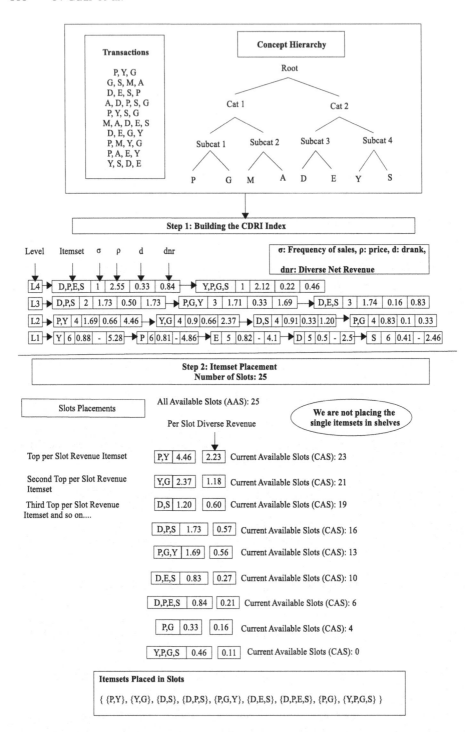

Fig. 4. Illustrative example of DRIP scheme

ratio, which represent the ratio of the portion of slots to be filled from RI and $CDRI$ index. For example, if RD_ratio is 0.3, 30% of given slots will be filled by itemsets from RI and 70% of given slots will be filled by itemsets from $CDRI$.

5   Performance Evaluation

In this section, we report the performance evaluation. The experiments were conducted on a Core i7-3537U CPU @ 2.00 GHz × 4 PC with 8 GB RAM running Ubuntu 18.04. We implemented the approaches in Python 3.7. The experiments are conducted on a real dataset, namely the *Instacart Market Basket Analysis* dataset [1]. The dataset is anonymous and contains over 3 million grocery orders from about 2,00,000 Instacart users. The dataset has 49,688 items and 1,31,208 transactions. The average transaction length is 10.55.

For conducting experiments, we require the hierarchy of the items in the dataset and the price value for each item. Regarding concept hierarchy, notably, the dataset has its own concept hierarchy in which the items are organized based on the department and the category information of the items. For all the experiments, we use this concept hierarchy. Regarding utility (price) values of the items, the dataset does not provide price related information. We generate the price for each item of the dataset randomly as follows. We consider a normalized price range of [0,0, 1.0] and divide it into the following six price brackets: [(0.01, 0.16), (0.17, 0.33), (0.34, 0.50), (0.51, 0.67), (0.67, 0.83), (0.84, 1.0)]. For each item in the dataset, we randomly select one of the afore-mentioned price brackets and then assign a random price within the range of that price bracket to the item.

Table 1. Parameters of performance evaluation

Parameters	Default	Variations
Total number of slots (10^3) (Ts)	10	(1, 2, 4, 8, 12)
RD_ratio	0.3	0.0, 0.2, 0.4, 0.6, 0.8, 1.0

Table 1 summarizes the parameters used in the evaluation. The *Total number of slots (Ts)* refers to the number of slots allocated. In the HIP approach, we fill a portion of slots from the index built for the RIP approach and the remaining portion from the index built for the proposed DRIP approach. To conduct experiments on HIP, we employ RD_ratio, which is the ratio of number of slots of the shelf to be filled from the index of the RIP approach and the number of slots of shelf to be filled from the index of the DRIP approach.

The performance metrics are *Total revenue (TR)* and *DRank*. *Total revenue (TR)* is the total retailer revenue for the test set. To calculate TR, for each transaction in the *test* set, we find the maximal set of items which are placed in the shelf, and add the revenue of that itemset to TR. Similarly, the $DRank$ is the mean $DRank$ of all the placed itemsets purchased by customers.

The experiments are performed on four approaches. They are Revenue based Itemset Placement (RIP), Diverse Revenue based Itemset Placement (DRIP), Hybrid Itemset Placement (HIP) and Diversity based Itemset Placement (DIP) approaches.

The RIP approach places high revenue itemsets in the given slots. The itemsets for RIP are picked from RI index. The proposed $DRIP$ approach places itemsets in $CDRI$ index. The HIP approach places a portion of high net revenue itemsets from the RI index and the remaining portion of itemsets from $CDRI$ index. The DIP approach places only high $DRank$ itemsets. For this we build index by considering $DRank$ value of itemsets, which we call DI index. For each of RI, $CDRI$, DI indexes, we fix the number of levels as 4 and the number of itemsets in each level as 5000.

For conducting the experiments, the dataset is divided into *training* set and *test* set in the ratio of 90:10. The index for the respective approaches such as RIP, DRIP and DIP approaches is built from the *training* set. We follow the placement approach, which was presented as a part of DRIP approach (Refer Sect. 4.2). We evaluate the performance considering the transactions in the *test* set in the following manner. We iterate over the transactions in the *test* set and check whether a subset (a set) of items of each transaction belongs to the itemsets placed in the given slots. We include the revenue of the corresponding subset towards TR. This helps in simulating a real life retail scenario where retailers place items on shelves and observe revenue only if the customer buys one of those items.

5.1 Effect of Variations in Number of Slots on Total Revenue

Figure 5a shows the variation of total revenue of RIP, DRIP, HIP and DIP. It can be observed that the revenue is increasing for all approaches with the number of slots. This is due to the fact that as the number of slots increases, more products will be available for purchase. As a result, increased number of customers can find the products in the retail store. Among the approaches, it can be observed that RIP's total revenue dominates other approaches. This is due to the fact that only high net revenue itemsets are placed in the RIP approach.

Also, as expected, it can be observed that the performance of DIP is low among four approaches, because DIP places only high $DRank$ itemsets by ignoring the net revenue value of the itemsets. The total revenue of DIP is not zero because, it also places some itemsets with high net revenue. It can be observed that the performance of DRIP is less than RIP, but significantly more than DIP. This is due to the fact that DRIP places the itemsets with both high net revenue as well as high diversity. The revenue of DRIP is less than RIP because, high net revenue itemsets with low $DRank$ values are not placed in the slots. This figure also shows the performance of HIP which places 30% of itemsets with top net revenue value from RIP and 70% of itemsets with both top net revenue as well as *DRank* values from DRIP. As expected, as we are also placing top net revenue itemsets, the total revenue of HIP is close to RIP.

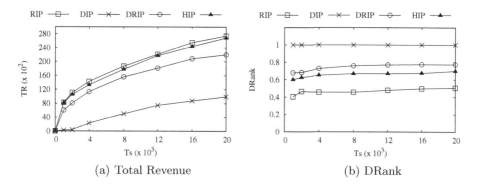

(a) Total Revenue   (b) DRank

Fig. 5. Effect of variations in number of slots

5.2 Effect of Variations in Number of Slots on DRank

Figure 5b shows the variation of *DRank* values of RIP, DRIP, HIP and DIP. It can be observed that except for DIP, the average *DRank* value is increasing initially and becomes stable for RIP, DRIP and HIP approaches. For DIP, it maintains a constant value with the number of slots. Among the approaches, as expected, it can be observed that the average *DRank* value is maximum for DIP. Also, DIP maintains the stable *DRank* value with the number of slots. This is due to the fact that as the number of slots increases, DIP places increased number of high *DRank* itemsets and the average *DRank* does not change significantly.

Also, as expected, it can be observed that the *DRank* performance of RIP is low among four approaches, because RIP places only high net revenue itemsets by ignoring the *DRank* value of the itemsets. The *DRank* value of RIP is not zero because, it also places some itemsets with high *DRank* value. Regarding the performance of the proposed DRIP approach, it can be observed that its *DRank* performance is less than DIP, but significantly more than RIP. This is due to the fact that DRIP contains the itemsets with both high *DRank* value and high net revenue value. The *DRank* value of DRIP is less than DIP because, high *DRank* itemsets with low net revenue values are not placed in the slots. This figure also shows the performance of HIP which places 30% of itemsets with top net revenue values from RIP and 70% of itemsets with both top net revenue as well as high *DRank* values from DRIP. Since, we exclude itemsets with top *DRank* values, the *DRank* value of HIP is less than DRIP.

From above results, it can be observed that DRIP gives high net revenue as well as high *DRank* values, its net revenue is less than RIP even though its *DRank* value is significantly higher than RIP. So, we can conclude that the HIP approach facilitates the placement of itemsets such that the total revenue can match with RIP, but trading the *DRank* value as compared to DRIP.

5.3   Effect of Variations in *RD_ratio*

For RIP, DIP, DRIP and HIP, Fig. 6a shows the variations of total revenue and Fig. 6b shows the variations of *DRank* value as we vary *RD_ratio*. In this experiment, we fix the number of slots at 12000. Notably, the *RD_ratio* does not affect RIP, DRIP and DIP due to the fact that the number of slots are fixed. Figure 6a shows the results of total revenue as we vary *RD_ratio*. It can be observed that HIP exhibits similar performance as that of RIP at *RD_ratio* = 0, and it is similar to DRIP at *RD_ratio* = 1. It can be observed that as we vary *RD_ratio* from 0 to 0.4, the total revenue of HIP is equal to RIP. It indicates that the top-revenue itemsets are giving the high revenue performance. Gradually, the performance of HIP starts declining from *RD_ratio*=0.5 and it becomes equal to DRIP at *RD_ratio* = 1. This indicates that after 0.4, the high *DRank* itemsets are added to the slots and hence, the performance of HIP becomes equal to DRIP.

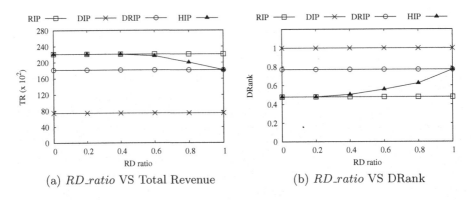

(a) *RD_ratio* VS Total Revenue   (b) *RD_ratio* VS DRank

Fig. 6. Effect of variations in *RD_ratio*

Figure 6b shows variations in *DRank* values as we vary *RD_ratio*. The behaviour HIP is similar to RIP at *RD_ratio*=0 and it is similar to DRIP at *RD_ratio* = 1. It can be observed that as we vary *RD_ratio* from 0 to 0.3, the *DRank* value of HIP is equal to RIP. Gradually, the *DRank* performance of HIP starts increasing from *RD_ratio* = 0.3 and it has reached to DRIP at *RD_ratio*=1. This indicates that after *RD_ratio* > 0.3, the high diversity itemsets are added to the slots and hence, the performance of HIP becomes equal to DRIP.

From Fig. 6a and Fig. 6b, we can conclude that HIP gives flexibility to trade total revenue versus *DRank*. HIP provides the option to realize the total revenue as that of RIP by increasing *DRank* to a reasonable extent. Moreover, HIP also provides the option to improve *DRank* significantly as that of DRIP, subject to trading of revenue.

6   Conclusion

Placement of items in a given retail store impacts the revenue of the retailer. Moreover, the placement of diverse items increases customer interest. In this paper, we have addressed the issue of itemset placement in a given retail store to improve diversity without compromising the revenue. In particular, we have proposed two itemset placement schemes, which take into account the top-revenue itemsets in conjunction with concept hierarchy based diversity. Our performance evaluation on a real dataset demonstrates the overall effectiveness of our proposed schemes in terms of both diversity as well as revenue. In the near future, we plan to investigate the notion of correlated patterns towards further improving itemset placement in retail stores.

References

1. Instacart market basket analysis - kaggle (2017). https://www.kaggle.com/c/instacart-market-basket-analysis/data. Accessed 30 June 2020
2. Bentalha, B., Hmioui, A., Alla, L.: The digitalization of the supply chain management of service companies: a prospective approach. In: Proceedings of the Conference on Smart City Applications (CSCA). ACM (2019)
3. Bradley, K., Smyth, B.: Improving recommendation diversity. In: Proceedings of the Irish Conference Artificial on Intelligence and Cognitive Science (ICAICS), pp. 75–84. Springer (2001)
4. Brijs, T., Swinnen, G., Vanhoof, K., Wets, G.: Using association rules for product assortment decisions: a case study. In: Proceedings of the Conference on Knowledge Discovery and Data Mining (CIKM), pp. 254–260. ACM (1999)
5. Brown, W., Tucker, W.: The marketing center: vanishing shelf space. Atlanta Econ. Rev. **11**, 9–13 (1961)
6. Chaudhary, P., Mondal, A., Reddy, P.K.: A flexible and efficient indexing scheme for placement of top-utility itemsets for different slot sizes. In: Reddy, P.K., Sureka, A., Chakravarthy, S., Bhalla, S. (eds.) BDA 2017. LNCS, vol. 10721, pp. 257–277. Springer, Cham (2017). https://doi.org/10.1007/978-3-319-72413-3_18
7. Chaudhary, P., Mondal, A., Reddy, P.K.: A diversification-aware itemset placement framework for long-term sustainability of retail businesses. In: Hartmann, S., Ma, H., Hameurlain, A., Pernul, G., Wagner, R.R. (eds.) DEXA 2018. LNCS, vol. 11029, pp. 103–118. Springer, Cham (2018). https://doi.org/10.1007/978-3-319-98809-2_7
8. Chaudhary, P., Mondal, A., Reddy, P.K.: An efficient premiumness and utility-based itemset placement scheme for retail stores. In: Hartmann, S., Küng, J., Chakravarthy, S., Anderst-Kotsis, G., Tjoa, A.M., Khalil, I. (eds.) DEXA 2019. LNCS, vol. 11706, pp. 287–303. Springer, Cham (2019). https://doi.org/10.1007/978-3-030-27615-7_22
9. Chaudhary, P., Mondal, A., Reddy, P.K.: An improved scheme for determining top-revenue itemsets for placement in retail businesses. Int. J. Data Sci. Analytics **10**(4), 359–375 (2020). https://doi.org/10.1007/s41060-020-00221-5
10. Chen, M., Lin, C.: A data mining approach to product assortment and shelf space allocation. Expert Syst. Appl. **32**, 976–986 (2007)

11. Chen, Y.L., Chen, J.M., Tung, C.W.: A data mining approach for retail knowledge discovery with consideration of the effect of shelf-space adjacency on sales. Decis. Support Syst. **42**(3), 1503–1520 (2006)

12. Coyle, M., Smyth, B.: On the Importance of Being Diverse. In: Shi, Z., He, Q. (eds.) IIP 2004. IIFIP, vol. 163, pp. 341–350. Springer, Boston, MA (2005). https://doi.org/10.1007/0-387-23152-8_43

13. Etgar, M., Rachman-Moore, D.: Market and product diversification: the evidence from retailing. J. Market. Channels **17**, 119–135 (2010)

14. Flamand, T., Ghoniem, A., Haouari, M., Maddah, B.: Integrated assortment planning and store-wide shelf space allocation: an optimization-based approach. Omega **81**(C), 134–149 (2018)

15. Han, J., Fu, Y.: Attribute-oriented induction in data mining. In: Fayyad, U.M., Piatetsky-Shapiro, G., Smyth, P., Uthurusamy, R. (eds.) Advances in Knowledge Discovery and Data Mining, pp. 399–421. AAAI (1996)

16. Han, J., Fu, Y.: Mining multiple-level association rules in large databases. Trans. Knowl. Data Eng. **11**(5), 798–805 (1999)

17. Hansen, P., Heinsbroek, H.: Product selection and space allocation in supermarkets. Eur. J. Oper. Res. **3**(6), 474–484 (1979)

18. Hart, C.: The retail accordion and assortment strategies: an exploratory study. Int. Rev. Retail Distrib. Consum. Res. **9**, 111–126 (1999)

19. Hurley, N., Zhang, M.: Novelty and diversity in Top-N recommendation - analysis and evaluation. Trans. Internet Technol. **10**(4), 14:1–14:30 (2011)

20. Kumara Swamy, M., Krishna Reddy, P.: Improving diversity performance of association rule based recommender systems. In: Chen, Q., Hameurlain, A., Toumani, F., Wagner, R., Decker, H. (eds.) DEXA 2015. LNCS, vol. 9261, pp. 499–508. Springer, Cham (2015). https://doi.org/10.1007/978-3-319-22849-5_34

21. Kumara Swamy, M., Krishna Reddy, P.: A model of concept hierarchy-based diverse patterns with applications to recommender system. Int. J. Data Sci. Analytics **10**(2), 177–191 (2020). https://doi.org/10.1007/s41060-019-00203-2

22. Kumara Swamy, M., Reddy, P.K., Srivastava, S.: Extracting diverse patterns with unbalanced concept hierarchy. In: Tseng, V.S., Ho, T.B., Zhou, Z.-H., Chen, A.L.P., Kao, H.-Y. (eds.) PAKDD 2014. LNCS (LNAI), vol. 8443, pp. 15–27. Springer, Cham (2014). https://doi.org/10.1007/978-3-319-06608-0_2

23. Liu, M., Qu, J.: Mining high utility itemsets without candidate generation. In: Proceedings of the Conference on Information and Knowledge Management (CIKM), pp. 55–64. ACM (2012)

24. Lu, Y.: Concept hierarchy in data mining: specification, generation and implementation. Master's thesis, School of Computer Science, Simon Fraser University, Canada (1997)

25. Kamber, M., Winstone, L., Gong, W., Cheng, S., Han, J.: Generalization and decision tree induction: efficient classification in data mining. In: Proceedings of the Workshop on Research Issues in Data Engineering, RIDE: High Performance Database Management for Large-Scale Applications, pp. 111–120. IEEE (1997)

26. Michalski, R., Stepp, R.: Automated construction of classifications: conceptual clustering versus numerical taxonomy. Trans. Pattern Anal. Mach. Intell. **5**(4), 396–410 (1983)

27. Srivastava, S., Kiran, R.U., Reddy, P.K.: Discovering diverse-frequent patterns in transactional databases. In: Proceedings of the Conference on Data Science and Management of Data (COMAD), pp. 69–78. CSI (2011)

28. Kumara Swamy, M., Krishna Reddy, P., Bhalla, S.: Association rule based approach to improve diversity of query recommendations. In: Benslimane, D., Damiani, E., Grosky, W.I., Hameurlain, A., Sheth, A., Wagner, R.R. (eds.) DEXA 2017. LNCS, vol. 10439, pp. 340–350. Springer, Cham (2017). https://doi.org/10.1007/978-3-319-64471-4_27

29. Tseng, V.S., Wu, C., Fournier-Viger, P., Yu, P.S.: Efficient algorithms for mining the concise and lossless representation of high utility itemsets. Trans. Knowl. Data Eng. (KDD) **27**(3), 726–739 (2015)

30. Tseng, V.S., Wu, C.W., Shie, B.E., Yu, P.S.: Up-growth: an efficient algorithm for high utility itemset mining. In: Proceedings of the Conference on Knowledge Discovery and Data Mining (KDD), pp. 253–262. ACM (2010)

31. Vargas, S., Castells, P.: Rank and relevance in novelty and diversity metrics for recommender systems. In: Proceedings of the Recommender Systems Conference (RecSys), pp. 109–116. ACM (2011)

32. Wang, M.T., Hsu, P.Y., Lin, K.C., Chen, S.S.: Clustering transactions with an unbalanced hierarchical product structure. In: Song, I.Y., Eder, J., Nguyen, T.M. (eds.) DaWaK 2007. LNCS, vol. 4654, pp. 251–261. Springer, Heidelberg (2007). https://doi.org/10.1007/978-3-540-74553-2_23

33. Wigley, S.M.: A conceptual model of diversification in apparel retailing: the case of Next plc. J. Text. Inst. **102**, 917–934 (2011)

34. Yang, M.H.: An efficient algorithm to allocate shelf space. Eur. J. Oper. Res. **131**(1), 107–118 (2001)

35. Yang, M.H., Chen, W.C.: A study on shelf space allocation and management. Int. J. Prod. Econ. **60**(1), 309–317 (1999)

36. Zheng, W., Fang, H., Yao, C.: Exploiting concept hierarchy for result diversification. In: Proceedings of the Conference on Information and Knowledge Management (CIKM), pp. 1844–1848. ACM (2012)

37. Ziegler, C.N., McNee, S.M., Konstan, J.A., Lausen, G.: Improving recommendation lists through topic diversification. In: Proceedings of the World Wide Web Conference (WWW), pp. 22–32. ACM (2005)

Recommending Question-Answers for Enriching Textbooks

Shobhan Kumar$^{(\boxtimes)}$ and Arun Chauhan

IIIT Dharwad, Dharwad, Karnataka, India
shobhank9@gmail.com, aruntakhur@gmail.com

Abstract. The students often use the community question-answering (cQA) systems along with textbooks to gain more knowledge. The millions of question-answers (QA) already accessible in these cQA forums. The huge amount of QA makes it hard for the students to go through all possible QA for a better understanding of the concepts. To address this issue, this paper provides a technological solution "Topic-based text enrichment process" for a textbook with relevant QA pairs from cQA. We used techniques of natural language processing, topic modeling, and data mining to extract the most relevant QA sets and corresponding links of cQA to enrich the textbooks. This work provides all the relevant QAs for the important topics of the textbook to the students, therefore it helps them to learn the concept more quickly. Experiments were carried out on a variety of textbooks such as Pattern Recognition and Machine Learning by Christopher M Bishop, Data Mining Concepts & techniques by Jiawei Han, Information Theory, Inference, & Learning Algorithms by David J.C. MacKay, and National Council of Educational Research and Training (NCERT). The results prove that we are effective in learning enhancement by enhancing the textbooks on various subjects and across different grades with the relevant QA pairs using automated techniques. We also present the results of quiz sessions which were conducted to evaluate the effectiveness of the proposed model in establishing relevance to learning among students.

Keywords: Community question-answers (cQA) · Relevance · Text-enrichment · Question-answers (QA) · Topic modeling · Data-preprocessing · e-learning

1 Introduction

Community Question Answer (cQA) forums like Quora[1], Stack Overflow[2], and Reddit[3] are getting more popular and it makes useful information available to the user (Daniele et al.) [4]. These forums are subject to a few restrictions, due

---

[1] https://www.quora.com/.

[2] https://stackoverflow.com/.

[3] https://www.reddit.com/.

© Springer Nature Switzerland AG 2020
L. Bellatreche et al. (Eds.): BDA 2020, LNCS 12581, pp. 308–328, 2020.
https://doi.org/10.1007/978-3-030-66665-1_20

to that any user can post and answer a question. On the positive side, a learner can ask any question and a variety of answers await. On the negative side, the effort is needed to get through all possible QA of varying quality and to make sense of them. This becomes the tiresome process that hampers the productivity of the learner. It would be very essential to have a functionality that helps users to retrieve the most appropriate cQA-QA pairs for all their topics in textbooks automatically for better insight. Here the main aim is to "fix" dilemmas or to give more clarity to the textbook concepts with relevant QAs, aiming "to enhance the textbook quality".

The textbook enrichment approach consists of the following steps: i) Identify key topics from a textbook. ii) Find relevant QA from cQA sites for those identified key topics. iii) Highlight the enriched topics to enhance the learning ability of the student. The main characteristics of cQA system describe how these educational community question answering systems can be used effectively by students, teachers to acquire more knowledge about their university courses.

Several studies were performed on various activities in cQA. For example, QA retrieval, question-question relatedness, and classification of comments (Moschitti et al.) [5], (Joty et al.) [12]. The conventional models of information retrieval (Chen et al.) [6], (Sewon et al.) [15], (Nakov et al.) [17], (Agrawal et al.) [1] can retrieve the QA sets from the old archived data. These existing methods have not addressed issues like how to use these large volumes of data from cQA forums to enrich textbook contents in a real-time manner to help the students fraternity with minimal effort. Though there is good content in the available textbooks, students tend to use cQA forums to gain more knowledge. There will be multiple answers that are attached to the same question in cQA, so it makes it hard for any user to select the relevant one. As a result, students use to spend more time collecting the necessary information from the web to get a better picture of a textbook concept.

The state-of-art techniques have not addressed the above-said problems. To address these issues this paper focuses on an automated model to enrich the given text document, with our *"topic-based text enrichment system"*. The proposed enrichment system selects the important topics (set of keywords) from the text document using LDA topic modeling. For each of these topics, it retrieves the most appropriate QA pairs form Quora cQA forums to enrich the textbook contents. The retrieved QA pairs are then present to the user in standard formats which helps them to understand concepts better.

Today, there exists an ecotones educational counterpart that acts as a bridge between conventional learning environments and the emergence of new knowledge environments that is truly composed around student-centered interaction. So this article mainly focusses on the enrichment of textbooks with digital content which offers new knowledge environments to the students. The rest of this paper has the following sections. Section 2 gives a brief overview of the related work.

Section 3 gives a detailed description of the proposed model. Section 4 reports the experimental results and discussion. Finally, we conclude the paper in Sect. 5.

2   Related Work

The open-domain QA was described as finding suitable QA in unstructured document collections. The QA retrieval tasks reflect subjects such as (Herrera J et al.) [10] semantic search, QA recommendation, and retrieval from cQA. Nowadays student frequently visits cQA site to find relevant QA when they find difficulty learning the concepts from the textbooks. The student community's viewpoint on the effects of cQA structures in education settings is emphasized (Ivan Srba et al.) [25]. After carrying out a large-scale quantitative and qualitative study, they highlighted major characteristics of QA systems that define the effect of educational cQA systems, and they specifically outline what kind of education cQA systems can be implemented in the academy to meet diversified student fraternities. In some studies, additional information is used to compensate for knowledge unbalance between the questions and the answers, some of the models are (Jiahui Wen et al.) [29], latent topic (Yoon et al.) [33], external knowledge (Shen et al.) [24] or question topic [30].

The traditional approaches tackled the question retrieval issues by converting the text of the questions with a Bag-of words representation, that uses TF-IDF (Chen et al.) [6] weighting methods to construct the suitable text representation. The retrieval can be carried out using dense representations alone (Vladimir et al.) [13], where embedding is learned by a simple dual encoder system from a limited number of questions and passages. The attention mechanism (Tan et al.) [26], (Wang et al.) [27] for attending important correlated information between QA pairs. These methods perform well when rating short answers, but the accuracy decreases with the increase in answers length (Gurevych et al.) [20]. Turn to end-to-end deep learning approaches (Chen Zheng et al.) [34] have shown promising results for QA retrieval tasks and the related task of question similarity ranking. The neural network models were once again used for Fast-QA retrieval tasks, a simple, context matching model (Dirk et al.) [28] for Fast-QA retrieval tasks that serve as the basis for FastQA invention. Few other options are made use of semantic relationships between questions and the available answers with the deep linguistic approach (Badri N. P. et al.) [19]. An interactive multitask and multi-head attention network for cQA (Min Yang et al.) [32] where the document modeling functionality uses the external knowledge to classify the background information and filter out long text noise information that has complicated semantic and syntactic structures.

Frequently Asked Question (FAQ) (Sakata et al.) [23] retrieval system that follows an unsupervised information retrieval system to calculate the similarity between the user query and an available question. The relevance between the query and answers are learned through the use of QA pairs in a FAQ database. The BERT model [8] is being used for the relevance computation. Detection of the duplicate question in cQA is closely related to QA retrieval and question-question similarity. A weak supervision method for (Andreas et al.) [21] duplicate question detection in cQA using the title and body of a question. The method shows that weak supervision with question title and body of the question is an effective approach to train cQA answer selection models without direct answer supervision. Recent research on coarse-to-fine QA for long texts, such as (Choi et al.) [7], (Wang et al.) [27] reading comprehension concentrate on the extraction of answer span in factoid QA, in which certain factoid questions can be answered with a certain word or a short sentence. A simple selector of sentences to choose a minimum set of sentences (Wu et al.) [30] used to fit the QA model to deal with most questions in datasets like SQuAD, NewsQA, etc. To increase the efficiency of retrieval (Sewon et al.) [16], the relationship between the subject and the body of the questions was used to condense the representation of the questions.

The text enrichment model (Agrawal et al.) [1] which augments the text documents with QAs from Wikipedia archives. A sentence-level text enrichment model (sk et al.) [14], it fetches the relevant cQA-QA pairs for the given text sentences. The textbook contains a large number of sentences so identifying the important sentence and fetching the relevant cQA-QAs is a tiresome process. An alternative is to have an abstract description using a clustering approach to narrow down the length of the document as it is not feasible to use full-text clustering data. The sparsity problem will occur during the clustering of short text, abstracts, different approaches [35], [31] to solve the problem were suggested. One of the most sought after approaches is the Latent Dirichlet Allocation (LDA) (Blei et al.) [3] model since LDA has the benefit of lowering the text dimension. A single topic is considered as a distribution of probabilities over terms so that each document can be presented as a mixture of different topics. Every word is created individually by a specific topic in the LDA model, and the corresponding topic is taken from its corresponding proportion distribution. Since LDA excels in extracting lateral topic information internally from the text document, in our approach we used the same for selecting the important topics from the given textbook.

The major issue current literature is that there is no special platform that enriches the textbook in a live manner with the relevant cQA-QA pairs for better insight. So this paper presents a new method that selects the important topics from the text document using topic modeling and then enriches the textbook contents by retrieving the relevant QA from cQA forums.

3 Methodology

In this section, we explain how our model works for a text enrichment with appropriate QA from cQA. Figure 1 illustrates the overall architecture of the proposed model where the model takes text documents as input and presents the appropriate number (based on the threshold values) of the respective Quora QA pairs for each of the important topics in the textbook. The work in this paper has the following steps. i) Data preprocessing ii) Selection of top K topics that occur in a given preprocessed data. iii) Text Enrichment-Finding appropriate cQA QA pairs and present the results (QA and the corresponding URLs) to the user in standard formats.

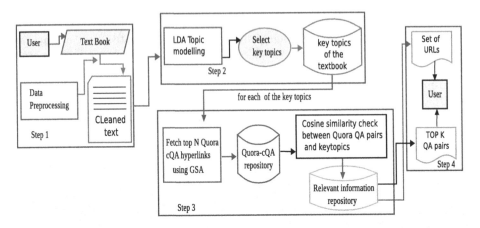

Fig. 1. The overall architecture of the proposed "Topic-based Text Enrichment" system.

Algorithm 1: Topic extraction from the text document

Input : Preprocessed text document
Output: The top K topics that occur in a given text document C

1 Identify the sentence list in the text document using sentence tokenizer
2 **for** *each sentence i* **do**
3 │ Use word tokeniser to compute the set C_i of keyphrases
4 │ Use predefined stop word list to remove non-keywords in C_i
5 │ Prepare a final C_i tokens list
6 **end**
7 Train an LDA model on a collection of X documents with Y topics
8 Save the model
9 Select the top K topics and its contributing words (collection of keywords)
10 Save the selected topics and corresponding words for the given text document.

3.1 Data Preprocessing

In general, the textbook contents have several special characters as well as HTML links, a duplicate copy of the sentence, etc. These details need to be cleaned because these details were not necessary for our work. In our data preprocessing work, the HTML tag, URLs, images, punctuation symbols, and redundant tuples were removed from the body of the document. In addition to that, the sentence compression method effectively reduces the duplication in a given text. Our method exploits various lexical and structural features for ensuring clean text and it becomes the facts for the subsequent topic modeling and QA retrieval process.

3.2   Top K Topics Selection

The main goal is to find important topics in a given text document, these topics are excellent means to provide a concise summary of textbook concepts. We used LDA topic modeling [3] to extract the important topics in a textbook, LDA excels at the internal extraction of the lateral topic information from the text. It learns a vector of a document that predicts words inside that document while neglecting any structure or how these words interact at a local level. The vectors of the text are often sparse, low-dimensional, and easily interpretable, highlighting the pattern and structure of documents. Here we need to determine a fair estimate of the number of topics that occur in the document collection. We enrich the textbooks, by providing these topics (set of words) to Google Search API (GSA) with a keyword "Quora". The GSA returns the relevant hyperlinks from which we considered only top N Quora cQA hyperlinks which can be used for further augmentation process (QA retrieval).

Algorithm 1 clearly describes how the relevant topics can be selected from the given text document. To prepare the cleaned data for LDA training it is preferred to traverse through the book line by line and save it as a list. The final list size becomes the total number of lines inside the book. Natural language processing (NLP) techniques such as Natural Language Toolkit (NLTK) word tokenizer and sentence tokenizer are used to generate token lists. Filtering useless data is the primary issue in this step, these words are so common in any document and may act as noise if we use them for model training and testing. The system scans the array of text documents extensively as shown in Fig. 2 and generates tokens for each of the text documents. Later important topics, its contributing words are extracted using the trained LDA model. We extract the relevant QA from the Quora cQA forum using GSA for each of these topics.

3.3   Text Enrichment-Finding Appropriate cQA-QA Pairs

In this work, we used the Quora cQA forum to enrich the textbooks. Figure 2 outlines the sequence of operations needed to produce the final copy of the augmented text (QA pairs and URLs) for the given text document. Algorithm 2, describes the operating sequence for finding relevant QA for the important topics. The system extracts top K topics and its contributing words from the text document using a trained LDA model. Since each topic has a collection of keywords we used different combinations of these words (unigram bigrams and trigrams) (let say 5 words in a topic and taking 3 keywords at a time, $^{5}C_3 = 10$ possible combinations) to extract Quora hyperlinks using GSA. We can find many QA pairs for every extracted Quora hyperlinks which may or may not be relevant for the given topics, hence we used Word2Vec [11] and cosine similarity method for checking the similarity between two text documents (retrieved QA pairs and contributing words of the topics). The documents (cQA QA pairs) can be represented in vector form as follows:

$$\boldsymbol{d} = (w_{d0}, w_{d1}, ..., w_{dn}) \qquad (1)$$

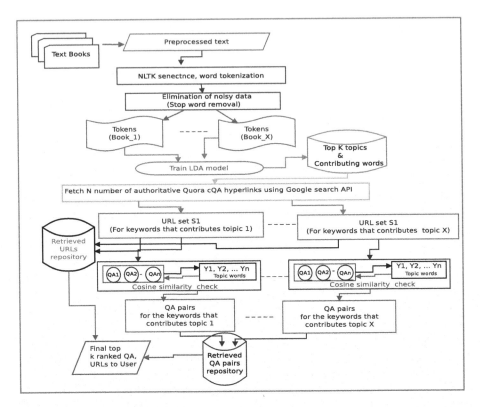

Fig. 2. Selection of top K topic from the given text document, for each topic the model fetches the relevant information (QAs, URLs) from the Quora cQA to enrich the text.

Similarly, the contributing words of the selected topic (say topic - x) can be represented in vector form as:

$$x = (w_{x0}, w_{x1}, ..., w_{xn}) \tag{2}$$

Where w_{di} and w_{xi} $(0 \leq i \leq n)$ represents the occurrence of each term in a given text documents. The cosine similarities between two vectors (QA pairs and collection of keywords (contributing words of the topics)) can be defined from Eq. 1 and 2 as:

$$Sim(d, x) = \frac{\sum_{i=1}^{n} x_i.d_i}{\sqrt{\sum_{i=1}^{n} x_i^2} \cdot \sqrt{\sum_{i=1}^{n} d_i^2}} \tag{3}$$

The cQA forums provide a voting system in both positive (upvote) and negative (downvote) directions. The difference between these two votes is often regarded as the question score [18]. Therefore we used the same metric to rank the available QAs and choose the top K QAs to enrich the given document. The enrichment system retrieves top k QA pairs, corresponding URLs based on the

---

Algorithm 2: Retrieval of relevant cQA-QA pairs for the important topics of the text document

---

    Input : A set of keywords y of the topics
    Output: A top k cQA-QA pairs and corresponding URLs.

1 **for** *each y_i elements* **do**
2   │ Select top N Quora cQA hyperlink $\mathbf{Q}_j$ using **GSA**
3   │ **for** *each link j* **do**
4   │   │ Perform cosine similarity check between topics keywords and cQA-QA pairs
5   │ **end**
6   │ Retrieve the top k relevant QA pairs, associated URLs based on upvote counts of QA pairs
7 **end**
8 Save the retrieved data onto the result-array $\mathbf{Z}_i$
9 **for** *each $\mathbf{Z}_i$ data* **do**
10 │ Integrate the retrieved information for all the topics to prepare a final enriched file
11 **end**
12 Present the enriched data file to user in suitable formats.

---

cosine similarity scores, and upvote counts where k is the threshold value set by the user. This process continues for all the important topics of the textbooks. The retrieved data are integrated to produce the final copy of the enriched information for the given text document to enhance the readability of the learners. In addition to the QA pairs, the enrichment model also provides a set of cQA forum URLs, offering a bunch of QA pairs apart from the retrieved one. All the retrieved data (QA pairs and URLs) are presented in a standard format to the user.

4 Experiment Results and Discussion

To enrich the text documents, the proposed model retrieves the relevant QAs, URLs from the Quora cQA platform. We tested our model using textbooks of various subjects and grades.

Data Set 1 (Training-Topic Modeling): To train the LDA model we used text documents such as National Council of Educational Research and Training (NCERT) Grade XI Computers and Communication Technology part I&II (nearly 12K words) and Pattern Recognition and Machine Learning [2] by Christopher M. Bishop (nearly 70K words). In total used 82K words to train the topic model.

Data Set 2 (Testing-Topic Modeling): To test our topic model we used a collection of text documents such as NCERT textbook (Science Grade X) and graduate levels textbooks - Data Mining Concepts & techniques [9] by Jiawei Han and Information Theory, Inference, & Learning Algorithms [15] by David J.C. MacKay. Besides that, We also extracted nearly 30 documents related to data mining, machine learning, etc. from Wikipedia. In total, we used 165K words to test our topic model.

Table 1. The details of the textbooks used for training and testing of the LDA topic model. (#words for training: 82K, #words for testing: 165K)

No	Textbook title	Author	#words (training)	#words (testing)
1	Science (Grade X)	NCERT	-	5K
2	Computers & Communication Technology part 1 & II	NCERT	12K	-
3	Data Mining Concepts & techniques	Jiawei Han	-	72K
4	Information Theory, Inference and Learning Algorithms	David J.C. MacKay	-	81K
5	Pattern Recognition and Machine Learning	Christopher M. Bishop	70K	-
6	Wiki document	Wikipedia	-	7K

LDA Model Training & Evaluation: We trained the LDA model in a system (intel i7 Desktop PC) which has the configuration Intel Core I7, 32 GB RAM, 1 Tb Hard Disk, and 2 GB graphics. Table 1 describes the document statistics used for training and testing purposes. The hyperparameters (alpha (document-topic density) and beta (word-topic density)) (ref. Fig. 3) affect the sparsity of the topics while training the model. The LDA model is trained for 10K iterations with chunk size 300.

Previously, the topics that originated from these LDA modeling were checked on their human interpretability by introducing them to humans and taking their feedback on them. It was not a quantitative measure, it was only qualitative. Quantitative approaches to assessing topic models are perplexity and [22] topic coherence but perplexity is a poor indicator of the quality of the topics. In contrast to that, the topic coherence measure is a convenient measure for judging a given topic model. Hence we used the same to gauze the quality of our trained topic model. The coherence score is computed as follows:

$$Coherence = \sum_{i<j} score(y_i, y_j) \tag{4}$$

where y_i, y_j are the keywords of the topic.

$$score_{\text{UCI}}(y_i, y_j) = \log \frac{p(y_i, x_j)}{p(y_i)p(y_j)} \tag{5}$$

Where p(y) reflects the probability of seeing y_i in a given text document, and $p(y_i, y_j)$ the probability that both y_i and y_j will co-occur in a given text document. We compute a topic's coherence (UCI coherence) by measuring the degree of semantic similarity among its high scoring keywords. The following sequence of operations is performed to measure the coherence value of our topic model. At first, we pick the top 10 keywords for each topic, then pairwise scores are calculated for each of the keywords and all pairwise scores are aggregated to determine the coherence score for a specific topic. Finally the mean of the coherence score for all topics is computed to obtain the final coherence score for the model. We trained our topic model by tuning the hyperparameter alpha: 0.01–1.0, beta: 0.01 and we obtain a good coherence score 0.702351 at alpha 0.03 which is a quite an acceptable value.

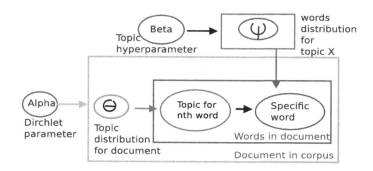

Fig. 3. Schematic representation of the flow of LDA topic modeling.

To verify the credentials of the proposed model we extracted 10–15 topics (each topic-10 keywords) from the above-mentioned textbooks and wiki documents based on the nature and complexity of the information which conveys. So we extracted nearly 77 topics since each topic has associated with multiple keywords (set of contributing words), we performed a different combination of the same (ngrams- where n: 1, 2, 3, 5). For each of these combinations, we extracted top N Quora cQA hyperlinks using GSA. Later we extracted top k QA pairs, URLs from cQA where the value of k ranges from 1 to 5. There are millions of QA pairs are available in cQA forums, due to that there will be high variance in the quality of the questions and available answers. Since we used different combinations of these words to retrieve QA, we find that there is plenty of cQA-QA which are closely related to the given set of keywords. So just to cut short the retrieval QA pairs we set the cosine similarity score limit as 0.3. Table 2 describes the statistics of retrieved QA pairs after handling the redundancy issues. We asked 30 students who have taken courses on "Machine Learning" for their undergraduate studies to assess our retrieval model efficiency. We took one-third of the total retrieved QA pairs corresponds to threshold values for a smoother evaluation process. The retrieved information, extracted topics (keywords) are presented to these students, and the evaluators are asked to mark it with the flags such as "relevant" or "irrelevant" based on the relevance of the QA pairs for the given topics of the textbook.

Retrieval Model Evaluation: We used rank-based measures such as Precision@k (P@k) as an evaluation metric for testing the efficiency of our retrieval model. (P@k) is defined as the proportion of the relevant k results. Where r is the number of relevant QA that were collected up to rank k.

$$P@k = \frac{r}{k} \tag{6}$$

Table 2. The statistics of the retrieved QA pairs from Quora (see footnote 1) cQA for important topics of the textbooks. There will be 10 keywords that contribute to each topic.

Text book	No of topics	Retrieved QA pairs @ different threshold value k				
		k = 1	k = 2	k = 3	k = 4	k = 5
Pattern Recognition and Machine Learning	15	150	220	291	335	372
Information Theory, Inference and Learning Algorithms	15	150	245	304	367	392
Data Mining Concepts & techniques	13	130	238	312	374	398
NCERT Science- Grade X	10	100	164	197	214	263
NCERT Computers and Communication Technology part 1 & II	12	120	189	228	253	279
Wiki document	07	70	126	173	194	211

Table 3. The retrieved top 3 questions from Quora (see footnote 1) cQA for each keyword "Bayesian" and "Probability".

Keyword: "Bayesian"	
Q1	What is the simplest explanation of Bayesian statistics?
Q2	What's the intuition behind Bayesian Inference?
Q3	What's the relationship between Bayesian statistics and machine learning?
Keyword: "Probability"	
Q1	What is probability?
Q2	What are the real life applications of probability in math?
Q3	How and where can I use probability in my daily life?

Table 4. The retrieved top 3 questions from Quora (see footnote 1) cQA for each keyword "Gaussian" and "k-means"

Keyword: "Gaussian"	
Q1	How would you explain the Gaussian distribution in layman's terms?
Q2	Why is the Gaussian distribution widely used in practice?
Q3	Why do we mostly use Gaussian processes in machine learning?
Keyword: "k means"	
Q1	What is the k-Means algorithm and how does it work?
Q2	What does K mean in clustering?
Q3	What is a k-means algorithm?

Table 5. The retrieved top 3 questions from Quora (see footnote 1) cQA for the keyword "conditional".

Keyword: "Conditional"	
Q1	What is a conditional sentence?
Q2	Grammar: What's wrong about using a mixed conditional?
Q3	What are the different conditional statements and what are the rules for using them?

Table 6. The retrieved top 3 questions from Quora (see footnote 1) cQA for the combination of three keywords.

Keywords: "Probability", "Gaussian", "k-means"	
Q1	What is the difference between K-means and the mixture model of Gaussian?
Q2	When do I use Gaussian Mixture Models v/s K-means for clustering data?
Q3	What are the advantages to using a Gaussian Mixture Model clustering algorithm?
Keywords: "Bayesian" "Conditional", "k-means"	
Q1	What is the difference between Naive Bayes and K-Means clustering?
Q2	What is the k-Means algorithm and how does it work?
Q3	How can we obtain the necessary conditions of K-means and K-median?

The retrieved questions are presented in Tables 3, 4, 5, 6 and 7 for the test data extracted from Wikipedia. Tables 3, 4 and 5 represents the retrieved question using the unigram approach, where for every keyword of the topics we extracted top k QA form cQA. We succeeded to retrieve the relevant for most of the keywords with few exceptions, (ref Table 5) where the intent is to retrieve the QA relates to conditional probability but with the single keyword, the model retrieves the questions which are related to a different conditional statement.

Table 7. The retrieved top 5 questions from Quora (see footnote 1) cQA for the combination of all five keywords.

	Keywords: "Probability", "Gaussian", "k-means", "Bayesian", "Conditional"
Q1	How is K-Means application of expectation maximization on Naive Bayes?
Q2	Why are unsupervised learning algorithms such as K-means clustering considered "machine" learning, when in fact, they have been explicitly programmed?
Q3	What are the advantages to using a Gaussian Mixture Model clustering algorithm?
Q4	What is the difference between K-means and the mixture model of Gaussian?
Q5	What is the k-Means algorithm and how does it work?

Table 8. The retrieved top 5 questions from Quora (see footnote 1) cQA for the combination of all five keywords.

	Keywords: "Multinomial", "Function","RVM", "Model", "Likelihood"
Q1	How do you explain maximum likelihood estimation intuitively
Q2	What is an intuitive explanation for why a likelihood function doesn't sum (or integrate) to 1?
Q3	What are the challenges of having Introverted Intuition (Ni) as the primary function?
Q4	What is an intuitive explanation of the Dirichlet distribution?
Q5	What is an intuitive explanation of the Dirichlet distribution as a probability distribution over the (k-1)-dimensional probability simplex?

Table 9. The retrieved top 5 questions from Quora (see footnote 1) cQA for the combination of all five keywords.

	Keywords: "Sequential", "Curve", "Distribution", "Fitting", "Bias"
Q1	How would you explain over-fitting issue to a non-technical user? Do you say something like training on 100% of the data doesn't make it useful for prediction and causes failure generalize the data?
Q2	What's ROC curve?
Q3	What is an intuitive explanation of over-fitting, particularly with a small sample set? What are you essentially doing by over-fitting? How does the over-promise of a high R, low standard error occur?
Q4	How does curve fitting work?
Q5	What is a statistical distribution?

Table 10. The obtained precision scores for the retrieved QA pairs, which describes whether the retrieved QA pairs are relevant to given topics or not.

Threshold value k	#QA for evaluation	Precision@k
1	245	0.497083
2	395	0.582173
3	510	**0.675852**
4	560	0.512148
5	590	0.437294

Hence we also used different combinations of the keywords (bigrams, trigrams) to extract the QA pairs. Table 6 and Table 7 represent the retrieved question for the trigram and five-gram approach where we used three and five keywords together. Similarly, Table 8 to Table 9 represent retrieved the top 5 questions when we used five keywords together from the "Machine learning" book for the given topic. The contents inside the text documents are not adequate to satisfy the curiosity of the students. The retrieved QA pairs have enough examples to clear all the doubts and it helps the students to learn the new concept in a timely manner.

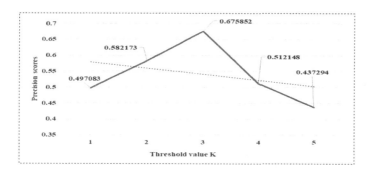

Fig. 4. The precision scores for the retrieved QA pairs. Though the retrieval system fetches many QAs for the given topics, in the majority of the cases evaluator concludes that out of 5 QA for each topic only 3 are most relevant and the rest are poor in terms of relevancy with reference to the given topic. Hence the retrieved QA pairs relevancy score reaches its peak point value when k = 3.

Table 10 describes the precision score comparison for the retrieved QA pairs for the given topics of the textbook. The precision scores state that the relevance percentage is reached to its peak rate (ref. Fig. 4) when the threshold value is 3 and for subsequent threshold value the precision score was gradually decreased. Through our model, the students can quickly access all the relevant QAs for important topics of the given textbook, so it definitely creates enhanced learning for the students.

Table 11. No. of quiz questions answered.

#Students: 24 (Group 1) Subject: Machine Learning [2]			
Week	1–4	5–8	9–10
Week 1	9	15	00
Week 2	5	17	2

Table 12. The obtained quiz scores.

q_score: Avg. quiz scores (Max 10) t_span: Avg. time spent (min:sec)			
Week	Students	q_score	t_span
Week 1	24	05.08	17:53
Week 2	24	06.17	15:49

Table 13. No. of quiz questions answered.

#Students: 21 (Group 2) Subject: Data Mining [9]			
Week	1–4	5–8	9–10
Week 1	6	14	1
Week 2	2	15	4

Table 14. The obtained quiz scores.

q_score: Avg. quiz scores (Max 10) t_span: Avg. time spent (min:sec)			
Week	Students	q_score	t_span
Week 1	21	05.21	16:11
Week 2	21	06.85	13:49

To ensure that the proposed system helped students to have better insight into the concepts a series of quiz sessions were conducted. A total of 45 students (23 Girls + 22 Boys) who have taken "Machine Learning" and "Data Mining" courses for their undergraduate studies took part in the evaluation process (two weeks). We made two groups (Group 1: 24 students who have taken "Machine Learning" courses for their undergraduate studies. Group 2: 21 students who have taken "Data Mining" courses for their undergraduate studies).

In the first week, we selected a text documents (For Group 1: Chap. 1-PRML [2], For Group 2: Chap. 8 & 9- Data Mining [9]). The students have read the text documents and then answered the quiz question. The quiz questions were prepared from the same text documents and asked students to answer the quiz questions. (please refer to Fig. 6a and Fig. 6c - Appendix) (Time - Max 20 min to answer the quiz, Max score: 10 points).

In the second week, we train the LDA topic model using the text documents (Chap. 2: from PRML and Chap. 10 & 11 from Data Mining) and selected the first N topics(each topic has a set of keywords that contribute the topic). Using combinations (n-grams where n: 1, 2, 3, 4, 5) of these keywords as queries, we selected the first N Quora cQA hyperlinks using GSA. Later we extracted the top K QA pairs from cQA. The students previewed both the retrieved QAs and text documents and then answered the second set of quiz questions (please refer to Fig. 6b and Fig. 6d - Appendix). In both the week the obtained quiz scores and total time spent by the students to answer the quiz are noted and analyzed.

Table 12 and Table 14 details the results obtained on short quiz questions. Table 11 and Table 13 describes the number of quiz questions answered by two groups of students during the evaluation process. The obtained results show that students who previewed the text document and relevant QA pairs completed quiz questions in a shorter time and obtained good scores.

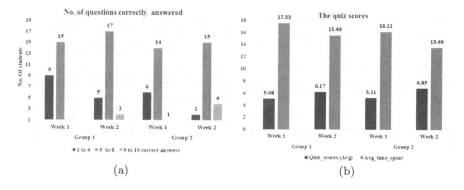

(a)

(b)

Fig. 5. (a) The number of quiz questions correctly answered by the students during the evaluation process. (b) The obtained quiz scores (avg) and time spent by the students to answer the quiz questions. When the students previewed both text documents and relevant QAs from Quora cQA more and more students correctly answered 5–8 and 9–10 quiz questions and also they completed the quiz in a shorter time compared to the first week. These results clearly state that the relevant cQA-QA pairs for the given text significantly enrich the textbook contents and improves the readability of the students.

Figure 5 is showing the results of our study (Quiz sessions for students). Group 1: We noticed that in week 1 nine students gave 1–4 and fifteen students gave 5–8 correct answers by learning text documents (Chap. 1- PRML). On the other hand in week 2, seventeen students gave 5–8 correct answers and 2 students gave 9–10 correct answers by previewing the text documents (Chap. 2- PRML) and relevant cQA-QAs.

Group 2: Fourteen students (Group 2) gave 5–8 correct answer and only one student gave 9–10 correct answers for the week 1 quiz by learning text documents (Chap. 8 & 9 - Data Mining). In the second week, fifteen students gave 5–8 correct answers, and four students 9–10 correct answers by previewing the text documents (Chap. 10 & 11- Data Mining) and relevant cQA-QAs. It can be easily noticed that when students previewed relevant QA pairs, maximum students gave 5–8 and 9–10 correct answers that are really high compared to week 1 performance (ref. Fig. 5a).

Group 1 students secured 5.08 points and they used 17.53 min to complete the quiz (week 1). They secured 6.17 points and used 15.49 min to complete the quiz (week 2). Group 2 students secured 5.21 points and they used 16.11 min to complete the quiz (week 1). They secured 6.85 points and used 13.49 min to complete the quiz (week 2) (ref. Fig. 5b).

The difficulty level of the week 2 quiz (quiz 2) (ref. Fig. 6b & Fig. 6d - Appendix) is high compared to the week 1 quiz (quiz 1). The difficulty level is high when we have to come up with our own answers for a given question and it is low when we need to choose out of options. Despite the high difficulty level they performed well for the week 2 quiz questions. One thing is very noticeable that in week 2 students have completed the quiz sessions in a shorter time compared to week 1 and they obtained good scores. These results prove that the relevant QA pairs from Quora cQA help students to improve their conceptual understanding ability.

5   Conclusion and Future Scope

The work in this paper mainly focusses on providing an enhanced learning environment to the students by recommending relevant QA from cQA forums. The textbook contains a huge volume of data, so reading and understanding each and every line is a tiresome process for the students. In addition to that, the textbooks in the markets are not adequate to satisfy student's curiosity, hence they frequently use cQA forums to learn more. Millions of QAs are available in cQA forums, due to that it requires an extra effort to go through all possible QAs and choose the right one for their better learning. To address this issue our enrichment model selects the required number of important topics of the textbook using topic modeling (LDA), these topics are enough to covey the concise summary of the textbook contents. Later for each of these selected topics, it fetches the top k relevant QA from Quora cQA. Here all the retrieved QAs, corresponding URLs are made available in a standard format for the given textbook, hence it definitely improves the productivity of the students.

The proposed model has been tested using the variety of text documents such as NCERT Grade X and XI textbooks, Pattern Recognition and Machine Learning by Christopher M Bishop, Data Mining Concepts & techniques by Jiawei Han, Information Theory, Inference, & Learning Algorithms by David J.C. MacKay and a few randomly selected Wikipedia documents show that our model is effective in enriching textbooks on different subjects across various grades by providing the relevant cQA-QA pairs in a real-time manner. We also conducted quiz sessions to evaluate the efficacy of the proposed model in establishing relevance to learning among students. The test results indicate that the relevant QAs from Quora cQA help the students develop their conceptual understanding ability. Currently, we are recommending relevant QA pairs form cQA for the important topics of the textbook, which has only subjective questions. Further, we can add various types of questions like multiple choice questions, fill in the blanks, true or false, etc., which can act as a personalized learning system for the students.

Appendix: Questionnaire for Quiz

See Fig. 6.

Questionnaires for Quiz 1 (Group 1) (Max time :20 Min, Max points:10)
(Refer PRML Chapter-1 : Introduction (Polynomial Curve Fitting, Probability Theory, Model Selection, The Curse of Dimensionality, Decision Theory and Information Theory)
....
1) For discrete random variables X and Y define
  a) The entropy H(X). b) The mutual information I(X; Y).
2.Compared to the variance of the Maximum Likelihood Estimate (MLE), the variance of the Maximum A Posteriori (MAP) estimate is _____
  a) higher b) same c) lower d) it could be any of the above
3.Answer the following
  i)The updated probabilities of an Event in light of newly-collected data are known as _____ probabilities
    a) Prior b) Posterior c) Conditional d)Simple
  ii) If P(A) = 0.20, P(B|A)=0.60 and P(B|A')=0.25, then P(A|B)=.............
    a) 0.3750 b)0.7060 c)0.6250 d) 0.2941
4.Fit a straight line y=a+bx into the given data. What is the value of y when x=8 ?. (x,y):(1,20)(2,21)(3,22)(4,23)(5,24)(6,25).
    a) 45.2 b) 26 c) 28   d) 37
5.Fit a straight line y=a+bx into the given data:
  (x,y):(5,12)(10,13)(15,14)(20,15)(25,16).
  a) y=11 b) y=0.2x c) y=11+0.2x   d) y=1.1+0.2x
6.Answer the following
  i)Which of the following is a good way to visualize Bayes' Theorem?
    a) Two-way table b)Weighted tree c) Candy-bar diagram d) All of these
  ii)A test says you have the flu, but you really don't. This is an example of a.......
    a)True Positive b)False Positive c)True Negative d) False Negative
7.Answer the following
  i)In Normal distribution, the highest value of ordinate occurs at
    a) Mean b) Variance c) Extremes d) Same value occurs at all points
  ii)Skewness of Normal distribution is _____
    a) Negative b) Positive c) 0 d) Undefined
8.Which of the following best describes the joint probability distribution P(X, Y, Z) for the given Bayes net.

$X \rightarrow Y \rightarrow Z$

a) P(X,Y,Z)=P(Y)*P(X|Y)*P(Z|Y).   b) P(X,Y,Z)=P(X)*P(Y|X)*P(Z|Y),
c) P(X,Y,Z) = P(Z)*P(X|Z)*P(Y|Z). d) P(X, Y, Z) = P(X) * P(Y) * P(Z)
9.True False Questions.
  i)We can use gradient descent to learn a Gaussian Mixture Model.
    a) True   b) False
  ii)When a decision tree is grown to full depth, it is more likely to fit the noise in the data. a) True   b) False
10. A virus has infected 1.8% of a population. A test detects this virus 95% of the time when it is actually present, but it returns a false positive 3% of the time when the virus is not present. If a person selected at random from this population tests positive for the virus, what is the probability that this person is actually infected? [Round to the nearest percent.]

    a) 37% b)63% c)34% d) 66%

(a)

Questionnaires for Quiz 2 (Group 1) (Max time :20 Min, Max points:10)
(Refer PRML Chapter 2 : Probability Distributions (Binary Variables, Multinomial Variable, The Gaussian Distribution, The Exponential Family, Nonparametric Methods)
....
1) Under what circumstances can you use the normal approximation when working with the beta distribution?. (Hint: w.r.t α and β).

2.Specify a conjugate prior when the likelihood is an exponential distribution with parameter θ > 0.

3.Consider a dataset of n points x_1 , . . . , x_n , where x_i is drawn from a Gaussian distribution with mean μ, and variance σ^2_i > 0, for i = 1, . . . , n. What is the ML estimate for μ, when the variances σ^2_1 , . . . , σ^2_n are known?

4.Let X|Y = −1 be distributed according to a mixture of two Gaussians given by 0.5N (3, 1) + 0.5N (9, 1). Let X|Y = +1 be distributed as N (6, 1). Give the Bayes classifier for the zero-one loss function and Bayes error (or the probability of mis-classification).

5.Consider two variables x and y with joint distribution p(x, y).
  Prove that E[x] = Ey [Ex[x|y]]

6.Show that the inverse of a symmetric matrix is itself symmetric.(Hint : $(M^{-1})' = M'$)
7.Answer the following
  i)At a certain university, 4% of men are over 6 feet tall and 1% of women are over 6 feet tall. The total student population is divided in the ratio 3:2 in favour of women. If a student is selected at random from among all those over six feet tall, what is the probability that the student is a woman?
  ii)A box of cartridges contains 30 cartridges, of which 6 are defective. If 3 of the cartridges are removed from the box in succession without replacement, what is the probability that all the 3 cartridges are defective?
8. Find the MLE of μ in the normal distribution, where the log likelihood function for the normal distribution is the given equation

$$\ln L(x_1, ..., x_n|\mu) = -\frac{n}{2} \ln(\sigma^2) - n \ln(\sqrt{2\pi}) - \frac{\sum(x_i - \mu)^2}{2\sigma^2}$$

9.i) What should the sample size be to use t distribution?
  ii) What should the sample size be to use t distribution if you know the data is normally distributed?
10. i) A coin is tossed 100 times and lands head 62 times. What is the maximum likelihood estimate for θ the probability of heads?
  ii) A coin is tossed n times and lands heads k times. What is the maximum likelihood estimate for θ the probability of heads?

(b)

Questionnaires for Quiz 1 (Group 2) (Max time :20 Min, Max points:10)
(Refer Data Mining concept & techniques Chapter 8 Classification: Basic Concepts, Decision Tree Induction, Bayes Classification Methods, Rule-Based Classification, Model Evaluation and Selection, Techniques to Improve Classification Accuracy and Chapter 9 Classification: Advanced Methods :Bayesian Belief Network, Classification by Backpropagation, Support Vector Machines, Lazy learners)
....
1. Which of the following is/are true about boosting trees?
  i)In boosting trees, individual weak learners are independent of each other.
  ii)It is the method for improving performance by aggregating the results of weak learners. a) i b) ii c) i and ii d) None of these

2. How the compactness of the Bayesian network can be described?
a) Locally structured b) Fully structured c) Partial structure d) All of the mentioned
3. What do you mean by a hard margin? a) The SVM allows a very low error in classification b) The SVM allows a high amount of error in classification c) None of the above
4.i) The cost parameter in the SVM means a) The number of cross-validations to be made b) The kernel to be used c) The tradeoff between misclassification and simplicity of the model d) None of the above.
  ii)Suppose you have trained an SVM classifier with a Gaussian kernel, and it learnt some function of the classes in the data. Now, say for training 1 time in one vs all method using the SVM is taking 10 seconds. How many seconds would it require to train one-vs-all method end to end?
    a) 20   b) 40   c) 60   d) 80

5.What is meant by generalized in the statement "backpropagation is a generalized delta rule" ?. a) because delta rule can be extended to hidden layer units b) because delta is applied to only input and output layers, thus making it more simple and generalized c) it has no significance d) none of the mentioned

6.For which of the following hyperparameters, higher value is better for the decision tree algorithm? i) A number of samples used for split ii) Depth of tree iii) Samples for leaf. a) i and ii   b)ii and iii
c) i, ii and iii   d)None
7._____ is the mathematical likelihood that something will occur. a) Classification. b) Probability. c) Naive Bayes Classifier. d) None of the other answers are correct.

8. To apply bagging to regression trees which of the following is/are true in such case? i) We build the N regression with N bootstrap sample ii) We take the average the of N regression tree iii) Each tree has a high variance with low bias. a) i and ii b) ii and iii c) i and iii d) i, ii and iii

9.Which of the following statement is true about the k-NN algorithm? i)k-NN performs much better if all of the data have the same scale. ii) k-NN works well with a small number of input variables (p), but struggles when the number of inputs is very large iii) k-NN makes no assumptions about the functional form of the problem is solved.
  a)i and ii   b)i and iii   c) i   d) All of the above
10. Write the Euclidean Distance and Manhattan Distance between the two data point A(1,3) and B(2,3)?.

(c)

Questionnaires for Quiz 2 (Group 2) (Max time :20 Min, Max points:10)
(Refer Data Mining concept & techniques Chapter 10: Cluster Analysis: Basic Concepts and Methods, Cluster Analysis, Partitioning Methods, Hierarchical Methods,Density-Based Methods, Grid-Based Methods, Evaluation of Clustering. Chapter-11:Advanced Cluster Analysis,Probabilistic Model-Based Clustering,Clustering High-Dimensional Data, Clustering Graph and Network Data,Clustering Graph and Network Data)
....
1. Which of the following metrics, do we have for finding dissimilarity between two clusters in hierarchical clustering?. i) Single-link ii) Complete-link iii) Average-link.
    a)i and ii   b)ii and iii   c)iii and iii   d)i, ii and iii

2.Assume, you want to cluster 7 observations into 3 clusters using K-Means clustering algorithm. After first iteration clusters, C1, C2, C3 has following observations: C1: {(2,2), (4,4), (6,6)}.   C2: {(0,4), (4,0)}.
C3: {(5,5), (9,9)}. Which will be the cluster centroids if you want to proceed for second iteration?

3.i) Name the clustering algorithm that follows a top to bottom approach?
  ii) Which clustering algorithm does not require a dendrogram?
4.i)Considering the K-median algorithm, if points (0, 3), (2, 1), & (-2, 2) are the only points which are assigned to the first cluster now, what is the new centroid for this cluster?.
  ii)Considering the K-means algorithm, after current iteration, we have 3 centroids (0, 1) (2, 1), (-1, 2). Will points (2, 3) and (2, 0.5) be assigned to the same cluster in the next iteration?.

5.i)When you break a group of items up in agglomerative hierarchical clustering, you want the deltas (changes) between items to be _____, and the deltas between the results to be _____.
    a) Small, Large b) Large, Large c) Small, Small d) Large, Small
  ii) The time complexity of the K-means is:

6. i)Name the data transformation technique that works well when minimum and maximum values for real-valued attributes are known
  ii)Find the odd man out: DBSCAN, K-mean, PAM, K-medoids

7. List the multilevel clustering techniques which are unable to correct erroneous merges or splits

8.i) What among the following is the intrinsic cluster evaluation parameter?. a) Cluster homogeneity b) Silhouette effect c) Cluster completeness d) BCubed precision and recall
  ii) List any two density based clustering techniques

9.What type of clusters can density-based clustering get?

10.i)What is the minimum no. of variables/ features required to perform clustering?
  ii)Which clustering algorithms suffer from the problem of convergence at local optima?

(d)

Fig. 6. (a) Questionnaire for Quiz 1 (Group 1). (b) Questionnaire for Quiz 2 (Group 1). (c) Questionnaire for Quiz 1 (Group 2). (d) Questionnaire for Quiz 2 (Group 2).

References

1. Agrawal, R., Gollapudi, S., Kenthapadi, K., Srivastava, N., Velu, R.: Enriching textbooks through data mining. In: Proceedings of the First ACM Symposium on Computing for Development, ACM DEV 2010. Association for Computing Machinery, New York (2010)
2. Bishop, C.M.: Pattern Recognition and Machine Learning. Information Science and Statistics. Springer, Heidelberg (2006)
3. Blei, D.M., Ng, A.Y., Jordan, M.I.: Latent Dirichlet allocation. J. Mach. Learn. Res. **3**, 993–1022 (2003)
4. Bonadiman, D., Uva, A., Moschitti, A.: Effective shared representations with multitask learning for community question answering. In: Proceedings of the 15th Conference of the European Chapter of the Association for Computational Linguistics (ACL): Volume 2, Short Papers, Vlencia, Spain, pp. 726–732, April 2017
5. Bonadiman, D., Uva, A., Moschitti, A.: Effective shared representations with multitask learning for community question answering. In: Proceedings of the 15th Conference of the European Chapter of the Association for Computational Linguistics (ACL): Volume 2, Short Papers, Valencia, Spain, pp. 726–732, April 2017
6. Chen, D., Fisch, A., Weston, J., Bordes, A.: Reading Wikipedia to answer open-domain questions. CoRR, abs/1704.00051 (2017)
7. Choi, E., Hewlett, D., Uszkoreit, J., Polosukhin, I., Lacoste, A., Berant, J.: Coarse-to-fine question answering for long documents. In: Proceedings of the 55th Annual Meeting of the Association for Computational Linguistics (Volume 1: Long Papers), Vancouver, Canada, pp. 209–220, July 2017
8. Devlin, J., Chang, M.-W., Lee, K., Toutanova, K.: BERT: pre-training of deep bidirectional transformers for language understanding. CoRR, abs/1810.04805 (2018)
9. Han, J., Kamber, M., Pei, J.: Data Mining: Concepts and Techniques, 3rd edn. Morgan Kaufmann Publishers Inc., San Francisco (2011)
10. Herrera, J., Poblete, B., Parra, D.: Learning to leverage microblog information for QA retrieval. In: Pasi, G., Piwowarski, B., Azzopardi, L., Hanbury, A. (eds.) ECIR 2018. LNCS, vol. 10772, pp. 507–520. Springer, Cham (2018). https://doi.org/10.1007/978-3-319-76941-7_38
11. Jatnika, D., Bijaksana, M.A., Suryani, A.A.: Word2Vec model analysis for semantic similarities in English words. Procedia Comput. Sci. **157**, 160–167 (2019). The 4th International Conference on Computer Science and Computational Intelligence (ICCSCI 2019): Enabling Collaboration to Escalate Impact of Research Results for Society
12. Joty, S., Màrquez, L., Nakov, P.: Joint multitask learning for community question answering using task-specific embeddings. In: Proceedings of the 2018 Conference on Empirical Methods in Natural Language Processing, pp. 4196–4207. Association for Computational Linguistics, Brussels (2018). https://doi.org/10.18653/v1/D18-1452
13. Karpukhin, V., et al.: Dense passage retrieval for open-domain question answering. arXiv preprint arXiv:2004.04906 [cs.CL] (2020)
14. Kumar, S., Chauhan, A.: Enriching textbooks by question-answers using cQA. In: TENCON 2019 – 2019 IEEE Region 10 Conference (TENCON), pp. 707–714 (2019)
15. MacKay, D.J.C.: Information Theory, Inference and Learning Algorithms. Cambridge University Press, New York (2002)

16. Min, S., Zhong, V., Socher, R., Xiong, C.: Efficient and robust question answering from minimal context over documents. CoRR, abs/1805.08092 (2018)
17. Nakov, P., et al.: SemEval-2017 task 3: community question answering. In: Proceedings of the 11th International Workshop on Semantic Evaluation (SemEval-2017), pp. 27–48. Association for Computational Linguistics, Vancouver (2017). https://doi.org/10.18653/v1/S17-2003
18. Nasehi, S.M., Sillito, J., Maurer, F., Burns, C.: What makes a good code example?: A study of programming Q&A in StackOverflow. In: 2012 28th IEEE International Conference on Software Maintenance (ICSM), pp. 25–34 (2012)
19. Patro, B.N., Kurmi, V.K., Kumar, S., Namboodiri, V.P.: Learning semantic sentence embeddings using sequential pair-wise discriminator. arXiv preprint arXiv:1806.00807 [cs.CL] (2019)
20. Rücklé, A., Moosavi, N., Gurevych, I.: COALA: a neural coverage-based approach for long answer selection with small data, vol. 33, July 2019
21. Rücklé, A., Moosavi, N.S., Gurevych, I.: Neural duplicate question detection without labeled training data. In: Proceedings of the 2019 Conference on Empirical Methods in Natural Language Processing and the 9th International Joint Conference on Natural Language Processing (EMNLP-IJCNLP), pp. 1607–1617. Association for Computational Linguistics, Hong Kong (2019). https://doi.org/10.18653/v1/D19-1171
22. Röder, M., Both, A., Hinneburg, A.: Exploring the space of topic coherence measures. In: Proceedings of the Eighth ACM International Conference on Web Search and Data Mining, WSDM 2015, New York, NY, USA, pp. 399–408 (2015)
23. Sakata, W., Shibata, T., Tanaka, R., Kurohashi, S.: FAQ retrieval using query-question similarity and BERT-based query-answer relevance. In: Proceedings of the 42nd International ACM SIGIR Conference on Research and Development in Information Retrieval, SIGIR 2019, pp. 1113–1116. Association for Computing Machinery, New York (2019)
24. Shen, Y., et al.: Knowledge-aware attentive neural network for ranking question answer pairs. In: The 41st International ACM SIGIR Conference on Research and Development in Information Retrieval, SIGIR 2018, pp. 901–904. Association for Computing Machinery, New York (2018)
25. Srba, I., Savic, M., Bielikova, M., Ivanovic, M., Pautasso, C.: Employing community question answering for online discussions in university courses: students' perspective. Comput. Educ. **135**, 75–90 (2019)
26. Tan, M., dos Santos, C., Xiang, B., Zhou, B.: Improved representation learning for question answer matching. In: Proceedings of the 54th Annual Meeting of the Association for Computational Linguistics (Volume 1: Long Papers), Berlin, Germany, pp. 464–473, August 2016
27. Wang, S., et al.: R3: reinforced ranker-reader for open-domain question answering. In: AAAI (2018)
28. Weissenborn, D., Wiese, G., Seiffe, L.: FastQA: a simple and efficient neural architecture for question answering. arXiv, abs/1703.04816 (2017)
29. Wen, J., Ma, J., Feng, Y., Zhong, M.: Hybrid attentive answer selection in CQA with deep users modelling. In: AAAI (2018)
30. Wu, W., Sun, X., Wang, H.: Question condensing networks for answer selection in community question answering. In: Proceedings of the 56th Annual Meeting of the Association for Computational Linguistics (Volume 1: Long Papers), Melbourne, Australia, pp. 1746–1755, July 2018

31. Xu, J., et al.: Short text clustering via convolutional neural networks. In: Proceedings of the 1st Workshop on Vector Space Modeling for Natural Language Processing, Association for Computational Linguistics, Denver, Colorado, pp. 62–69, June 2015

32. Yang, M., Tu, W., Qu, Q., Zhou, W., Liu, Q., Zhu, J.: Advanced community question answering by leveraging external knowledge and multi-task learning. Knowl.-Based Syst. **171**, 106–119 (2019)

33. Yoon, S., Shin, J., Jung, K.: Learning to rank question-answer pairs using hierarchical recurrent encoder with latent topic clustering. In: Proceedings of the 2018 Conference of the North American Chapter of the Association for Computational Linguistics: Human Language Technologies, Volume 1 (Long Papers), New Orleans, Louisiana, pp. 1575–1584, June 2018

34. Zheng, C., Zhai, S., Zhang, Z.: A deep learning approach for expert identification in question answering communities. CoRR, abs/1711.05350 (2017)

35. Zheng, C.T., Liu, C., San Wong, H.: Corpus-based topic diffusion for short text clustering. Neurocomputing **275**, 2444–2458 (2018)

OWI: Open-World Intent Identification Framework for Dialog Based System

Jitendra Parmar$^{(\boxtimes)}$, Sanskar Soni, and Satyendra Singh Chouhan

Department of CSE, MNIT, Jaipur, India
{2019rcp9044,2018ucp1265,sschouhan.cse}@mnit.ac.in

Abstract. Automated task-oriented dialog based system, generally stated as *Chatbot*, is widely used nowadays by service-oriented platforms such as banking, mobile service providers and travel management firms. The most imperative part of the task-oriented dialog system is to distinguish the intent of the queries asked. If the system erroneously identifies the intent of the query, then the given answer is either incorrect or not related to the query asked. This raises the risk of deteriorating the reliability of the entire system and the organization. Such kind of systems struggle when a user asks queries that contain words for which training classes are not available. These classes may be termed as unseen classes. Our aim is to find the unseen classes in an automated task-oriented dialog system. This paper focuses on open-world learning. Specifically, we propose a deep learning-based Intent Identification framework, OWI, to identify unseen classes for an automated dialog-based system. The OWI framework is based on convolutional neural network with a 1-vs-rest output layer to identify the unseen classes. The proposed model is evaluated on various performance matrices. In addition, we compare OWI with an existing state-of-the-art model. The experimental results show that the OWI outperforms the existing model with respect to identifying unseen classes.

Keywords: Intent classification · Unseen class discovery · Open-world learning · Text classification · Deep learning

1 Introduction

For every system which accepts input as a text or voice, to solve or answer a user's query, it is very important to understand the intention of the user. In artificial intelligence, intent classification is a method to recognize the purpose or intention of a user by evaluating the text language of input. It refers to as a intent classification or intent identification. Nowadays, there are many establishments that are using text-based chat system to solve their customers' queries without any human interactions [1]. These systems are also used for physically challenged people [2], a few service-oriented firms such as booking hotels and flights [3], automated customer support [4] and providing information about tourists [5].

© Springer Nature Switzerland AG 2020
L. Bellatreche et al. (Eds.): BDA 2020, LNCS 12581, pp. 329–343, 2020.
https://doi.org/10.1007/978-3-030-66665-1_21

Such systems can not provide a correct response to any query without identifying the intent of the query. Hence intent classification is the most imperative part of such systems.

In a real-world scenario, new intents emerge regularly. It is essential to identify such unseen intents which have not appeared before. Intents that are not appearing in any training set can be termed as unseen intents/classes. If unseen intents are identified incorrectly by any system, it may damage the credibility of the entire system. For example, Fig. 1 shows customer queries with a dialog-based system. The domain of this system is banking and it is specifically designed for the balance and transaction related queries.

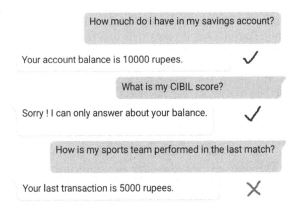

Fig. 1. Message exchange between the user and XYZ bank dialog based system

In this example, there are three queries of a user in which the intent of the two queries are identified correctly while the intent of one query is identified incorrectly. In query 1, a user asks about the account balance; the banking system identifies it correctly and gives the relevant response. In query 2, the user asks about his CIBIL score which is rejected by the system as it finds the query out-of-scope. In query 3, a user asks about the sport team's performance in the last match and the system replies about the last payment details as the system finds it in-scope and gives an irrelevant response.

There are numerous research works that have been done in intent classification for various platforms such as web search, dialogue system and cross-domain. In [6], authors proposed a system for multi-class intent classification using a bag-of-token method which exploits hybrid features representation and classifies intent such as seeking help, offering help, and advising. In [7], Naive Bayes and logistic regression algorithms were used to classify intent. There is a work on close domain data to identify intent in chatbots [8]. It uses RNN (Recurrent neural network) and LSTM (Long short-term memory) to classify the intent. In [9] authors used Semantic hashing as an embedding to classify the intent of a particular task. BERT (Bidirectional Encoder Representations from

Transformers) [10] is also used in many tasks for embedding the text to understand the talk oriented dialog and as an intent detector [11,12].

Above mentioned related works follow the closed-world assumption i.e., they work with known intents (seen classes) only. A very few efforts have been done so far in a discourse to identify unknown intent. Some of the researches proposed solutions to address open-world intent classification. In [13], authors proposed, SMDN (SofterMax and deep novelty detection), which uses softmax as a classifier that does not have rejection capability for the unseen classes. In [14], researchers used SEG (semantic-enhanced Gaussian mixture model), an approach of LOF (local outlier factor), to classify new intent from the data. In their work, enough labeled data was available throughout the model training and such mechanisms are unable to address novel or earlier unseen intent classes.

In the light of above works, we propose a solution to handle the unseen intents for a dialog-based system. We propose, OWI, an Open-World Intent Identification Framework to identify unseen intents in the data. The OWI framework is based on convolutional neural network. In addition, OWI has a 1-vs-rest layer [15] to identify unseen classes. We evaluate the performance of OWI Framework on intent classification and out-of-scope dataset [16]. The performance evaluation of model is done by using various performance matrices such as accuracy, precision, recall, F1-score, specificity and MCC (Matthews correlation coefficient). We also compare the performance of OWI with an adopted version existing state-of-the-art model given in Ref. [17].

The rest of the paper is organised as follows. Section 2 presents the detailed literature review. Section 3 presents the proposed OWI. Section 4 discusses the performance evaluation of OWI. Conclusions and future directions are given in Sect. 5.

2   Related Works

In this section, first we discuss the related works in intent classification problem. Next, we discuss relevant works in open-world learning followed by discussions on related works that address open-world learning in intent classification problem.

2.1   Intent Identification Approaches

In [18], authors used two combinations in LSTM network former being Naïve Bayes grouping with Bag-of-Words and latter being Bag-of-Words with SVM (Support vector machine) to classify the intents for dialog utterances uses. This work clearly states that SVM models are superior to the LSTM model in terms of performance. It is because Hierarchical structure in intent and word embedding improves the performance of SVM. Use of word embedding is significantly better than other methods such as count of words.

In [19], researchers adopted Word2Vec (Word to Vector) [20] with nearest neighbor, Siamese network, LSTM and feed forward neural network with an end to end memory network to identify the user intent. They applied many multi-label classification approaches on their dataset and suggested that the machine learning based models under-perform than the simpler matching architectures. It was also found that Word2Vec with nearest neighbor perform better than the Siamese, LSTM and feed forward neural network. It is because the extra training data is not required. Memory based architectures are sustainably prefered when classes are out-sized but training data is inadequate.

Some works have also been done to identify the specific intent in social media. In [21], authors used feature set to identify domestic violence. They found that the psycho-linguistic evidence has strong suggestive powers in the estimation of any social media article than the documentary features. Some of the works done on NLP (Natural Language Processing), to classify documents to understand neural language, are based on deep learning techniques such as CNN (Convolutional Neural Network) based approaches [22] and hierarchical attention-based networks [23].

In [24], authors used Pre-trained models to train large data such as Wikipedia. In [25], authors adapted Code-mix methods for task-oriented systems, Which used a combination of vector representation and various models to detect intent in multiple languages. In this work deep-recurrent network models significantly perform better than the classical machine learning models. However performance of traditional methods of vector representation is underachieved than the self-trained word embedding methods. This work found that the combination of self-trained word embedding and deep- recurrent network performed better to find intent from multi-language statements.

In [26], authors proposed a framework of joint intent classification and spot labelling for the goal-oriented intent prediction purposes. It is a convolution-based model that decomposes the task into components for analysis and then apply word contextualization and determine label recurrent model.

2.2 Open-World Learning Techniques

Traditional machine learning functions in two parts, training and testing, and for each instance of testing we must have a training instance to identify that class. This is a closed-world learning assumption. For example, if we have a system to identify the diseases of crops by analyzing the pattern of rust on leaves. For each kind of rust pattern, we have a training class/seen class. The system will easily predict the disease of crops. However, if a pattern of rust is not available in the training class, it will not be able to predict a disease and it will predict something from the seen class, leading to the incorrect classification of disease.

In [27], researchers adapted multi-class classifier for open set recognition that has a rejection ability like 1-vs-1 and 1-vs-Rest. It can reject data which is not available during the training. In [28], authors used probability-based multi classes open set for the normalized posterior likelihood of class insertion. They used multi-class EVT (Extreme Value Theory) based likelihood modelling for PI-SVM training and multi-class likelihood estimation for PI-SVM testing. In [29], authors proposed DOC (Deep Open Classification) to identify new documents or tasks which may not belong to any of the training class. The ideal classifier should be able to identify both the classes i.e., seen and unseen classes correctly. DOC builds a multi-class classifier with the 1-vs-rest final layer of sigmoid instead of soft max. It reduced open space risk. To lighten the decision boundaries of sigmoid function and reduce the open space risk they used Gaussian function and 1-vs-rest layer was used to enable rejection capability of the proposed network.

In [30], authors proposed a learning without forgetting method which improved open-world outcomes. They used joint training by giving algorithm "learning without forgetting" which works on shared parameters. Learning without forgetting used only examples of a new task and acquire both extraordinary accuracy and optimization. The algorithm preserves responses in the present task from the original network. The network is conserved in such a way that previously learned tasks produce the same output for all related inputs. The main improvement of learning without forgetting is classification performance which is improved through feature extraction and fine-tuning. In terms of training time, computational effectiveness of learning without forgetting is better than the joint training method.

In [31], authors aim to identify seen classes and reject unseen classes. Their proposed approach is quite different from the knowledge transfer because in case of knowledge transfer, knowledge is sent from supervised to supervised or unsupervised to unsupervised, but here knowledge is shared from supervised to unsupervised. For this objective, they developed a model which consisted of Combination of two networks that have been used for open classification network (OCN) and pairwise classification network (PCN). Both networks shared the same components for learning.

2.3 Unseen Classes Discovery in Intent Identification

The research of identifying unseen intent classes in dialog systems is still in its early stage. Here, we discuss some of the works done in unknown class discovery in intent classification. In [13], authors used SMDN (Softmax and deep novelty) architecture to detect an unknown intent. They calculated decision boundaries to pick unknown intent. It used softmax function for the output layer. SMDN can be applied to any of the existing architectures without changing the model structure. 1-vs-Rest [15] multi-class classification is one of the used mechanism for outputs that has the ability to detect unseen classes.

In [14], authors proposed Gaussian mixture model (GMM) that used SEG (Semantic-enhanced Gaussian mixture) classifier for training and testing and LOF [32] to detect outliers. Some of the models emphasize on intent extraction from text, such as Open intent extraction model. In [33], authors used open intent extraction model to detect an open intent from neural language interaction. They used bidirectional LSTM to discover more than one common intent type from text expression in order to understand the neural language better for the discovery of unseen intent. In [34], authors proposed ADVIN (Automatic discovery of novel intents and domains) framework; which could detect unique unseen intents and domains from text expressions.

3   Proposed OWI Framework

The architecture of OWI is shown in Fig. 2. It consists of three major components. First is the text embedding that uses a Word2Vec [20] method. Second is sequential convolutions and max-pooling operation and third is a fully connected 1-vs-rest layer. In [17,35], authors suggested that the Convolutional neural network had outstanding outcomes in both image processing as well as in text classification. Therefore, we select CNN for OWI. OWI consists of two convolution layers (Sect. 3.1) with intermediate ReLu (Rectified Linear Unit) function to normalize the output of convolution. Next, Max-Pooling layer (Sect. 3.2) used to reduce the size of convolution layer's output. It is followed by two fully connected layers (Sect. 3.3) with one having in-between ReLU function. The last output layer of OWI is 1-vs-rest sigmoid layer (Sect. 3.4). The detailed description is as follows.

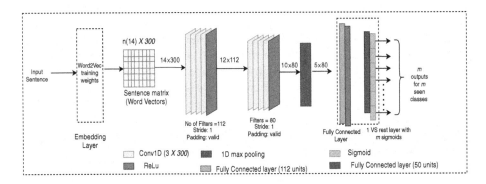

Fig. 2. OWI framework

3.1   Convolution Layer of OWI

Let d dimensional word vector $a_i \in \mathbb{R}^d$, which is corresponding to the ith word in the sentence. Assume that the length of a sentence is l (including padding). The concatenation all the word vectors will be,

$$a_{i:l} = a_1||a_2||a_3||, \ldots, a_l \tag{1}$$

Here, "$||$" symbol denotes the concatenation. General representation of concatenation is $a_{i:i+j}$ which is a concatenation of words: a_i, $a_i + 1, \ldots a_{i+j}$. Applying convolution filters $w \in \mathbb{R}^{sd}$ on word window size of s to generate feature map v_i from the window of word $a_{i:1+s-1}$ is represented by,

$$v_i = f(w.a_{i:1+s-1} + b) \tag{2}$$

Here f is nonlinear function and $b \in \mathbb{R}$ is a bias term. All possible windows of word in sentence are $a_{1:s}, a_{2:s}, \ldots a_{l-s+1:l}$, We apply filters f on these windows. It produces a Feature Map v ,Where $v \in \mathbb{R}^{l-s+1}$

$$v = (v_1, v_2, \ldots, v_{l-s+1}) \tag{3}$$

ReLu activation layer has been used after each convolution layer to normalize the output. In OWI, we use two convolution layers. For the first convolution layer, 112 filters are applied with filter size 3 and stride length of 1 to each convolution operation. For second convolution layer 80 filters are applied with filter size 3 and stride length of 1 to each convolution operation.

3.2   MaxPooling Layer

OWI uses 1D Max-pooling layer to reduce the feature vector v. The output of convolution layer is reduced through max-poling layer by taking the maximum value $\hat{v} = Max(v)$ for each corresponding filter.

3.3   Fully Connected Layer

The outcome of pooling layer, from a d-dimension feature vector v is fed into a fully connected layer with 112 units whose primary purpose is to reduce the dimensionality of the vector to 112. We applied one in-between ReLu activation layer.

3.4   1-Vs-Rest Sigmoid Layer [15]

Classical machine learning uses a softmax as an output layer for multi-class classification. However, softmax does not have rejection capability. For this, the output of softmax needs to be normalized for each class in all the training classes.

In OWI we use 1-vs-rest layer with m sigmoid functions, where m is seen classes. We have i^{th} sigmoid function for x_i class. OWI designates all classes as positive and negative for x_i class such that $y = x_i$ is positive class and rest of all $y \neq x_i$ classes are negative. We calculate log loss for all m sigmoid functions for the training data.

$$loss = \sum_{i=1}^{m} \sum_{j=1}^{n} -I(y_j = x_i)logp(y_j = x_i) - I(y_j \neq x_i)(1 - logp(y_j = x_i)) \tag{4}$$

Where I is the Indicator function, j= 1 to n (n = Number of instances) and probability output of m sigmoids for j^{th} input of ith dimension of r is $p(y_j = x_i) = sigmoid(r_{j,i})$ reject unseen intent through the sigmoid such that

$$\hat{y} = \left\{ \begin{array}{l} reject, \; if sigmoid(r_i) < z_i, \; \forall \; x_i \; in \; y_i \\ argmax_{x_i \in y} \; sigmoid(r_i), \; Otherwise \end{array} \right\} \tag{5}$$

3.5   Analysis of OWI Framework

Since most of the chatbot based conversations are primarily very short therefore, after cleaning the data, we were left with sentences with a maximum length of 14. Thus, we have to choose the hyperparameters in such a way that there is minimal loss of the important information. In order to achieve it, we use a hyperparameter optimization technique - '*Keras tuner*' and run random Search on it. The method somehow gives similar results pertaining to our assumption of using a lower Kernel size i.e., 3 which yields the best results. For deciding the number of filters and learning rate, we again use the same technique and get the best results with filter sizes of 112 and 80 for two convolution layers respectively. The learning rate is selected to be 0.001. The number of convolution layers were also limited to 2, as it was enough to detect the main features from the data and nonetheless adding more layers led to overfitting and a decrease in computational speed. The detailed configuration of OWI is shown in Table 1.

Table 1. OWI configuration details

Type	Sub-layer	Number of filters	Kernel size	Stride	Output shape	#Parameters
Conv 1D	3.1	112	3	1	(None, 12, 112)	100912
Conv 1D		80	3	1	(None, 10, 80)	26960
Max pooling 1D	3.2	–	2(pool)	–	(None, 5, 80)	0
Flatten		–	–	–	None, 400)	0
Dense	3.3	112 units	–	–	(None, 112)	44912
1 Vs Rest	3.4	50 (seen classes)			(None, 50)	5650

4   Experiments and Results

To validate the performance of proposed OWI framework, we have performed experiments on publicly obtainable Intent classification and out-of-scope prediction dataset [16]. In this section, we discuss the dataset specifications, implementation details, comparative performance of proposed framework with the adopted version of an existing state-of-the-art text classification framework [17].

4.1   Dataset Used

We perform the experiments on the publicly available Intent classification and out-of-scope prediction dataset [16]. Other existing datasets have limited intents/classes, whereas Intent classification and out-of-scope prediction dataset has 150 intent from about 10 different domains. This recently constructed data set has 23700 queries which are short and unstructured in nature. Out of 23700 queries, 22500 queries are in-scope and 1200 queries are out-of-scope from all 10 domains. For the extent of our experiments, we used 50 classes out of the available 150 classes and included all of the out-of-scope (OOS) data (1200 queries) into the test dataset along with the 300 queries from the validation set belonging to the untrained classes. The test data includes 1500 samples from the seen class along with 1500 samples from OOS class.

4.2 Experimental Setup

The experimental environment of OWI is a computer equipped with 1.80 GHz Dual-Core Intel core i5 and 8 GB DDR3 RAM, running MacOS 10.15.3, 64 bit processor. All experiments are implemented in python 3.0.

The proposed OWI Model implemented on dataset uses Word2Vec embedding. OWI uses Google's pre-trained Word2Vec[1] model which includes word vectors for a vocabulary of 3 million words and phrases which were trained on roughly 100 billion words from a Google News dataset with the dimension of each vector as 300. For experimental validation, we have also evaluated OWI with GloVe (Global Vector) [36] embeddings. We evaluated the OWI framework on difference performance matrices as discussed below.

1. *Accuracy:* The ratio of appropriately classified intents to the total number of intents:

$$Accuracy = \frac{(T_{positive} + T_{negative})}{(T_{positive} + T_{negative} + F_{positive} + F_{negative})} \tag{6}$$

2. *Precision:* Precision shows the positive predicted intents. Precision states as the proportion of intents anticipated as out-of-scope/unseen which truly are out-of-scope/unseen.

$$Precision = \frac{T_{positive}}{T_{positive} + F_{positive}} \tag{7}$$

3. *Recall:* It is rate of true positive. Recall is a proportion of detected positives intent foreseen as positives through a classifier. It also called Sensitivity.

$$Recall = \frac{T_{positive}}{T_{positive} + F_{negative}} \tag{8}$$

4. *F1 Score:* F1-Score gives an accuracy metrics to analyze erroneously classified out-of-scope/unseen classes. It also refers as F-Score. F1-Score is the harmonic mean of Precision and Recall.

$$F1 - score = 2 \times \frac{precision \times recall}{precision + recall} \tag{9}$$

5. *Specificity:* It is an opposite of Recall. Specificity is Proportion of detected negatives intents foreseen as negatives through a classifier.

$$Specificity = \frac{T_{negative}}{F_{positive} + T_{negative}} \tag{10}$$

---

[1] http://code.google.com/archive/p/word2vec/.

6. *MCC:* Matthews Correlation Coefficient is used where predictions are taking place. It is established on the base theory of Chi-Square test and generally uses where outcome of a methods are based on predictions.

$$MCC = \frac{T_{positive} \times T_{negative} - F_{positive} \times F_{negative}}{\sqrt{(F_{positive} + T_{positive})(T_{positive} + F_{negative})(T_{negative} + F_{positive})(T_{negative} + F_{negative})}}$$

4.3 Unseen Class Identification

To discover out-of-scope/unseen classes we have used 1-vs-rest sigmoid layer (Sect. 3.4). Sigmoid layer typically uses the default value of the threshold ($z_i = 0.5$) for classification of the targeted class C_i. However, if we fix the value of threshold $z_i = 0.5$, it will not provide a probable result for rejection of unseen classes. To improve a boundary of threshold we can check it for all the possible values of z_i. We have selected the range of threshold values, $z_i = 0.1$ to 0.9, for better recall of unseen classes. OWI tested on all the performance parameters for each threshold value for both the models with different embeddings.

4.4 OWI Results

Table 2 and Table 3 show the results of OWI framework on intent classification dataset discussed in Sect. 4.1. Table 2 shows the results of OWI with Word2Vec embedding and Table 3 shows the results of OWI with GloVe word embedding techniques.

Table 2. OWI results with Word2Vec Embeddings

Threshold	Accuracy_oos	Precision	Recall	F1_score	Specificity	MCC
0.2	0.68756	0.71171	0.10533	0.18351	0.97867	0.18285
0.3	0.71689	0.74249	0.23067	0.35198	0.96	0.295
0.4	**0.75733**	**0.75**	**0.408**	**0.5285**	**0.932**	**0.41599**
0.5	**0.78444**	**0.73619**	**0.55067**	**0.63005**	**0.90133**	**0.49251**
0.6	**0.79644**	**0.70739**	**0.664**	**0.68501**	**0.86267**	**0.53545**
0.7	0.79111	0.66627	0.748	0.70477	0.81267	0.54616
0.8	0.78222	0.63374	0.82133	0.71545	0.76267	0.55576
0.9	0.76578	0.60109	0.884	0.7156	0.70667	0.55699

Table 3. OWI results with GloVe Embeddings

Threshold	Accuracy_oos	Precision	Recall	F1_score	Specificity	MCC
0.2	0.67022	0.7	0.01866	0.03636	0.996	0.07366
0.3	0.70044	0.82203	0.12933	0.22350	0.986	0.24389
0.4	**0.74977**	**0.75900**	**0.36533**	**0.49324**	**0.942**	**0.39474**
0.5	**0.78444**	**0.74311**	**0.54**	**0.62548**	**0.90666**	**0.49147**
0.6	**0.78666**	**0.69014**	**0.65333**	**0.67123**	**0.85333**	**0.51393**
0.7	0.784888	0.65573	0.74666	0.69825	0.804	0.53492
0.8	0.77733	0.62717	0.81866	0.71023	0.75666	0.54705
0.9	0.75555	0.58896	0.88266	0.70651	0.692	0.54180

The following inferences can be drawn from Tables 2 and 3.

- In view of all the performance measures for OWI with Word2Vec embedding, we are having better results on threshold $z_i = 0.4$, $z_i = 0.5$ and $z_i = 0.6$. Specifically, we obtain highest value for accuracy is 0.79644, for recall 0.6664, for F1-Score 0.6850 and for MCC 0.5354 at threshold $z_i = 0.6$, and for precision and specificity we obtain highest value at threshold $z_i = 0.5$, that is 0.7361 and 0.90133 respectively.
- Similarly, with GloVe word embedding the results obtained on $z_i = 0.4$, $z_i = 0.5$ and $z_i = 0.6$ are better in comparison to the other threshold values. Specifically, we obtain highest value for accuracy which is 0.78666, for recall 0.6533, for F1-Score 0.67123 and MCC 0.5139 at threshold $z_i = 0.6$. We obtained the highest value of precision 0.75900 at threshold $z_i = 0.4$ and Specificity 0.90666 at threshold $z_i = 0.5$.
- The overall performance measures' values with Word2Vec are significantly higher than the values produced by using GloVe word embedding with OWI.
- In addition, the performance measures clearly state that the OWI performance is suggestively higher for threshold values $z_i = 0.4$, $z_i = 0.5$ and $z_i = 0.6$. Thus we take only these 3 threshold values for better results.

4.5 Comparison and Discussion

We have compared the performance of OWI with an adopted model of Yoon kim's text classification model given in [17]. Though, original Kim model uses softmax layer at the end for classification purpose. Here, we have replaced it with 1-vs-rest layer for adding rejection capability. For simplicity, we are referring this model as *YKTC* model. The *YKTC* is having 3 kernel size (3, 4, 5) each having 100 filters. After performing max-pooling, each output vector is concatenated to produce a single vector. The details are given in [17]. The resultant vector is then passed through the 1-vs-rest layer. For a fair comparison of the OWI framework with the *YKTC* model we have compared these models on threshold values $z_i = 0.4$, $z_i = 0.5$ and $z_i = 0.6$ with accuracy.

Figure 3 and 4 shows the Comparative results of OWI with *YKTC* model. We have compared OWI and *YKTC* with both the word embeddings. From the results we observe that:

- the OWI with Word2Vec outperform the *YKTC* at threshold $z_i = 0.5$ and $z_i = 0.6$. Figure 3 illustrates that even at threshold $z_i = 0.4$ OWI performed better than *YKTC* on the performance parameters.
- Similarly, with GloVe word embedding Fig. 3 clearly states that OWI performed excellently over the *YKTC* at threshold $z_i = 0.5$ and $z_i = 0.6$ and performed slightly better at threshold $z_i = 0.4$.
- Thus we conclude that the overall performance of OWI is better than that of *YKTC* for defined parameters.

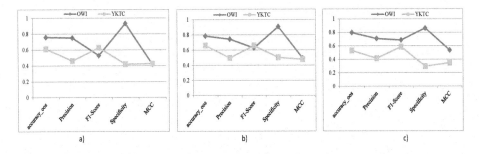

Fig. 3. Comparison of OWI with YKTC with Word2Vec embeddings with threshold values (z_i): a) 0.4 b) 0.5 and c) 0.6

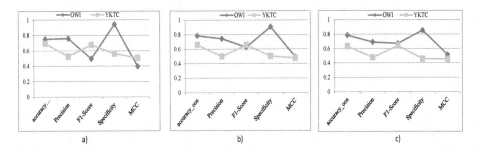

Fig. 4. Comparison of OWI with YKTC with GloVe embeddings with threshold values (z_i): a) 0.4 b) 0.5 and c) 0.6

5 Conclusion

The world is progressively moving towards the dynamic environment therefore closed-world learning has the least significance in such an environment, Whereas, open-world learning is useful to address dynamic environment. This paper proposed an open-world learning method and presented a framework to identify the unseen intent in the dialog-based system. It can classify seen intents as well as reject unseen intents. We have performed experiments on a dataset with two different word embeddings methods. We observed that OWI is appropriate and reliable to reject unseen intents. In Sect. 4 we have shown that the OWI has significantly improved the correctness in unseen class discovery as compared to the state-of-art method.

In near future, we plan to improve OWI by an accumulation of continual learning capability. With an aim to learn about out-of-scope classes using previous knowledge, thereby finding new in scope classes by categorising intents.

References

1. Thomas, N.T.: An e-business chatbot using AIML and LSA. In: International Conference on Advances in Computing, Communications and Informatics (ICACCI), pp. 2740–2742 (2016)
2. Naveen Kumar, M., Linga Chandar, P.C., Venkatesh Prasad, A., Sumangali, K.: Android based educational chatbot for visually impaired people. In: IEEE International Conference on Computational Intelligence and Computing Research (ICCIC), pp. 1–4 (2016)
3. Braun, D., Hernandez Mendez, A., Matthes, F., Langen, M.: Evaluating natural language understanding services for conversational question answering systems. In: Proceedings of the 18th Annual SIGdial Meeting on Discourse and Dialogue, pp. 174–185 (2017)
4. Xu, A., Liu, Z., Guo, Y., Sinha, V., Akkiraju, R.: A new chatbot for customer service on social media. In: Proceedings of the CHI Conference on Human Factors in Computing Systems, pp. 3506–3510 (2017)
5. Paweł Budzianowski, T.-H.W., et al.: Multiwoz-a large-scale multi-domain wizard-of-oz dataset for task-oriented dialogue modelling. In Proceedings of the Conference on Empirical Methods in Natural Language Processing, pp. 5016–5026 (2018)
6. Purohit, H., Dong, G., Shalin, V., Thirunarayan, K., Sheth, A.: Intent classification of short-text on social media. In: Proceedings of the IEEE International Conference on Smart City/Socialcom/Sustaincom (Smartcity), pp. 222–228 (2015)
7. Helmi Setyawan, M.Y., Awangga, R.M., Efendi, S.R.: Comparison of multinomial naive bayes algorithm and logistic regression for intent classification in chatbot. In: Proceedings of the International Conference on Applied Engineering (ICAE), pp. 1–5 (2018)
8. Nigam, A., Sahare, P., Pandya, K.: Intent detection and slots prompt in a closed-domain chatbot. In: Proceedings of the IEEE 13th International Conference on Semantic Computing (ICSC), pp. 340–343 (2019)
9. Shridhar, K., et al.: Subword semantic hashing for intent classification on small datasets. In: Proceedings of the International Joint Conference on Neural Networks (IJCNN), pp. 1–6 (2019)

10. Devlin, J., Chang, M.-W., Lee, K., Toutanova, K.: Bert: pre-training of deep bidirectional transformers for language understanding. In: Proceedings of the Conference of the North American Chapter of the Association for Computational Linguistics: Human Language Technologies, Volume 1 (Long and Short Papers), pp. 4171–4186 (2019)
11. Wu, C.-S., Hoi, S., Socher, R., Xiong, C.: TOD-BERT: pre-trained natural language understanding for task-oriented dialogues. arXiv preprint arXiv:2004.06871 (2020)
12. Casanueva, I., Temčinas, T., Gerz, D., Henderson, M., Vulić, I.: Efficient intent detection with dual sentence encoders. arXiv preprint arXiv:2003.04807 (2020)
13. Lin, T.-E., Hua, X.: A post-processing method for detecting unknown intent of dialogue system via pre-trained deep neural network classifier. Knowl. Based Syst. **186**, 104979 (2019)
14. Yan, G., et al.: Unknown intent detection using Gaussian mixture model with an application to zero-shot intent classification. In: Proceedings of the 58th Annual Meeting of the Association for Computational Linguistics, pp. 1050–1060 (2020)
15. Rifkin, R., Klautau, A.: In defense of One-Vs-all classification. J. Mach. Learn. Res. **5**, 101–141 (2004)
16. Larson, S., et al.: An evaluation dataset for intent classification and out-of-scope prediction. In: Proceedings of the Conference on Empirical Methods in Natural Language Processing and the 9th International Joint Conference on Natural Language Processing (EMNLP-IJCNLP), pp. 1311–1316 (2019)
17. Kim, Y.: Convolutional neural networks for sentence classification. In: Proceedings of the Conference on Empirical Methods in Natural Language Processing (EMNLP), pp. 1746–1751 (2014)
18. Schuurmans, J., Frasincar, F.: Intent classification for dialogue utterances. IEEE Intell. Syst. **35**(1), 82–88 (2019)
19. Bhardwaj, A., Rudnicky, A.: User intent classification using memory networks: a comparative analysis for a limited data scenario. arXiv preprint arXiv:1706.06160 (2017)
20. Mikolov, T., Chen, K., Corrado, G., Dean, J.: Efficient estimation of word representations in vector space. arXiv preprint arXiv:1301.3781 (2013)
21. Subramani, S., Vu, H.Q., Wang, H.: Intent classification using feature sets for domestic violence discourse on social media. In: Proceedings of the 4th Asia-Pacific World Congress on Computer Science and Engineering (APWC on CSE), pp. 129–136 (2017)
22. Zhou, J., Xu, W.: End-to-end learning of semantic role labeling using recurrent neural networks. In: Proceedings of the 53rd Annual Meeting of the Association for Computational Linguistics and the 7th International Joint Conference on Natural Language Processing, Volume 1: Long Papers, pp. 1127–1137 (2015)
23. Yang, Z., Yang, D., Dyer, C., He, X., Smola, A., Hovy, E.: Hierarchical attention networks for document classification. In: Proceedings of the conference of the North American Chapter of the Association for Computational Linguistics: Human Language Technologies, pp. 1480–1489 (2016)
24. Radford, A., Narasimhan, K., Salimans, T., Sutskever, I.: Improving language understanding by generative pre-training (2018)
25. Gupta, A., Hewitt, J., Kirchhoff, K.: Simple, fast, accurate intent classification and slot labeling for goal-oriented dialogue systems. In: Proceedings of the 20th Annual SIGdial Meeting on Discourse and Dialogue, pp. 46–55 (2019)

26. Jayarao, P., Srivastava, A.: Intent detection for code-mix utterances in task oriented dialogue systems. In: Proceedings of the International Conference on Electrical, Electronics, Communication, Computer, and Optimization Techniques (ICEECCOT), pp. 583–587 (2018)
27. Scheirer, W.J., de Rezende Rocha, A., Sapkota, A., Boult, T.E.: Toward open set recognition. IEEE Trans. Patt. Anal. Mach. Intell. **35**(7), 1757–1772 (2012)
28. Jain, L.P., Scheirer, W.J., Boult, T.E.: Multi-class open set recognition using probability of inclusion. In: Fleet, D., Pajdla, T., Schiele, B., Tuytelaars, T. (eds.) ECCV 2014. LNCS, vol. 8691, pp. 393–409. Springer, Cham (2014). https://doi.org/10.1007/978-3-319-10578-9_26
29. Shu, L., Xu, H., Liu, B.: DOC: deep open classification of text documents. In: Proceedings of the Conference on Empirical Methods in Natural Language Processing, pp. 2911–2916 (2017)
30. Li, Z., Hoiem, D.: Learning without forgetting. IEEE Trans. Pattern Anal. Mach. Intell. **40**(12), 2935–2947 (2017)
31. Shu, L., Xu, H., Liu, B.: Unseen class discovery in open-world classification. arXiv preprint arXiv:1801.05609 (2018)
32. Breunig, M.M., Kriegel, H.-P., Ng, R.T., Sander, J.: LOF: identifying density-based local outliers. In: Proceedings of the ACM SIGMOD International Conference on Management of Data, pp. 93–104 (2000)
33. Vedula, N., Lipka, N., Maneriker, P., Parthasarathy, S.: Open intent extraction from natural language interactions. In: Proceedings of The Web Conference, pp. 2009–2020 (2020)
34. Vedula, N., Gupta, R., Alok, A., Sridhar, M.: Automatic discovery of novel intents & domains from text utterances. arXiv preprint arXiv:2006.01208 (2020)
35. Hinton, G.E., Srivastava, N., Krizhevsky, A., Sutskever, I., Salakhutdinov, R.R.: Improving neural networks by preventing co-adaptation of feature detectors. arXiv preprint arXiv:1207.0580 (2012)
36. Pennington, J., Socher, R., Manning, C.D.: Glove: global vectors for word representation. In: Proceedings of the Conference on Empirical Methods in Natural Language Processing (EMNLP), pp. 1532–1543 (2014)

Author Index

Printed in the United States
By Bookmasters